Progress in Horticultural Crops - from Genotype to Phenotype

Progress in Horticultural Crops - from Genotype to Phenotype

Editors

Guanglong Wang
Lijun Ou
Aisheng Xiong

MDPI

Basel • Beijing • Wuhan • Barcelona • Belgrade • Novi Sad • Cluj • Manchester

Editors
Guanglong Wang
Huaiyin Institute of Technology
Huai'an
China

Lijun Ou
Hunan Agricultural University
Changsha
China

Aisheng Xiong
Nanjing Agricultural University
Nanjing
China

Editorial Office
MDPI
St. Alban-Anlage 66
4052 Basel, Switzerland

This is a reprint of articles from the Special Issue published online in the open access journal *Agronomy* (ISSN 2073-4395) (available at: https://www.mdpi.com/journal/agronomy/special_issues/R2ZCWQ40C3).

For citation purposes, cite each article independently as indicated on the article page online and as indicated below:

Lastname, A.A.; Lastname, B.B. Article Title. *Journal Name* **Year**, *Volume Number*, Page Range.

ISBN 978-3-7258-0931-8 (Hbk)
ISBN 978-3-7258-0932-5 (PDF)
doi.org/10.3390/books978-3-7258-0932-5

Contents

agronomy

Article

Nitric Oxide Promoted the Seed Germination of *Cynanchum auriculatum* under Cadmium Stress

Fang-Fang Liu [1,†], Xuan-Huan Qiao [1,†], Tao Yang [1], Peng Zhao [1], Zhi-Peng Zhu [1], Jun-Hao Zhao [1], Jia-Ming Luo [1], Ai-Sheng Xiong [2,*] and Miao Sun [1,2,*]

1 College of Marine and Biological Engineering, Yancheng Teachers University, Yancheng 224002, China; liuff@yctu.edu.cn (F.-F.L.); 18936320875@163.com (X.-H.Q.); 13382156654@163.com (T.Y.); 18862736573@163.com (P.Z.); zhuzp@stu.sci.edu.rs (Z.-P.Z.); 13915760321@163.com (J.-H.Z.); 18862272510@163.com (J.-M.L.)

2 State Key Laboratory of Crop Genetics & Germplasm Enhancement and Utilization, College of Horticulture, Nanjing Agricultural University, Nanjing 210095, China

* Correspondence: xiongaisheng@njau.edu.cn (A.-S.X.); sunm@yctu.edu.cn (M.S.)

† These authors contributed equally to this work.

Abstract: *Cynanchum auriculatum*, an early food-medicine homologous plant native to Asia, possesses significant nutritional and health benefits. However, the presence of cadmium (Cd) in the soil poses a hazard to the germination and growth of *C. auriculatum*. As nitric oxide (NO) plays a vital role in plant resistance to heavy metal stress, we used three different concentrations of SNP treatment during the germination phase, aiming to alleviate the inhibitory effects of Cd stress on the seed germination of *C. auriculatum*. The results indicated that when compared to seeds treated with SNP concentrations of 0.2 mM and 0.8 mM, *C. auriculatum* seeds treated with 0.4 mM SNP exhibited an improved germination rate and germination index, as well as longer hypocotyl. Furthermore, in comparison to NOS-like, the SNP application stimulated the production of endogenous NO through NR catalysis. Additional investigations showed that the ABA level decreased while the GA level increased under normal conditions, while the SNP application enhanced the accumulation of both ABA and GA in *C. auriculatum* seeds under Cd stress. Histochemical staining and biochemical indicators demonstrated that SNP treatment enhanced the enzymatic activity of SOD, POD, and CAT, while inhibiting the production of hydrogen peroxide and superoxide anion. Moreover, SNP treatment resulted in increased α-amylase activity, which facilitated starch hydrolysis and the generation of soluble sugar. Ultimately, the seed vitality of *C. auriculatum* under Cd stress was promoted. Our findings present a theoretical framework for the application of SNP in the seed germination mechanism of *C. auriculatum* and establish the groundwork for comprehending the physiological role of NO under Cd stress.

Keywords: *Cynanchum auriculatum*; cadmium stress; seed germination; phytohormone; reactive oxygen species

Citation: Liu, F.-F.; Qiao, X.-H.; Yang, T.; Zhao, P.; Zhu, Z.-P.; Zhao, J.-H.; Luo, J.-M.; Xiong, A.-S.; Sun, M. Nitric Oxide Promoted the Seed Germination of *Cynanchum auriculatum* under Cadmium Stress. *Agronomy* **2024**, *14*, 86. https://doi.org/10.3390/agronomy14010086

Academic Editor: Jose Maria Barrero

Received: 8 December 2023
Revised: 26 December 2023
Accepted: 28 December 2023
Published: 29 December 2023

1. Introduction

The global rise of environmental pollution problems, particularly the concern regarding cadmium (Cd) contamination in soil, has become increasingly acknowledged as a consequence of the rapid advancements in industry and agriculture [1]. Cd possesses a prolonged biological half-life, allowing it to persist in the human kidney for a considerable period of 10–30 years, causing significant harm to human health [2]. The majority of Cd pollution is assimilated into the human body through dietary sources like cereals, produce, and root crops, accounting for a significant ninety percent [3]. In plants, Cd pollution also affects a variety of life activities, including the absorption of essential elements, photosynthesis, respiration, and plant growth [4]. Recently, there has been a noticeable correlation between the growing human demand and the increasing necessity for medicinal plants [5]. However, the presence of Cd pollution within the soil has had a detrimental effect on the process of seed germination and subsequent seedling development in these medicinal plants due to the enrichment, accumulation, and transportation of Cd [6].

Cynanchum auriculatum, a member of the *Cynanchum* family, is a medicinal plant commonly found in China, Korea, and India [7]. It is widely recognized as a traditional Chinese medicinal food homologous herb, renowned for its medicinal properties and significant levels of steroids, acetophenones, terpenes, and alkaloids [8]. Recent studies have shown that *C. auriculatum* exhibits various therapeutic effects, such as enhancing immunity by improving the phagocytosis ability of macrophages [9], treating breast cancer by controlling the formation of breast CSCs [10], and the inhibition of neuron damage by reducing apoptosis and reactive oxygen species (ROS) [11]. However, our previous study revealed that soil contamination with Cd negatively affected the germination process of *C. auriculatum*, leading to a decreased germination rate. Therefore, it would be valuable to develop an effective method for germinating *C. auriculatum* seeds under Cd stress.

Recent research has demonstrated that nitric oxide (NO), an influential gaseous signaling molecule with the unique capability to freely permeate biofilms, possesses multiple biological functions in various plant physiological processes, including the regulation of plant growth and development [12], stomatal movement [13], respiration [14], and the biosynthesis and transduction of phytohormone [15]. In the seeds of *Oryza sativa* [16], *Brassica chinensis* [17], and *Cicer arietinum* [18], NO has been found to enhance the germination rate through various regulatory mechanisms, such as the combined actions of NO and photosensitive pigments [19], the interaction between NO and phytohormones [20], and the regulation of the "oxidative window" model by controlling ROS generation [21]. More interestingly, in the defense mechanism of the plant exposed to Cd pollution, NO is also involved in the response mechanism and plays a significant role as a signaling molecule in regulating the resistance. Through S-nitrosylation, NO plays a crucial role in regulating the activity and functionality of Cd response proteins [22], as well as interacting with superoxide anions to modify various biomolecules [23]. In addition, the reaction between NO and hydrogen sulfide leads to the production of thionirous acid, which in turn influences physiological processes and stress responses [24]. Furthermore, NO has the ability to bind to heme, resulting in the inhibition of bis-(3′-5′)-cyclic dimeric guanosine monophosphate (c-di-GMP) production, and the inhibition of soil pathogens through the NosP/NahK signaling pathway [25].

In *Arabidopsis thaliana* [26], *Hordeum vulgare* [27], and *Capsicum annuum* [28], it was found that increasing the endogenous NO level or administering exogenous NO enhanced the resistance of plants against Cd stress, suggesting that NO is a potential gas component to improve the resistance of plants to Cd stress. However, it remains unclear whether NO is effective for the resistance of *C. auriculatum* against Cd stress, or the underlying mechanism by which NO regulates the Cd resistance network of *C. auriculatum*. Thus, to investigate the role of NO in regulating the resistance of *C. auriculatum* against Cd stress, we treated *C. auriculatum* seeds with sodium nitroprusside (SNP, exogenous NO donor) and cadmium chloride ($CdCl_2$, Cd stress) for 48 h, followed by the photography of the germination phenotype, as well as measured the germination rate, hypocotyl length, biomass, and seed viability. To explore whether exogenous NO affected the production of endogenous NO, we determined the endogenous NO level, the enzymatic activity of nitric oxide synthase-like (NOS-like) and nitrate reductase (NR), and the expression level of the NO biosynthesis gene. To explore the interaction between the NO and phytohormone of *C. auriculatum* seeds under Cd stress, the contents of abscisic acid (ABA) and gibberellin (GA) and the expression of phytohormone biosynthesis and signaling genes were measured. In addition, we determined the α-amylase activity, starch content, soluble sugar content, and the level of superoxide anion (O^{2-}) and hydrogen peroxide (H_2O_2). We also visualized the accumulation of ROS through histochemical staining. To summarize, the current study has substantiated the impact of NO on the seed germination of *C. auriculatum* under Cd stress. Furthermore, these results provide preliminary findings regarding the most effective concentration of SNP that can be employed to enhance the seed biology of *C. auriculatum*.

2. Materials and Methods

2.1. Plant Materials and Treatments

The seeds of *C. auriculatum* variety 'Binwu No. 1' were collected from the *C. auriculatum* Planting Base located at Yancheng Teachers University (120°12′20″ E, 34°11′48″ N), Yancheng, China. Full, pest-free seeds were screened out and washed with ethanol (75%), sodium hypochlorite (10%), mercuric chloride (0.1%), and sterile water. Then, the *C. auriculatum* seeds were immersed in distilled water for a duration of 48 h at 30 °C. After that, the seed germination of *C. auriculatum* was carried out with between two layers of filter paper lined in sterile petri dishes.

To examine the influence of Cd stress on the seed germination of *C. auriculatum*, $CdCl_2$ (Catalog No. 202908, purchased from Shanghai Sigma-Aldrich Co., Ltd., Shanghai, China) was added into the filter paper [29]. Considering the Cd exposure level in the soil, air, and water [30], the $CdCl_2$ solution was used at the final concentration of 50 μM. Meanwhile, filter papers and the $CdCl_2$ solution were regularly replaced every 48 h in the incubation medium to ensure a constant water potential [31]. To investigate the role of SNP in regulating the resistance of *C. auriculatum* against Cd stress, SNP solutions (Catalog No. PHR1423, purchased from Shanghai Sigma-Aldrich Co., Ltd., Shanghai, China) at various concentrations (0.2, 0.4, and 0.6 mM) were added into the filter paper, while ddH_2O-treated seeds were set as the control group. According to previous research [32], the 0.1 mM SNP solution released an average of 0.65 $\mu M\ min^{-1}$ of NO within 3 h. Each petri dish contained a total of 50 seeds and was placed in the thermostatic incubator at (25 ± 1) °C for 6 d, and the determination and photographic documentation of the germination rate occurred at 2-day intervals. The indication of germination was observed by subjecting the lower radicle to a measurement of 2 mm. The germination index of *C. auriculatum* seeds under different treatments was measured using the method [33].

2.2. Endogenous NO Content and the Activity of NO Biosynthesis Enzyme

To investigate the role of endogenous NO in regulating the resistance of *C. auriculatum* against Cd stress, the compound 4-(4-carboxyphenyl)-4,4,5,5-tetramethylimidazoline-1-oxyl-3-oxide potassium salt (cPTIO, Catalog No. C221, purchased from Shanghai Sigma-Aldrich Co., Ltd., Shanghai, China) was used as a specific NO scavenger [34]. Given that the clearance of one molecule of NO is facilitated by one molecule of cPTIO, and cPTIO is concentration-dependent and time-sensitive [35], we applied a final concentration of 10 mM for the cPTIO solution. Following germination, the seeds of *C. auriculatum* underwent grinding in liquid nitrogen and subsequent homogenization. The resultant homogenized powder, weighing 200 mg, was utilized for the subsequent analysis. Additionally, in this study, each treatment included 3 biological replicates consisting of 20 seeds. Triphenyl tetrazolium chloride (TTC) was used to evaluate seed viability in petri dishes in addition to the germination experiment [36].

The endogenous NO content of *C. auriculatum* seeds was determined using the Total NO Assay kit (Catalog No. A012–1). In brief, the procedure involved combining the mixture with coenzyme, reaction enhancer, and nitrate reductase, and incubating it at 37 °C for 30 min. Following that, lactate dehydrogenase (LDH, 10 μL) and 10 μL of LDH buffer were added and incubated at 37 °C for an additional cycle. Griess reagent I and Griess reagent II were mixed in, and the OD_{540} value of the mixture was determined after incubating at 30 °C for 10 min. Furthermore, sodium nitrite with various concentrations (0.0031, 0.0063, 0.0125, 0.025, 0.05, 0.1, and 1 μM) was used to construct the standard curve. Additionally, the assessment of NOS-like and NR enzymatic activities was conducted using the Plant NOS ELISA (Catalog No. A014–2–2) and Plant NR ELISA (Catalog No. A096–1–1) kits, respectively. The kits were all procured from Nanjing Jiancheng Biological Engineering Co., Ltd. (Nanjing, China).

2.3. Endogenous ABA and GA Content

The preparation method for the determination of the endogenous phytohormone was referred to in the previous study [37]. The *C. auriculatum* seeds (1.0 g) were mixed with 2.0 mL of the extraction solution containing 100% methanol and ground using liquid nitrogen. The resulting mixture was left to stand at 4 °C for a duration of 12 h. Afterward, it was centrifugated at 4000 rpm for 15 min. The supernatant was then collected, and the aforementioned steps were repeated once. The resulting supernatant was combined and subjected to solid phase extraction using C18 and activated carbon. The supernatant was eluted using 100% methanol and purified using a 0.45 μm microporous filter to enable the determination of the phytohormone. Then, the contents of endogenous ABA and GA in *C. auriculatum* seeds were detected using the Plant ABA ELISA kit (Catalog No. 077235) and Plant GA ELISA kit (Catalog No. 072782), respectively. The above kits were all purchased from Shanghai Mlbio Biotech Co., Ltd. (Shanghai, China).

2.4. The Determination of Starch and Soluble Sugar and the Activity of α-amylase

Briefly, the initial step involved drying the seeds in an oven, followed by the grinding process with a grinder, and ultimately screening the resulting powder for the extraction. Sterile water (400 μL) was added to the powder, which was then gelatinized in a boiling water bath for 15 min. After cooling, a mixture of 400 μL perchloric acid solution (9.2 mM) and 600 μL sterile water was added to the powder. The mixture was thoroughly mixed and centrifuged at 4000 rpm for 10 min, and the extract was used to determine the starch content [38]. Meanwhile, the soluble sugar of *C. auriculatum* seeds was extracted with 80% ethanol [39]. The starch content and soluble sugar level of *C. auriculatum* seeds under different treatments were measured according to the phenol sulfuric acid method [40].

The α-amylase activity of *C. auriculatum* seeds was measured using the dinitrosalicylic (DNS) acid method [41] with minor modifications. *C. auriculatum* seeds (1.0 g) were meticulously ground and homogenized using liquid nitrogen. The mixture was then subjected to centrifugation at 3000 rpm for 10 min. Following this, the supernatant was heated to 70 °C for 15 min. The amylase liquid was subsequently cooled, and then combined with 2 mL of DNS and 1 mL of starch solution (10%). The resulting mixture was agitated thoroughly and subsequently boiled for 10 min. After being cooled in an ice bath, the colorimetry of the mixture at 540 nm was measured.

2.5. Histochemical Staining and Quantification of H_2O_2 and O^{2-}

To detect the distribution of H_2O_2 in *C. auriculatum* seeds under different treatments, we used the diaminobenzidine (DAB) histochemical staining method as previously described [42]. Following the removal of the seed coat, the seeds were longitudinally incised into two segments with the sterilized scalpel. Then, these segments were submerged in the DAB solution (1.0%) at 30 °C for 1 h. Afterwards, the resultant stained phenotype of the seeds was recorded with a camera (EOS 200D, Canon). At the same time, the H_2O_2 content of *C. auriculatum* seeds was quantified using the Plant H_2O_2 ELISA kit (Catalog No. A064-1-1, purchased from Nanjing Jiancheng Biological Engineering Co., Ltd., Nanjing, China).

The O^{2-} staining of *C. auriculatum* seeds was performed according to the method [43]. Following the removal of the seed coat using tweezers, the seeds were longitudinally divided into two sections and then submerged in the nitroblue tetrazolium (NBT) solution (0.02%) at 30 °C in darkness for 10 min. The stained seeds were then treated with 95% alcohol for several decolorization steps and captured using a camera. Meanwhile, the O^{2-} generation rate of *C. auriculatum* seeds was measured using the previous method [44].

2.6. The Antioxidant Enzyme Activity of C. auriculatum Seeds

In brief, *C. auriculatum* seeds (1.0 g) were carefully sterilized and washed and meticulously transferred into a pre-cooled mortar. Subsequently, 50 mM of phosphate buffer (2.0 mL, pH = 7.8) was added to the homogenate at 4 °C. The mixture was then centrifuged at 12,000 rpm for 20 min, and the resulting supernatant was carefully preserved for further analysis of antioxidant enzyme activity. The enzymatic activities of superoxide dismutase (SOD), peroxidase (POD), and catalase (CAT) in *C. auriculatum* seeds were determined using the SOD assay kit (Catalog No. A001-1-1), POD assay kit (Catalog No. A084-3-1), and CAT assay kit (Catalog No. A007-1-1), respectively. Furthermore, these kits were procured from Nanjing Jiancheng Biological Engineering Co., Ltd., Nanjing, China.

2.7. Quantitative Real-Time PCR (qRT-PCR) Analysis

Based on our previous report, *CaActin7* was used as the reference gene for quantifying the expression profiles of target genes [37]. In this study, NO synthesis genes (*CaNOS-like*, *CaNR1*) [45], ABA synthesis (xanthoxin dehydrogenase, *CaABA2*) [46] and signaling gene (SNF1-related protein kinase 2, *CaSnRK2*) [47], and GA synthesis (G protein alpha subunit 1, *CaGα1*) and signaling gene (Ga20 oxidase, *CaGA20ox1*) [48], whose expression profiles were determined using qRT-PCR (primer information in Supplementary Table S1) were used. The titanium One-Step RT-PCR kit (Catalog No. 639503, purchased from Beijing Takara Co., Ltd., Beijing China) was used for qRT-PCR amplification. Additionally, the relative expression levels of the target genes were calculated using the $2^{-\Delta\Delta Ct}$ method [49] with *CaActin7* as the reference gene.

2.8. Statistical Analysis and Data Visualization

In this study, we performed three biological and three technical repeats for each treatment to eliminate the accumulative error. To identify the statistical significance between the ddH_2O, Cd, Cd + SNP, and Cd + cPTIO treatment groups, a statistical analysis was performed with a t-test and one-way ANOVA, while the data were visualized using the Graphpad Prism 8.0 software. In addition, the p values of ≤ 0.05 and ≤ 0.01 were considered to be significant (*) and very significant (**), respectively.

3. Results

3.1. SNP Treatment Promoted the Germination of C. auriculatum Seeds under Cd Stress

To investigate the effects of Cd stress on the seed germination of *C. auriculatum*, we treated sterilized *C. auriculatum* seeds with Cd stress in the $CdCl_2$ (50 μM)-containing plates. As shown in Figure 1, compared to the control group (ddH_2O treatment), Cd stress severely reduced the germination of *C. auriculatum* seeds during the 6-day culture cycle. However, under Cd stress, SNP treatments significantly promoted the germination of *C. auriculatum* seeds, of which 0.4 mM SNP-treated *C. auriculatum* seeds exhibited the best results.

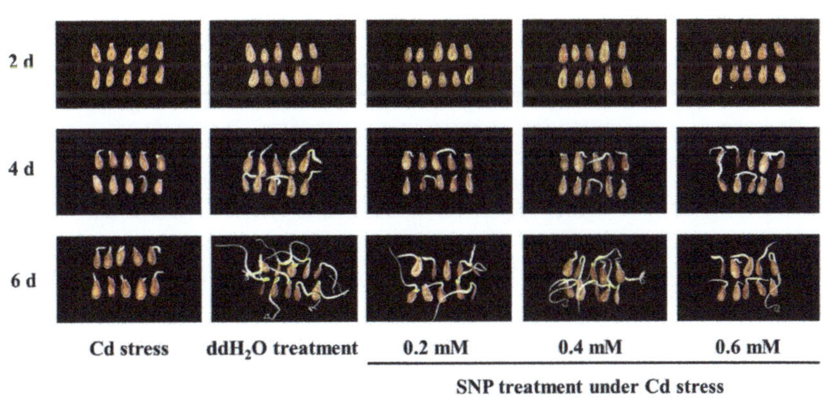

Figure 1. The effects of Cd stress and multiple concentrations of SNP treatment on the germination phenotype of *C. auriculatum* seeds.

At the same time, we measured the effects of Cd stress and SNP treatments on the germination rate, germination index, hypocotyl length, and biomass of *C. auriculatum* seeds. The data showed that during the 6-day culture cycle of the control group, the seed germination was initiated on the 4th day of sowing, with a germination rate of 92.67% and a germination index of 45.34%, respectively. In contrast, Cd stress significantly reduced the germination rate and germination index throughout the entire culture cycle. Meanwhile, under Cd stress, the SNP treatment groups showed a higher germination rate and germination index. Among these groups, the 0.4 mM SNP-treated *C. auriculatum* group showed the highest germination rate (80.67%) and germination index (41.74%) compared to the 0.2 mM and 0.6 SNP treatment groups. Additionally, the hypocotyl length of different treatment groups showed a similar trend (Table 1). These results indicated that the SNP treatment effectively alleviated the inhibitory effect of Cd stress on the germination of *C. auriculatum* seeds, with 0.4 mM being the most suitable concentration compared to other concentrations of SNP solutions.

Table 1. The effect of Cd stress and multiple concentrations of SNP treatment on germination rate, germination index, and hypocotyl length of *C. auriculatum* seeds.

Day	ddH_2O Treatment	Cd Stress	SNP Treatment under Cd Stress		
			Cd + 0.2 mM SNP	Cd + 0.4 mM SNP	Cd + 0.6 mM SNP
			Germination rate		
2	0	0	0	0	0
4	82.67 ± 1.62	6.67 ± 0.13 **	9.33 ± 0.45 **	12.67 ± 0.34 **	8.56 ± 0.29 **
6	92.67 ± 5.21	14.11 ± 0.31 **	75.33 ± 4.62 **	80.67 ± 4.96 **	71.25 ± 3.75 **
			Germination index		
2	0	0	0	0	0
4	41.34 ± 1.13	3.18 ± 0.06 **	4.42 ± 0.07 **	6.40 ± 0.04 **	4.17 ± 0.07 **
6	46.36 ± 1.21	6.67 ± 0.25 **	35.70 ± 1.06 **	40.74 ± 1.65 **	34.28 ± 1.25 **
			Hypocotyl length		
2	0	0	0	0	0
4	3.44 ± 0.05	1.10 ±0.01 **	2.51 ± 0.02 **	2.90 ± 0.02 **	2.32 ± 0.01 **
6	7.74 ± 0.12	1.81 ± 0.02 **	5.85 ± 0.11 **	6.33 ± 0.18 **	5.32 ± 0.15 **

The values were expressed as the means ± standard error of triplicate assays, and the significance level for differences between groups was denoted by asterisks, with ** indicating a very significant level of $p < 0.01$.

3.2. The Role of NO on the Germination of C. auriculatum Seeds under Cd Stress

To verify the impact of the 0.4 mM SNP treatment on the germination of *C. auriculatum* seeds under Cd stress, we determined the seed viability of *C. auriculatum* under ddH_2O treatment, Cd treatment, Cd + 0.4 mM SNP treatment, and Cd + 10 mM cPTIO treatment using the TTC staining method. The results showed that the highest staining degree was observed in ddH_2O-treated seeds, followed by Cd + 0.4 mM SNP-treated seeds, Cd-treated seeds, and Cd + 10 mM cPTIO-treated seeds (Figure 2A). This indicated that the presence of 0.4 mM SNP not only enhanced the seed viability of *C. auriculatum* under Cd stress, but also highlighted the significant role of endogenous NO in the seed germination process of *C. auriculatum*.

To investigate the role of endogenous NO in the seed germination process of *C. auriculatum* under Cd stress, we determined the content of endogenous NO at four different time points. Throughout the 6-day culture cycle, the endogenous NO content of the control group increased from 0.22 nM kg^{-1} (0 d) to 1.52 nM kg^{-1} (4 d) and then decreased slowly, indicating that NO was positively involved in the germination process of the *C. auriculatum* seed. However, the NO content of *C. auriculatum* seeds under Cd stress was significantly higher than that of the control group at 4 d (increased by 15.13%) and 6 d (increased by 34.03%). At the same time, the highest NO content was noted in the Cd + 0.4 mM SNP treatment group, while the Cd + 10 mM cPTIO treatment group had the lowest level of NO content among all the groups (Figure 2B).

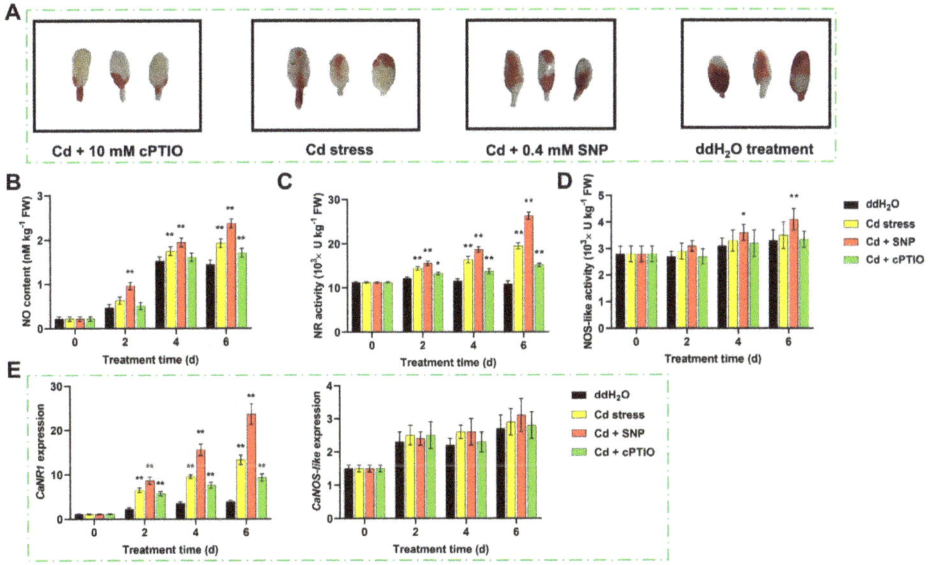

Figure 2. The effects of 0.4 mM SNP treatment on the biosynthesis of endogenous NO in *C. auriculatum* seeds under Cd stress. (**A**) Seed viability of *C. auriculatum* under the treatments of Cd + 10 mM cPTIO, Cd stress, Cd + 0.4 mM SNP, and ddH_2O at 6 d using the TTC staining method. (**B**) Endogenous NO content, (**C**) NR activity, (**D**) NOS-like activity, and (**E**) the expression profiles of *CaNR1* and *CaNOS-like* of *C. auriculatum* seeds under different treatments during the culture cycle. The values were expressed as the means ± standard error of triplicate assays, and the significance level for differences between groups was denoted by asterisks, with * indicating a significance level of $p \leq 0.05$, and ** indicating a very significant level of $p \leq 0.01$.

We also measured the enzymatic activities of NOS-like and NR to evaluate the impact of SNP treatment on endogenous NO biosynthesis in *C. auriculatum* seeds. Throughout the culture cycle, there were no significant differences in NR activity observed in the control group at four time points. However, the NR activity in *C. auriculatum* seeds under Cd stress increased by 74.11% at 6 d compared to the control group. Furthermore, treatment with Cd + 0.4 mM SNP resulted in a significant 1.44-fold increase in NR activity, while the NR activity in the Cd + 10 mM cPTIO treatment group was significantly lower than that of the Cd-treated group and Cd + 0.4 mM SNP-treated group (Figure 2C). In contrast, the activity of NOS-like showed no significant difference between different groups during the culture cycle, except for the Cd + 0.4 mM SNP treatment group (Figure 2D). Meanwhile, the expression profiles of *CaNOS-like* and *CaNR1* were also evaluated at four time points and showed similar trends as the enzymatic activity (Figure 2E).

3.3. The Role of NO on the Accumulation of ABA and GA in C. auriculatum Seeds under Cd Stress

Since ABA and GA are involved in the germination process and abiotic stress resistance of plants, we determined the content of ABA and GA in *C. auriculatum* seeds under Cd stress. In the control group, the ABA content of *C. auriculatum* seeds gradually declined from 23.33 μg kg^{-1} (0 d) to 11.65 μg kg^{-1} (6 d). However, the ABA content of *C. auriculatum* seeds in other treatment groups were higher compared to the control group, with the Cd + 0.4 mM SNP treatment group exhibiting the highest ABA content (Figure 3A). At the same time, the GA content of *C. auriculatum* seeds under ddH_2O treatment showed a gradual increasing trend, rising from 0.47 μg kg^{-1} (0 d) to 2.03 μg kg^{-1} (6 d). Compared to the control group, the GA content in other treatment groups was lower, with the Cd + 10 mM cPTIO treatment group having the lowest GA content (Figure 3B).

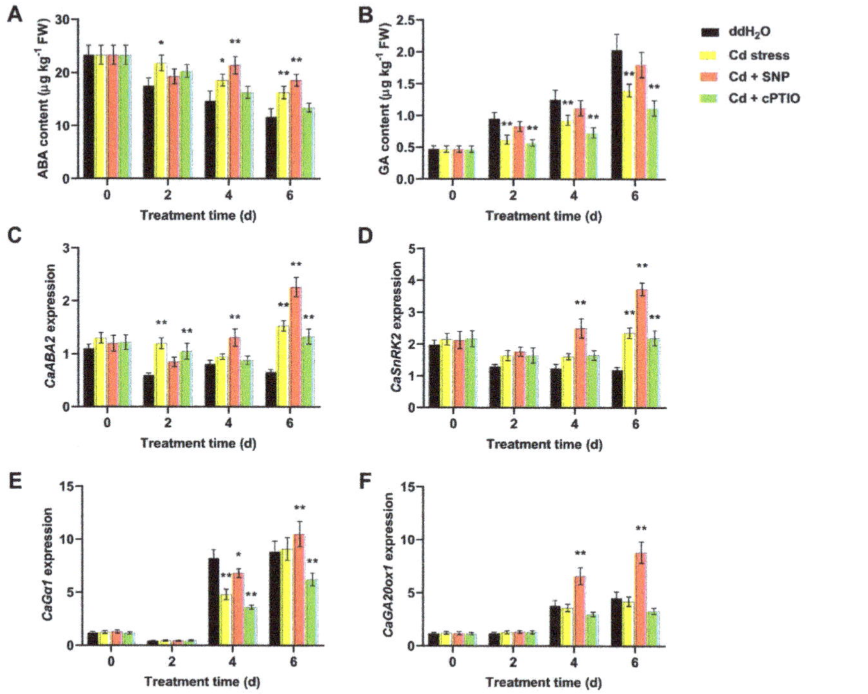

Figure 3. The effects of 0.4 mM SNP treatment on the biosynthesis and signaling of ABA and GA of *C. auriculatum* seeds under Cd stress. (**A**) ABA content, (**B**) GA content, as well as the expression levels of (**C**) *CaABA2*, (**D**) *CaSnRK2*, (**E**) *CaGα1*, and (**F**) *CaGA20ox1* of *C. auriculatum* seeds under different treatments during the culture cycle. The values were expressed as the means ± standard error of triplicate assays, and the significance level for differences between groups was denoted by asterisks, with * indicating a significance level of $p \leq 0.05$, and ** indicating a very significant level of $p \leq 0.01$.

To further investigate the critical role of NO in regulating the biosynthesis and signal transduction of phytohormones, we measured the relative expression levels of genes related to the biosynthesis and signaling of ABA and GA. Compared to the control group, the expression of *CaABA2* under Cd stress showed an overall upward trend, particularly in the Cd + 0.4 mM SNP treatment group (Figure 3C). A similar pattern was observed in the expression of *CaSnRK2* (Figure 3D). In the biosynthesis and signaling of GA, it was found that the expression of *CaGα1* (Figure 3E) and *CaGA20ox1* (Figure 3F) under different treatments increased at 4 d and peaked at 6 d, among which the Cd + 0.4 mM SNP-treated samples showed the highest expression profiles.

3.4. The Role of NO in Starch Degradation of C. auriculatum Seeds under Cd Stress

To investigate the role of NO in starch degradation during the culture cycle, we determined the levels of starch and soluble sugar in *C. auriculatum* seeds under different treatments. The results showed that in the control group, the decrease in starch content was accompanied by an increase in soluble sugar content. Under Cd stress, the hydrolysis of starch was slower in the Cd + 10 mM cPTIO treatment group, whereas the Cd + 0.4 mM SNP treatment group exhibited lower starch content (Figure 4A). Meanwhile, at 4 d and 6 d, it was found that the soluble sugar content of Cd + 0.4 mM SNP-treated samples was higher compared to the Cd treatment group, while the Cd + 10 mM cPTIO-treated samples had the lowest soluble sugar content, particularly at 6 d (39.54 g kg^{-1}) (Figure 4B).

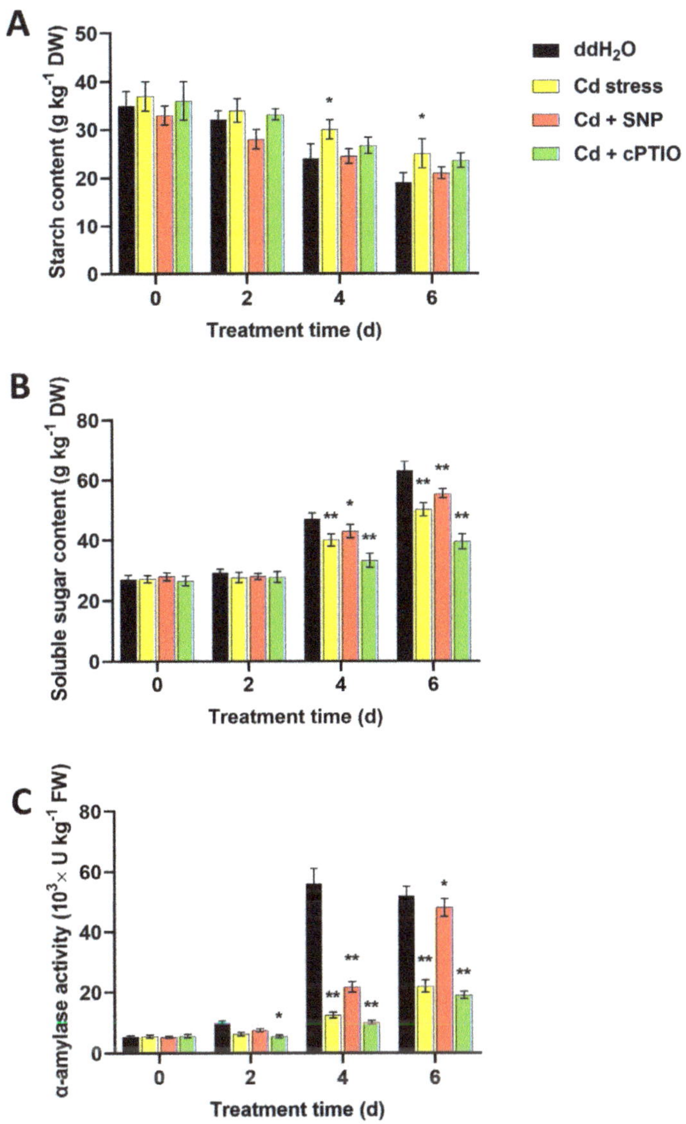

Figure 4. The effects of 0.4 mM SNP treatment on starch hydrolysis of *C. auriculatum* seeds under Cd stress. (**A**) Starch content, (**B**) soluble sugar content, and (**C**) α–amylase activity of *C. auriculatum* seeds under different treatments during the culture cycle. The values were expressed as the means ± standard error of triplicate assays, and the significance level for differences between groups was denoted by asterisks, with * indicating a significance level of $p \leq 0.05$, and ** indicating a very significant level of $p \leq 0.01$.

Meanwhile, to further analyze the impact of Cd stress on the starch hydrolysis of *C. auriculatum* seeds, we measured the enzymatic activity of α-amylase. In the control group, α-amylase activity peaked at 4 d (56.17×10^3 U kg^{-1}) and slightly decreased at 6 d. In other groups, the trend of changes in α-amylase activity was similar to the soluble sugar content. α-amylase activity was found to be the highest in the Cd + 0.4 mM SNP treatment group, whereas the lowest was assayed in the Cd + 10 mM cPTIO treatment group (Figure 4C).

3.5. The Role of NO on ROS Scavenging of *C. auriculatum* Seeds under Cd Stress

To explore the effects of NO on the antioxidant enzyme activity of *C. auriculatum* seeds under Cd stress, we measured the enzymatic activities of SOD, POD, and CAT throughout the culture cycle. As shown in Figure 5A, the SOD activity of the control group displayed a gradual increase, with the highest activity observed in the Cd + 0.4 mM SNP-treated samples at 6 d (57.17×10^3 U kg^{-1}), which was 1.76 times higher than the control group. Overall, the POD activities of all groups peaked at 4 d after sowing, followed by a slight decline at 6 d. The Cd + 0.4 mM SNP treatment group showed the highest POD activity during the 6-day culture cycle (Figure 5B). Additionally, the CAT activities of all groups showed a notable upward trend, with changes in the CAT activity aligning with the variations of SOD and POD. The Cd + 0.4 mM SNP treatment group presented the highest CAT activity (Figure 5C).

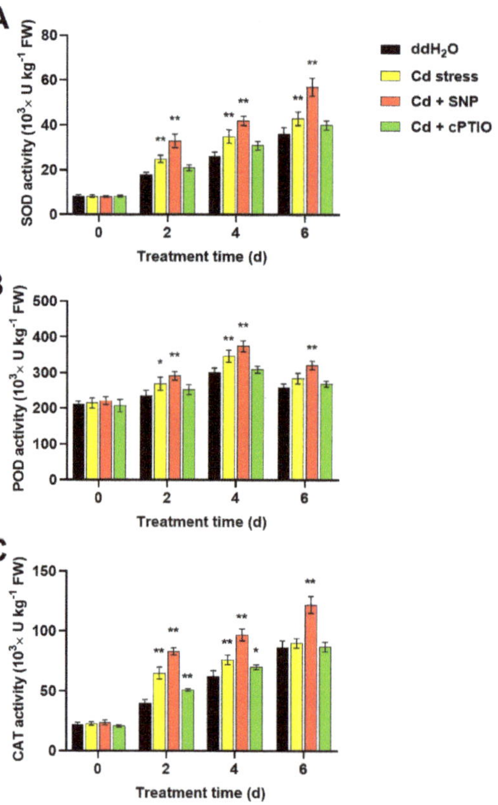

Figure 5. The effects of 0.4 mM SNP treatment on the activity of antioxidant enzyme of *C. auriculatum* seeds under Cd stress. (**A**) SOD activity, (**B**) POD activity, and (**C**) CAT activity of *C. auriculatum* seeds under different treatments during the culture cycle. The values were expressed as the means ± standard error of triplicate assays, and the significance level for differences between groups was denoted by asterisks, with * indicating a significance level of $p \leq 0.05$, and ** indicating a very significant level of $p \leq 0.01$.

To assess the impact of the increased activity of the antioxidant enzyme on the ROS generation of *C. auriculatum* seeds under Cd stress, we analyzed the distribution and the content of H_2O_2 and O^{2-} throughout the culture cycle. As shown in Figure 6A, after 6 d of cultivation, the Cd + 0.4 mM SNP group exhibited a lower degree of brown and blue staining compared to the control group, while a greater degree of staining was observed in the Cd + 10 mM cPTIO group. Moreover, we quantified the production of H_2O_2 and O^{2-} to evaluate the ROS-scavenging ability of NO. Compared to the Cd stress group, the Cd + 0.4 mM SNP treatment group exhibited a decrease in H_2O_2 content at 2 d (reduced by 25.05%), 4 d (reduced by 18.09%), and 6 d (reduced by 18.33%) (Figure 6B). Additionally, the O^{2-} generation rate of the Cd + 0.4 mM SNP treatment group was slower, with a reduction of 0.91-fold at 2 d, 0.73-fold at 4 d, and 0.67-fold at 6 d (Figure 6C).

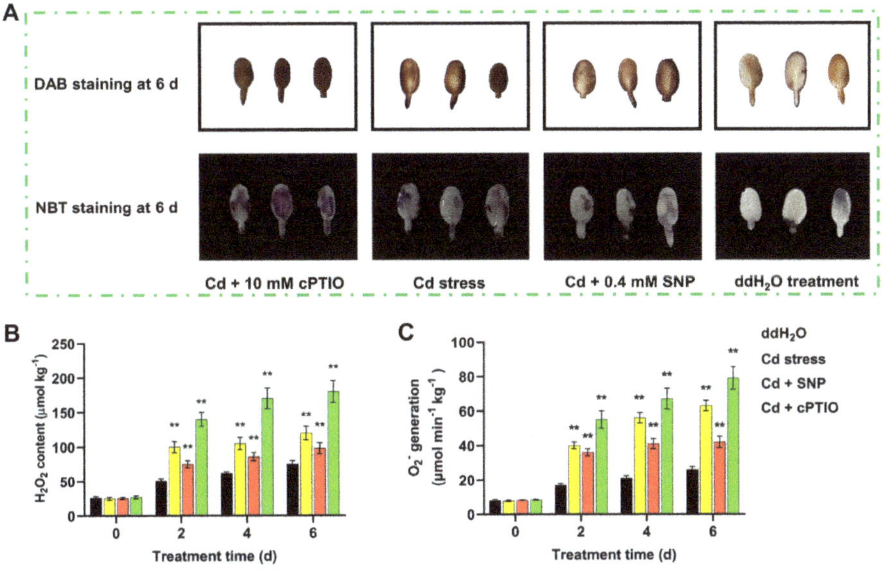

Figure 6. The effects of 0.4 mM SNP treatment on ROS scavenging of *C. auriculatum* seeds under Cd stress. (**A**) DAB and NBT staining seeds at 6 d, (**B**) H_2O_2 content, and (**C**) O^{2-} generation rate of *C. auriculatum* seeds under different treatments during the culture cycle. The values were expressed as the means ± standard error of triplicate assays, and the significance level for differences between groups was denoted by asterisks, with ** indicating a very significant level of $p \leq 0.01$.

4. Discussion

The germination of crops like *Solanum lycopersicum* [50], *Triticum aestivum* [51], and *Oryza sativa* [52] is severely hindered by Cd contamination in soil. However, there is little research about the impact of Cd stress on the germination of *C. auriculatum*. The present study found that Cd stress induced the accumulation of ABA and GA, activated the production of H_2O_2 and O^{2-}, inhibited the hydrolysis of starch, and decreased the seed viability, thereby inhibiting the germination of *C. auriculatum* seeds. Therefore, to stimulate seeds' germination and establishment of *C. auriculatum*, we treated *C. auriculatum* seeds with the nitric oxide donor SNP. SNP has been used for alleviating the inhibitory effect of Cd on the seed germination and seedling growth of *Oryza sativa* [53] and *Brassica oleracea* [54], but the suitable concentration of SNP treatment varies among different plant species and developmental stages [34,45]. *Nitraria tangutorum* treated with 70 μM SNP had the best growth rate, while 100 μM SNP significantly inhibited the growth rate [55]. *Brassica chinensis* seeds pre-soaked with 10 μM SNP showed a higher germination rate, germination index, and vitality index under NaCl stress, while 200 μM SNP had no significant effect on the germination [17]. In this study, compared to concentrations of 0.2 mM and 0.6 mM, *C. auriculatum* seeds treated with 0.4 mM SNP exhibited a higher germination rate, germination index, and longer hypocotyl length, suggesting that the concentration of 0.4 mM SNP is more effective in promoting seed germination of *C. auriculatum* under Cd stress. Furthermore, excessively high concentrations of SNP in the environment can lead to nitrate poisoning in vertebrates [56], suggesting that using low concentrations of SNP avoids negative effects on other organisms in the ecological environment.

In plants, the production of NO depends on NR and the mitochondrial electron transport chain, or through the arginine pathway, similar to the NOS activity present in animals [57]. In *Solanum tuberosum*, SNP and cPTIO affected the metabolism and signal transduction of ABA by up-regulating and down-regulating NR activity, and ultimately controlled the dormancy mechanism of a potato tuber [58]. In *Glycine max*, NO was the key signaling molecule in static magnetic field-stimulated tolerance towards UV-B stress, and NOS may possibly be accountable for the triggered NO production in seedlings [59]. Our study revealed that the change in NR activity and *CaNR1* expression corresponded to the variation in NO content, indicating that the promotion of endogenous NO synthesis by exogenous SNP was mainly dependent on the NR pathway, which was similar to findings in *Daucus carota* [45], *Fragaria annassa* [60], and *Cucumis sativus* [61]. In addition, NO is a crucial player in the leaf senescence of plants under abiotic stress [62], and it was observed that compared to young leaves, the NO level was higher in senescent leaves [63]. Given the association between nitrogen content and cardiovascular diseases [64], it also suggested that controlling for NO production helps to regulate early senescence in plants, while prolonged SNP treatment on *C. auriculatum* may even have a negative impact.

Moreover, further studies have shown that the intricate balance between ABA and GA plays a significant role in overcoming dormancy and facilitating the germination process [20]. During normal germination conditions, NO acts prior to ABA and GA to enhance the production of GA and suppress the production of ABA, thereby promoting optimal seed germination while preserving essential nutrients [65]. However, under heavy meal stress, the increased expression of *GA20ox1* and accelerated GA accumulation counteracted the inhibition of abiotic stress in seed germination [66], leading to stomatal closure and enhanced plant resistance to abiotic stress [67]. Our data also showed that under normal conditions, the biosynthesis and signaling of the ABA pathway were repressed during the germination process. However, the increase in NO induced by SNP plays a protective role in ABA accumulation under Cd stress, while the removal of NO induced by cPTIO inhibits ABA accumulation. Furthermore, under Cd stress, the treatment of SNP significantly enhances the biosynthesis and signaling of GA in *C. auriculatum* seeds, which has also been confirmed in *Oryza sativa* [68,69] and *Cucumis sativus* [70].

Additionally, during the seed germination process, heavy metal stress is accompanied by ROS generation, which decreases the activity of antioxidant enzymes and increases oxidative damage [71]. While a moderate level of ROS is required for the seed germination process of plants, excess ROS directly harms the mitochondrial membrane structure and has detrimental effects [72]. During the germination process of *Oryza sativa* seeds under Cd stress, the application of SNP enhanced the activities of SOD, POD, CAT, and APX, while reducing the accumulation of H_2O_2 and MDA [53]. In this study, the results of the histochemical staining and biochemical indicators demonstrated that under Cd stress, the increase in the endogenous NO level by SNP treatment enhanced the activity of the antioxidant enzyme and inhibited the excessive generation of H_2O_2 and O^{2-}. On the other hand, the decrease in endogenous NO content through cPTIO treatment had the opposite effect, indicating the significant role of NO in protecting the ROS scavenging system during the germination of *C. auriculatum* under Cd stress. Besides the ROS scavenging system, Cd stress also impeded the starch degradation by reducing the α-amylase activity of seeds, which led to the immobilization of starch and lack of nutrients [71]. Recent reports have shown that NO accelerated the starch hydrolysis to form soluble sugar by increasing the activity of α-amylase and β-amylase [73]. In *Vigna radiata*, drought stress stimulated the production of endogenous NO, while SNP treatment significantly up-regulated α-amylase activity and accelerated the formation of reducing sugar and soluble sugar, which alleviated the inhibition of seed germination [74]. Our findings showed that Cd stress inhibited the hydrolysis of starch and the production of soluble sugar, and cPTIO treatment further exacerbated this problem by weakening the α-amylase activity of *C. auriculatum* seeds. However, the utilization of SNP to enhance the endogenous NO level enhanced the α-amylase activity, facilitated starch hydrolysis and soluble sugar accumulation, consequently promoting seed germination under Cd stress, and similar results were also obtained in *Triticum aestivum* [75] and *Glycine max* [76]. Furthermore, it has been reported that SNP treatment on seeds significantly reduced the accumulation of Cd and stress markers in seedlings during the development process [53,77]. Therefore, the resistance of SNP-treated *C. auriculatum* plants against Cd stress during seedling development will be one of our future research directions. Furthermore, NO metabolism related genes and transcription factors can be further functionally characterized using genome sequencing [78], overexpression [79,80], and knockout [81] strategies.

5. Conclusions

Overall, the importance of NO in enhancing the resistance of *C. auriculatum* to Cd stress during germination was demonstrated. Among different concentrations, a 0.4 mM treatment of SNP was found to be particularly effective in increasing the germination rate and germination index and promoting hypocotyl growth. A further study showed that 0.4 mM SNP treatment promoted endogenous NO accumulation by enhancing the NR activity. This, in turn, affected the biosynthesis and signaling of ABA and GA, as well as the activities of SOD, POD, and CAT. Moreover, it inhibited the generation of H_2O_2 and O_2^- during germination. Together, these combined effects led to starch hydrolysis and the production of soluble sugar, resulting in the improved seed germination of *C. auriculatum* under Cd stress (Figure 7).

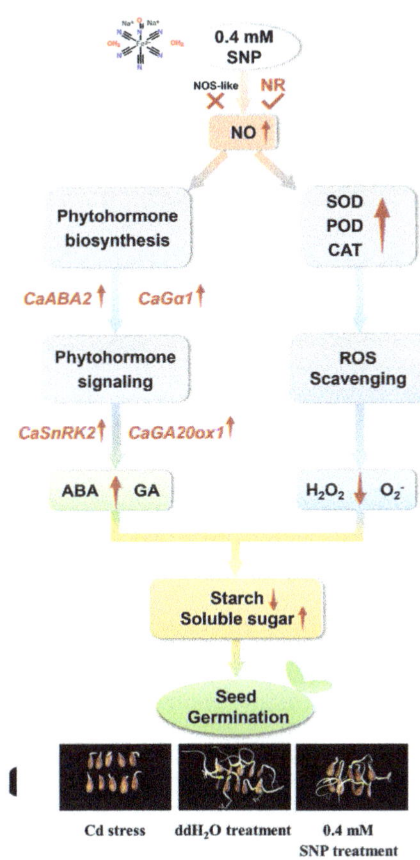

Figure 7. The potential mechanism of 0.4 mM SNP treatment enhancing the germination of *C. auriculatum* seeds under Cd stress.

Supplementary Materials: The following supporting information can be downloaded at: https://www.mdpi.com/article/10.3390/agronomy14010086/s1.

Author Contributions: F.-F.L. and X.-H.Q.: experimental design, writing—original draft, data analysis, investigation, methodology. T.Y., P.Z., Z.-P.Z., J.-H.Z. and J.-M.L.: investigation, data curation. A.-S.X. and M.S.: experimental design, writing—review and editing, methodology, funding acquisition. All authors have read and agreed to the published version of the manuscript.

Funding: This study was financed by the Natural Science Foundation for Higher Education Institutions in Jiangsu Province (21KJB210007), the Key Research and Development Program of Jiangsu (BE2022386), and the Priority Academic Program Development of Jiangsu Higher Education Institutions Project (PAPD).

Data Availability Statement: The data presented in this study are available on request from the corresponding author.

Conflicts of Interest: The authors declare that they have no known competing financial interests or personal relationships that could have appeared to influence the work reported in this paper.

References

1. Yang, S.; Yang, D.; Taylor, D.; He, M.; Liu, X.; Xu, J. Tracking cadmium pollution from source to receptor: A health-risk focused transfer continuum approach. *Sci. Total Environ.* **2023**, *867*, 161574. [CrossRef] [PubMed]
2. Clemens, S.; Aarts, M.G.; Thomine, S.; Verbruggen, N. Plant science: The key to preventing slow cadmium poisoning. *Trends Plant Sci.* **2013**, *18*, 92–99. [CrossRef]
3. Yan, J.; Wu, X.; Li, T.; Fan, W.; Abbas, M.; Qin, M.; Li, R.; Liu, Z.; Liu, P. Effect and mechanism of nano-materials on plant resistance to cadmium toxicity: A review. *Ecotoxicol. Environ. Saf.* **2023**, *266*, 115576. [CrossRef]
4. Li, Y.; Ding, L.; Zhou, M.; Chen, Z.; Ding, Y.; Zhu, C. Transcriptional Regulatory Network of Plant Cadmium Stress Response. *Int. J. Mol. Sci.* **2023**, *24*, 4378. [CrossRef] [PubMed]
5. Siadjeu, C.; Pucker, B. Medicinal plant genomics. *BMC Genom.* **2023**, *24*, 429. [CrossRef] [PubMed]
6. Al-Khayri, J.M.; Banadka, A.; Rashmi, R.; Nagella, P.; Alessa, F.M.; Almaghasla, M.I. Cadmium toxicity in medicinal plants: An overview of the tolerance strategies, biotechnological and omics approaches to alleviate metal stress. *Front. Plant Sci.* **2022**, *13*, 1047410. [CrossRef] [PubMed]
7. Jung, D.H.; Kim, H.Y.; Won, J.H.; Park, S.H. Development of a classification model for *Cynanchum wilfordii* and *Cynanchum auriculatum* using convolutional neural network and local interpretable model-agnostic explanation technology. *Front. Plant Sci.* **2023**, *14*, 1169709. [CrossRef]
8. Wang, L.; Cai, F.; Zhao, W.; Tian, J.; Kong, D.; Sun, X.; Liu, Q.; Chen, Y.; An, Y.; Wang, F.; et al. *Cynanchum auriculatum* Royle ex Wight., *Cynanchum bungei* Decne. and *Cynanchum wilfordii* (Maxim.) Hemsl.: Current Research and Prospects. *Molecules* **2021**, *26*, 7065. [CrossRef]
9. Wang, Y.; Han, J.; Yue, Y.; Wu, Y.; Zhang, W.; Xia, W.; Wu, M. Purification, structure identification and immune activity of a neutral polysaccharide from *Cynanchum auriculatum*. *Int. J. Biol. Macromol.* **2023**, *237*, 124142. [CrossRef]
10. Zhen, X.; Choi, H.S.; Kim, J.H.; Kim, S.L.; Liu, R.; Ko, Y.C.; Yun, B.S.; Lee, D.S. Caudatin Isolated from *Cynanchum auriculatum* Inhibits Breast Cancer Stem Cell Formation via a GR/YAP Signaling. *Biomolecules* **2020**, *10*, 925. [CrossRef]
11. Zhang, X.; Gao, M.; Rao, Z.; Lei, Z.; Zeng, J.; Huang, Z.; Shen, C.; Zeng, N. The antitumour activity of C(21) steroidal glycosides and their derivatives of Baishouwu: A review. *J. Ethnopharmacol.* **2022**, *293*, 115300. [CrossRef] [PubMed]
12. Gupta, K.J.; Kaladhar, V.C.; Fitzpatrick, T.B.; Fernie, A.R.; Møller, I.M.; Loake, G.J. Nitric oxide regulation of plant metabolism. *Mol. Plant* **2022**, *15*, 228–242. [CrossRef] [PubMed]
13. Van Meeteren, U.; Kaiser, E.; Malcolm Matamoros, P.; Verdonk, J.C.; Aliniaeifard, S. Is nitric oxide a critical key factor in ABA-induced stomatal closure? *J. Exp. Bot.* **2020**, *71*, 399–410. [CrossRef] [PubMed]
14. Poderoso, J.J.; Helfenberger, K.; Poderoso, C. The effect of nitric oxide on mitochondrial respiration. *Nitric Oxide* **2019**, *88*, 61–72. [CrossRef] [PubMed]
15. Xie, Z.; Yang, C.; Li, M.; Zhang, Z.; Wu, Y.; Gu, L.; Peng, X. Nitric Oxide Crosstalk With Phytohormone Is Involved in Enhancing Photosynthesis of *Tetrastigma hemsleyanum* for Photovoltaic Adaptation. *Front. Plant Sci.* **2022**, *13*, 852956. [CrossRef] [PubMed]
16. Kumari, A.; Singh, P.; Kaladhar, V.C.; Manbir; Paul, D.; Pathak, P.K.; Gupta, K.J. Phytoglobin-NO cycle and AOX pathway play a role in anaerobic germination and growth of deepwater rice. *Plant Cell Environ.* **2022**, *45*, 178–190. [CrossRef]
17. Ren, Y.; Wang, W.; He, J.; Zhang, L.; Wei, Y.; Yang, M. Nitric oxide alleviates salt stress in seed germination and early seedling growth of pakchoi (*Brassica chinensis* L.) by enhancing physiological and biochemical parameters. *Ecotoxicol. Environ. Saf.* **2020**, *187*, 109785. [CrossRef]
18. Pandey, S.; Kumari, A.; Shree, M.; Kumar, V.; Singh, P.; Bharadwaj, C.; Loake, G.J.; Parida, S.K.; Masakapalli, S.K.; Gupta, K.J. Nitric oxide accelerates germination via the regulation of respiration in chickpea. *J. Exp. Bot.* **2019**, *70*, 4539–4555. [CrossRef]
19. Beligni, M.V.; Lamattina, L. Nitric oxide stimulates seed germination and de-etiolation, and inhibits hypocotyl elongation, three light-inducible responses in plants. *Planta* **2000**, *210*, 215–221. [CrossRef]

20. Zhang, Y.; Wang, R.; Wang, X.; Zhao, C.; Shen, H.; Yang, L. Nitric Oxide Regulates Seed Germination by Integrating Multiple Signalling Pathways. *Int. J. Mol. Sci.* **2023**, *24*, 9052. [CrossRef]
21. Bailly, C.; Merendino, L. Oxidative signalling in seed germination and early seedling growth: An emerging role for ROS trafficking and inter-organelle communication. *Biochem. J.* **2021**, *478*, 1977–1984. [CrossRef] [PubMed]
22. Feng, J.; Chen, L.; Zuo, J. Protein S-Nitrosylation in plants: Current progresses and challenges. *J. Integr. Plant Biol.* **2019**, *61*, 1206–1223. [CrossRef]
23. Domingos, P.; Prado, A.M.; Wong, A.; Gehring, C.; Feijo, J.A. Nitric oxide: A multitasked signaling gas in plants. *Mol. Plant* **2015**, *8*, 506–520. [CrossRef] [PubMed]
24. Kolupaev, Y.E.; Yemets, A.I.; Yastreb, T.O.; Blume, Y.B. The role of nitric oxide and hydrogen sulfide in regulation of redox homeostasis at extreme temperatures in plants. *Front. Plant Sci.* **2023**, *14*, 1128439. [CrossRef] [PubMed]
25. Anantharaman, S.; Guercio, D.; Mendoza, A.G.; Withorn, J.M.; Boon, E.M. Negative regulation of biofilm formation by nitric oxide sensing proteins. *Biochem. Soc. Trans.* **2023**, *51*, 1447–1458. [CrossRef]
26. Besson-Bard, A.; Gravot, A.; Richaud, P.; Auroy, P.; Duc, C.; Gaymard, F.; Taconnat, L.; Renou, J.P.; Pugin, A.; Wendehenne, D. Nitric oxide contributes to cadmium toxicity in Arabidopsis by promoting cadmium accumulation in roots and by up-regulating genes related to iron uptake. *Plant Physiol.* **2009**, *149*, 1302–1315. [CrossRef]
27. Valentovicová, K.; Halusková, L.; Huttová, J.; Mistrík, I.; Tamás, L. Effect of cadmium on diaphorase activity and nitric oxide production in barley root tips. *J. Plant Physiol.* **2010**, *167*, 10–14. [CrossRef]
28. Kaya, C.; Ashraf, M.; Alyemeni, M.N.; Ahmad, P. The role of nitrate reductase in brassinosteroid-induced endogenous nitric oxide generation to improve cadmium stress tolerance of pepper plants by upregulating the ascorbate-glutathione cycle. *Ecotoxicol. Environ. Saf.* **2020**, *196*, 110483. [CrossRef]
29. Zhang, Z.X.; Wang, S.L.; Wang, J.; Zhang, C.; Liu, D.; Wang, C.; Xu, F.S. The overexpression of LOW PHOSPHATE ROOT 1 (LPR1) negatively regulates Arabidopsis growth in response to Cadmium (Cd) stress. *Plant Physiol. Biochem.* **2023**, *196*, 556–566. [CrossRef]
30. Li, G.Y.; Yan, L.J.; Chen, X.M.; Lam, S.S.; Rinklebe, J.; Yu, Q.; Yang, Y.F.; Peng, W.X.; Sonne, C. Phytoremediation of cadmium from soil, air and water. *Chemosphere* **2023**, *320*, 138058. [CrossRef]
31. Pegler, J.L.; Oultram, J.M.J.; Nguyen, D.Q.; Grof, C.P.L.; Eamens, A.L. MicroRNA-Mediated Responses to Cadmium Stress in Arabidopsis thaliana. *Plants* **2021**, *10*, 130. [CrossRef] [PubMed]
32. Ederli, L.; Reale, L.; Madeo, L.; Ferranti, F.; Gehring, C.; Fornaciari, M.; Romano, B.; Pasqualini, S. NO release by nitric oxide donors in vitro and in planta. *Plant Physiol. Biochem.* **2009**, *47*, 42–48. [CrossRef] [PubMed]
33. Zhang, Z.; Fan, J.; Wu, J.; Zhang, L.; Wang, J.; Zhang, B.; Wang-Pruski, G. Alleviating effect of silicon on melon seed germination under autotoxicity stress. *Ecotoxicol. Environ. Saf.* **2020**, *188*, 109901. [CrossRef] [PubMed]
34. Sun, M.; Yang, X.L.; Zhu, Z.P.; Xu, Q.Y.; Wu, K.X.; Kang, Y.J.; Wang, H.; Xiong, A.S. Comparative transcriptome analysis provides insight into nitric oxide suppressing lignin accumulation of postharvest okra (*Abelmoschus esculentus* L.) during cold storage. *Plant Physiol. Biochem.* **2021**, *167*, 49–67. [CrossRef]
35. D'Alessandro, S.; Posocco, B.; Costa, A.; Zahariou, G.; Schiavo, F.L.; Carbonera, D.; Zottini, M. Limits in the use of cPTIO as nitric oxide scavenger and EPR probe in plant cells and seedlings. *Front. Plant Sci.* **2013**, *4*, 340. [CrossRef]
36. Qiao, J.; Liao, Y.; Yin, C.; Yang, X.; Tú, H.M.; Wang, W.; Liu, Y. Vigour testing for the rice seed with computer vision-based techniques. *Front. Plant Sci.* **2023**, *14*, 1194701. [CrossRef]
37. Sun, M.; Zhu, Z.P.; Yu, J.X.; Wu, K.X.; Guo, Y.X.; Shen, M.; Liu, F.F.; Tang, X.H.; Kang, Y.J. Transcriptomic and physiological analysis reveal phytohormone and phenylpropanoid biosynthesis in root of *Cynanchum auriculatum*. *Plant Growth Regul.* **2023**, *101*, 67–85. [CrossRef]
38. Maniglia, B.C.; Silveira, T.M.G.; Tapia-Blácido, D.R. Starch isolation from turmeric dye extraction residue and its application in active film production. *Int. J. Biol. Macromol.* **2022**, *202*, 508–519. [CrossRef]
39. Zhu, Z.P.; Yu, J.X.; Qiao, X.H.; Yu, Z.F.; Xiong, A.S.; Sun, M. Hydrogen sulfide delays yellowing and softening, inhibits nutrient loss in postharvest celery. *Sci. Hortic.* **2023**, *315*, 111991. [CrossRef]
40. Paraginski, R.T.; Colussi, R.; Dias, A.R.G.; da Rosa Zavareze, E.; Elias, M.C.; Vanier, N.L. Physicochemical, pasting, crystallinity, and morphological properties of starches isolated from maize kernels exhibiting different types of defects. *Food Chem.* **2019**, *274*, 330–336. [CrossRef]
41. Visvanathan, R.; Qader, M.; Jayathilake, C.; Jayawardana, B.C.; Liyanage, R.; Sivakanesan, R. Critical review on conventional spectroscopic α-amylase activity detection methods: Merits, demerits, and future prospects. *J. Sci. Food Agric.* **2020**, *100*, 2836–2847. [CrossRef] [PubMed]
42. Zhu, Z.P.; Yu, J.X.; Liu, F.F.; Zhu, D.W.; Xiong, A.S.; Sun, M. AeWRKY32 from okra regulates anthocyanin accumulation and cold tolerance in Arabidopsis. *J. Plant Physiol.* **2023**, *287*, 154062. [CrossRef]
43. Jambunathan, N. Determination and detection of reactive oxygen species (ROS), lipid peroxidation, and electrolyte leakage in plants. *Methods Mol. Biol.* **2010**, *639*, 292–298. [CrossRef] [PubMed]
44. Poór, P.; Kovács, J.; Borbély, P.; Takács, Z.; Szepesi, Á.; Tari, I. Salt stress-induced production of reactive oxygen- and nitrogen species and cell death in the ethylene receptor mutant Never ripe and wild type tomato roots. *Plant Physiol. Biochem.* **2015**, *97*, 313–322. [CrossRef] [PubMed]

45. Sun, M.; Yang, T.; Qiao, X.H.; Zhao, P.; Zhu, Z.P.; Wang, G.L.; Xu, L.L.; Xiong, A.S. Nitric oxide regulates the lignification and carotenoid biosynthesis of postharvest carrot (*Daucus carota* L.). *Postharvest Biol. Technol.* **2024**, *207*, 112593. [CrossRef]
46. Sun, M.; Yang, T.; Qiao, X.H.; Zhao, P.; Zhu, Z.P.; Su, K.Y.; Sun, C.W.; Xie, Z.B.; Zhang, S.S.; Xu, M.; et al. Hydrogen sulfide delays postharvest ripening of white fleshed strawberry by regulating phytohormone accumulation and ROS generation. *Postharvest Biol. Technol.* **2023**, *205*, 112536. [CrossRef]
47. Belda-Palazón, B.; Adamo, M.; Valerio, C.; Ferreira, L.J.; Confraria, A.; Reis-Barata, D.; Rodrigues, A.; Meyer, C.; Rodriguez, P.L.; Baena-González, E. A dual function of SnRK2 kinases in the regulation of SnRK1 and plant growth. *Nat. Plants* **2020**, *6*, 1345–1353. [CrossRef]
48. Hwang, O.J.; Back, K. Exogenous Gibberellin Treatment Enhances Melatonin Synthesis for Melatonin-Enriched Rice Production. *Biomolecules* **2022**, *12*, 198. [CrossRef]
49. Schmittgen, T.D.; Livak, K.J. Analyzing real-time PCR data by the comparative CT method. *Nat. Protoc.* **2008**, *3*, 1101–1108. [CrossRef]
50. Lv, Y.; Zhao, Y.; He, Y.; Wang, J.; Zheng, Y.; Chen, X.; Huang, F.; Liu, J.; Yu, L. Synergistic effects of gamma-aminobutyric acid and melatonin on seed germination and cadmium tolerance in tomato. *Plant Signal Behav.* **2023**, *18*, 2216001. [CrossRef]
51. Zhang, L.; Gao, B. Effect of Isosteviol on Wheat Seed Germination and Seedling Growth under Cadmium Stress. *Plants* **2021**, *10*, 1779. [CrossRef] [PubMed]
52. Rizwan, M.; Ali, S.; Adrees, M.; Rizvi, H.; Zia-Ur-Rehman, M.; Hannan, F.; Qayyum, M.F.; Hafeez, F.; Ok, Y.S. Cadmium stress in rice: Toxic effects, tolerance mechanisms, and management: A critical review. *Environ. Sci. Pollut. Res. Int.* **2016**, *23*, 17859–17879. [CrossRef] [PubMed]
53. He, J.; Ren, Y.; Chen, X.; Chen, H. Protective roles of nitric oxide on seed germination and seedling growth of rice (*Oryza sativa* L.) under cadmium stress. *Ecotoxicol. Environ. Saf.* **2014**, *108*, 114–119. [CrossRef] [PubMed]
54. Ma, J.; Saleem, M.H.; Alsafran, M.; Jabri, H.A.; Mehwish; Rizwan, M.; Nawaz, M.; Ali, S.; Usman, K. Response of cauliflower (*Brassica oleracea* L.) to nitric oxide application under cadmium stress. *Ecotoxicol. Environ. Saf.* **2022**, *243*, 113969. [CrossRef]
55. Gao, Z.; Zhang, J.; Zhang, J.; Zhang, W.; Zheng, L.; Borjigin, T.; Wang, Y. Nitric oxide alleviates salt-induced stress damage by regulating the ascorbate-glutathione cycle and Na(+)/K(+) homeostasis in *Nitraria tangutorum* Bobr. *Plant Physiol. Biochem.* **2022**, *173*, 46–58. [CrossRef]
56. Chen, H.; Pang, Y.; Wei, Y.; He, X.; Zhang, Y.; Xie, L. Nitrate and sodium nitroprusside alter the development of Asian black-spined toads' embryos by inducing nitric oxide production. *Environ. Sci. Pollut. Res. Int.* **2023**, *30*, 23060–23069. [CrossRef]
57. Astier, J.; Gross, I.; Durner, J. Nitric oxide production in plants: An update. *J. Exp. Bot.* **2018**, *69*, 3401–3411. [CrossRef]
58. Wang, Z.; Ma, R.; Zhao, M.; Wang, F.; Zhang, N.; Si, H. NO and ABA Interaction Regulates Tuber Dormancy and Sprouting in Potato. *Front. Plant Sci.* **2020**, *11*, 311. [CrossRef]
59. Raipuria, R.K.; Kataria, S.; Watts, A.; Jain, M. Magneto-priming promotes nitric oxide via nitric oxide synthase to ameliorate the UV-B stress during germination of soybean seedlings. *J. Photochem. Photobiol. B* **2021**, *220*, 112211. [CrossRef]
60. Kaya, C.; Ashraf, M.; Alyemeni, M.N.; Ahmad, P. Nitrate reductase rather than nitric oxide synthase activity is involved in 24-epibrassinolide-induced nitric oxide synthesis to improve tolerance to iron deficiency in strawberry (*Fragaria* × *annassa*) by up-regulating the ascorbate-glutathione cycle. *Plant Physiol. Biochem.* **2020**, *151*, 486–499. [CrossRef]
61. Feng, Y.; Fu, X.; Han, L.; Xu, C.; Liu, C.; Bi, H.; Ai, X. Nitric Oxide Functions as a Downstream Signal for Melatonin-Induced Cold Tolerance in Cucumber Seedlings. *Front. Plant Sci.* **2021**, *12*, 686545. [CrossRef] [PubMed]
62. Procházková, D.; Wilhelmová, N. Nitric oxide, reactive nitrogen species and associated enzymes during plant senescence. *Nitric Oxide* **2011**, *24*, 61–65. [CrossRef] [PubMed]
63. Corpas, F.J.; Barroso, J.B.; Carreras, A.; Quirós, M.; León, A.M.; Romero-Puertas, M.C.; Esteban, F.J.; Valderrama, R.; Palma, J.M.; Sandalio, L.M.; et al. Cellular and subcellular localization of endogenous nitric oxide in young and senescent pea plants. *Plant Physiol.* **2004**, *136*, 2722–2733. [CrossRef] [PubMed]
64. Hong, C.; Zhu, H.; Zhou, X.; Zhai, X.; Li, S.; Ma, W.; Liu, K.; Shirai, K.; Sheerah, H.A.; Cao, J. Association of Blood Urea Nitrogen with Cardiovascular Diseases and All-Cause Mortality in USA Adults: Results from NHANES 1999–2006. *Nutrients* **2023**, *15*, 461. [CrossRef] [PubMed]
65. Bethke, P.C.; Libourel, I.G.; Aoyama, N.; Chung, Y.Y.; Still, D.W.; Jones, R.L. The Arabidopsis aleurone layer responds to nitric oxide, gibberellin, and abscisic acid and is sufficient and necessary for seed dormancy. *Plant Physiol.* **2007**, *143*, 1173–1188. [CrossRef] [PubMed]
66. Bouteraa, M.T.; Mishra, A.; Romdhane, W.B.; Hsouna, A.B.; Siddique, K.H.M.; Saad, R.B. Bio-Stimulating Effect of Natural Polysaccharides from Lobularia maritima on Durum Wheat Seedlings: Improved Plant Growth, Salt Stress Tolerance by Modulating Biochemical Responses and Ion Homeostasis. *Plants* **2022**, *11*, 1991. [CrossRef]
67. Malcheska, F.; Ahmad, A.; Batool, S.; Müller, H.M.; Ludwig-Müller, J.; Kreuzwieser, J.; Randewig, D.; Hänsch, R.; Mendel, R.R.; Hell, R.; et al. Drought-Enhanced Xylem Sap Sulfate Closes Stomata by Affecting ALMT12 and Guard Cell ABA Synthesis. *Plant Physiol.* **2017**, *174*, 798–814. [CrossRef]
68. Liu, Y.S.; Tao, Y.; Yang, X.Z.; Liu, Y.N.; Shen, R.F.; Zhu, X.F. Gibberellic acid alleviates cadmium toxicity in rice by regulating NO accumulation and cell wall fixation capacity of cadmium. *J. Hazard. Mater.* **2022**, *439*, 129597. [CrossRef]
69. Singh, P.K.; Chakrabarty, D.; Dwivedi, S.; Kumar, A.; Singh, S.P.; Sinam, G.; Niranjan, A.; Singh, P.C.; Chatterjee, S.; Majumdar, D.; et al. Nitric oxide-mediated alleviation of arsenic stress involving metalloid detoxification and physiological responses in rice (*Oryza sativa* L.). *Environ. Pollut.* **2022**, *297*, 118694. [CrossRef]

70. Wu, P.; Kong, Q.; Bian, J.; Ahammed, G.J.; Cui, H.; Xu, W.; Yang, Z.; Cui, J.; Liu, H. Unveiling Molecular Mechanisms of Nitric Oxide-Induced Low-Temperature Tolerance in Cucumber by Transcriptome Profiling. *Int. J. Mol. Sci.* **2022**, *23*, 5615. [CrossRef]
71. Seneviratne, M.; Rajakaruna, N.; Rizwan, M.; Madawala, H.; Ok, Y.S.; Vithanage, M. Heavy metal-induced oxidative stress on seed germination and seedling development: A critical review. *Environ. Geochem. Health* **2019**, *41*, 1813–1831. [CrossRef] [PubMed]
72. Oracz, K.; Karpiński, S. Phytohormones Signaling Pathways and ROS Involvement in Seed Germination. *Front. Plant Sci.* **2016**, *7*, 864. [CrossRef] [PubMed]
73. Hajihashemi, S.; Skalicky, M.; Brestic, M.; Pavla, V. Cross-talk between nitric oxide, hydrogen peroxide and calcium in salt-stressed Chenopodium quinoa Willd. At seed germination stage. *Plant Physiol. Biochem.* **2020**, *154*, 657–664. [CrossRef] [PubMed]
74. Zhang, Y.; Su, J.; Cheng, D.; Wang, R.; Mei, Y.; Hu, H.; Shen, W.; Zhang, Y. Nitric oxide contributes to methane-induced osmotic stress tolerance in mung bean. *BMC Plant Biol.* **2018**, *18*, 207. [CrossRef] [PubMed]
75. Zheng, C.F.; Jiang, D.; Liu, F.L.; Dai, T.B.; Liu, W.C.; Jing, Q.; Cao, W.X. Exogenous nitric oxide improves seed germination in wheat against mitochondrial oxidative damage induced by high salinity. *Environ. Exp. Bot.* **2009**, *67*, 222–227. [CrossRef]
76. Basit, F.; Bhat, J.A.; Guan, Y.J.; Jan, B.L.; Tyagi, A.; Ahmad, P. Nitric oxide and spermine revealed positive defense interplay for the regulation of the chromium toxicity in soybean (*Glycine max* L.). *Environ. Pollut.* **2022**, *308*, 119602. [CrossRef]
77. Kumar, D.; Manhas, R.K.; Ohri, P. Deciphering the growth promoting and stress curtailing role of nitric oxide in Meloidogyne incognita infested *Solanum lycopersicum* seedlings. *Sci. Hortic.* **2023**, *319*, 112147. [CrossRef]
78. Li, M.Y.; Feng, K.; Hou, X.L.; Jiang, Q.; Xu, Z.S.; Wang, G.L.; Liu, J.X.; Wang, F.; Xiong, A.S. The genome sequence of celery (*Apium graveolens* L.), an important leaf vegetable crop rich in apigenin in the *Apiaceae* family. *Hortic. Res.* **2020**, *7*, 9. [CrossRef]
79. Li, T.; Liu, J.X.; Deng, Y.J.; Duan, A.Q.; Liu, H.; Zhuang, F.Y.; Xiong, A.S. Differential hydroxylation efficiency of the two non-heme carotene hydroxylases: *DcBCH1*, rather than *DcBCH2*, plays a major role in carrot taproot. *Hortic. Res.* **2022**, *9*, uhac193. [CrossRef]
80. Feng, K.; Xing, G.M.; Liu, J.X.; Wang, H.; Tan, G.F.; Wang, G.L.; Xu, Z.S.; Xiong, A.S. *AgMYB1*, an R2R3-MYB factor, plays a role in anthocyanin production and enhancement of antioxidant capacity in celery. *Veg. Res.* **2021**, *1*, 2. [CrossRef]
81. Duan, A.Q.; Liu, H.; Shu, S.; Xiong, A.S. CRISPR/Cas9-mediated precise targeted mutagenesis of phytoene desaturase in celery. *Hortic. Res.* **2022**, *9*, uhac162. [CrossRef]

agronomy

MDPI

Article

Exogenous GR24 Inhibits Strawberry Tillering by Affecting the Phytohormone Signaling and Sugar Metabolism Pathways

Yuting Peng †, Yuyan Jiang †, Caixia He, Musha She, Mengyao Li, Qing Chen, Yong Zhang, Yuanxiu Lin, Yunting Zhang, Yan Wang, Wen He, Xiaorong Wang, Haoru Tang and Ya Luo *

College of Horticulture, Sichuan Agricultural University, Chengdu 611130, China; 2022205007@stu.sicau.edu.cn (Y.P.); jiangyuyan@stu.sicau.edu.cn (Y.J.); 2022305087@stu.sicau.edu.cn (C.H.); 2022205078@stu.sicau.edu.cn (M.S.); limy@sicau.edu.cn (M.L.); supnovel@sicau.edu.cn (Q.C.); zhyong@sicau.edu.cn (Y.Z.); linyx@sicau.edu.cn (Y.L.); asyunting@sicau.edu.cn (Y.Z.); wangyanwxy@sicau.edu.cn (Y.W.); hewen0724@gmail.com (W.H.); wangxr@sicau.edu.cn (X.W.); htang@sicau.edu.cn (H.T.)

* Correspondence: luoya945@sicau.edu.cn; Tel.: +86-28-86291741

† These authors contributed equally to this work.

Abstract: Tillering is an important part in strawberry growth, and strawberries can reproduce nutritionally through stolons to generate genetically stable offspring. However, excessive tillering during the fruit-growing stage can negatively impact fruit yield and quality. In this study, different concentrations of exogenous rac-GR24 (GR24) are used to treat the strawberry plants. It was found that GR24 effectively inhibited the sprouting of strawberry stolons, while promoting the growth of the stems and leaves. Among the treatments, the most effective concentration was found to be 5 μmol/L GR24. This treatment resulted in a decrease in the glucose content in the strawberry crowns and also caused changes in the contents of two endogenous phytohormones, gibberellic acid (GA_3) and trans-zeatin riboside (tZR). Transcriptome data further suggested that exogenous GR24 may inhibit strawberry plant tillering by affecting various phytohormone signaling pathways and the sugar metabolism pathway. In 5 μmol/L GR24-treated plants, the expression level of type-B response regulator (*B-ARR*) was down-regulated and the expression level of CYTOKININ RESPONSE 1 (*CRE1*), histidine-containing phosphotransfer protein (*AHP*), and type-A response regulator (*A-ARR*) were up-regulated, suggesting the inhibition of the cytokinin (CTK) signaling pathway. The down-regulation of auxin (*AUX*) and auxin response factor (*ARF*), as well as the up-regulation of auxin/indole-3-acetic acid (*AUX/IAA*), led to the inhibition of the indole-3-acetic acid (IAA) signaling pathway. Additionally, the up-regulation of pyrabactin resistance 1/ pyrabactin resistance 1-like (*PYR/PYL*), non-fermenting 1-related protein kinase 2 (*SnRK2*), and ABRE binding factors (*ABF*) and the down-regulation of protein phosphatase 2C (*PP2C*) were observed in the up-regulated abscisic acid (ABA) signaling pathways. In the sugar metabolism pathway, the up-regulation of invertase (*INV*), hexokinase (*HK*), and fructokinase (*FRK*) and the down-regulation of trehalase (*TREH*) and beta-amylase (*BMY*) led to a decreased glucose synthesis and an increased glucose consumption. Therefore, GR24 can effectively inhibit strawberry plant tillering through these pathways, making it an effective reagent for tillering inhibition.

Keywords: strigolactone; stolon; cytokinin; auxin; abscisic acid

Citation: Peng, Y.; Jiang, Y.; He, C.; She, M.; Li, M.; Chen, Q.; Zhang, Y.; Lin, Y.; Zhang, Y.; Wang, Y.; et al. Exogenous GR24 Inhibits Strawberry Tillering by Affecting the Phytohormone Signaling and Sugar Metabolism Pathways. *Agronomy* **2023**, *13*, 3078. https://doi.org/10.3390/agronomy13123078

Academic Editor: Samir C. Debnath

Received: 31 October 2023
Revised: 4 December 2023
Accepted: 13 December 2023
Published: 17 December 2023

1. Introduction

Strawberries (*Fragaria* × *ananassa* Duch.) are perennial herbaceous plants in the Rosaceae family, known for their high nutritional and economic value [1]. In the production of cultivated strawberries, the propagation of stolons is a commonly used breeding method to preserve the plants' excellent traits. However, in China, during mid-October every year, favorable external environmental conditions lead to a significant sprouting of stolons. This excessive growth can disrupt the balance between the reproductive and nutritional growth processes of strawberries, resulting in yield reduction and loss of fruit quality [2]. Therefore, it is critical to effectively control the formation of stolons during the cultivation process of strawberries.

As perennial rosette-forming herbaceous plants, strawberries have dense verticillate leaves with long petioles. Axillary buds, which are located in the leaf axils, can germinate into branches under low-temperature and short-day conditions [3]. These branches can develop into long branches called stolons [4]. The occurrence of stolons is influenced by various factors, such as the development and location of axillary buds, phytohormones, as well as environmental conditions. It has been found that the duration of sunlight can affect the asexual differentiation of strawberry axillary buds [5]. Long-day conditions promote the formation of stolons [6]. Gibberellin (GA) has been found to enhance the ability of strawberry plants to sprout stolons [7]. The photoperiod also plays a role in regulating stolon formation by affecting the GA biosynthesis or balancing asexual and sexual reproduction patterns in axillary meristems [8]. The *GIBBERELLIN20-oxidase 4* gene (*FvGA20ox4*) is essential for stolon growth in woodland strawberries (*Fragaria vesca* L.) [9]. In addition, 6-Benzylaminopurine (6-BA) can break the dormancy of axillary buds and promote stolon sprouting by increasing the soluble sugar content in strawberries [10]. In other plant species, such as creeping bentgrass (*Agrostis stolonifera* L.), an elevated carbon dioxide (CO_2) concentration promotes stolon growth and increases the accumulation of glucose, sucrose, fructose, and endogenous IAA at the stolon nodes and internodes, resulting in longer stolons and larger stolon shoot biomass [11]. Kentucky bluegrass (*Poa pratensis* L.) with more ramets also exhibits a significant increase in the soluble sugar content in the rhizome [12]. In addition, studies on strawberries have revealed that auxin and CTK play antagonistic roles in controlling the development of axillary buds. Reduced auxin accumulation promotes stolon formation, while exogenous CTK promotes the formation of flower buds [13]. In wheat (*Triticum aestivum* L.), a lower red light/far-red light ratio up-regulated CTK degradation genes (*TaCKX5* and *TaCKX11*) and GA biosynthesis genes (*TaGA20ox1* and *TaGA3ox2*) in the tillering nodes, leading to a decreased CTK level and an increased GA level. This promotes CTK degradation and inhibits tillering in wheat [14]. The temperature mainly affects the dormancy of buds or the germination of buds. In experiments with roses, it has been shown that a low temperature can promote the signaling pathway of strigolactone, which in turn affects the gradient germination rate of lateral branches [15]. Additionally, phosphorus (P) and nitrogen (N) have important effects on the development of plant lateral branches, which can be influenced by the synthesis and transportation changes of IAA, CTK, and SLs, thereby affecting the development of lateral branches [16].

Strigolactone (SL) is a carotenoid-derived phytohormone that plays a crucial role in regulating various plant activities. It was initially discovered as a germinating agent [17,18] and later found to be an effective inhibitor of branching [19]. The function of SL in plants is mainly achieved through its complex interaction with other phytohormones. Research has shown that reducing the activity of IAA4/5, which is a downstream component of the auxin signaling pathway in the pea (*Pisum sativum* L.) plant, can promote auxin signaling transduction. This, in turn, leads to higher content of SL and a lower content of CTK, resulting in the inhibition of axillary bud branching [20]. Furthermore, SL and CTK have antagonistic effects on shoot growth regulation [21]. SL also affects root development through the transduction of GA signaling [22].

GR3, GR7, GR5, and GR24 are common artificially synthesized SL analogues; among them, GR24 is the most effective [23]. GR24 ($C_{17}H_{14}O_5$) is a triterpene lactone analogue of SL, which has been found to promote seed germination, inhibit plant branching, and promote anthocyanin accumulation in various plant species [24,25]. In addition, it has been found that GR24 can affect the cell size of cherry rootstocks and inhibit their stem growth [26]. GR24 treatment can also reduce the susceptibility of tobacco and grapevine plantlets to *Botrytis cinerea* infection [27]. However, studies on SL in strawberries are limited to its effects on strawberry fruit quality and endogenous SL affecting branching. It was found that exogenous SL alleviated the oxidative damage of strawberries by improving the defense capability of the antioxidant system [28]. And it was reported that the expression of most SL biosynthetic genes was high in developing carpel, anther, and style, while that of SL signaling genes was high in carpel and style, but low in anther. This indicated that SL plays a role in the early stages of woodland strawberry fruit development [29]. Meanwhile, research has also shown that branching features or the number of branches are dependent on the content of SL in strawberry plants [30]. However, the role of exogenous SL in regulating strawberry stolons has not been reported. Therefore, this study intends to analyze the physiological and transcriptome changes in strawberry plants treated with different concentrations of GR24. We also examine the interaction between endogenous phytohormones and sugar in response to exogenous SL. By studying the effects of exogenous SL on strawberry plants, we hope to gain insights into the methods of controlling plant tillering in strawberry cultivation.

2. Materials and Methods

2.1. Plant Materials and Treatments

Cultivated strawberries (*Fragaria* × *ananassa* cv. 'Benihoppe') were used as the experimental materials at the strawberry picking base in Jinggang Village, Pidu District, Chengdu City, Sichuan Province. Strawberry seedlings were planted in September 2022. One month after planting, uniform and robust strawberry plants were selected and sprayed with 50 mL of the GR24 (Coolabar, Beijing, China) solution at concentrations of 0 μmol/L, 5 μmol/L, 10 μmol/L, and 20 μmol/L on their crowns. Each treatment consisted of 15 plants, and the spraying was conducted every 9 days for a total of 3 times, with 3 replicates. One functional leaf (15 days old) with a uniform growth state from each plant was marked for the subsequent measurement of leaf physiological indicators. Stem diameter, plant height, petiole length, leaf thickness, and the number of stolons were measured 9 days after each spraying. The crowns, 9 days after the 3rd spraying of 5 μmol/L GR24, were quickly frozen in liquid nitrogen, ground into a uniform powder, and stored in an ultra-low temperature refrigerator at −80 °C for the subsequent determination of plant endogenous phytohormones and transcriptome analysis.

2.2. Measurement of Plant Growth Indexes

The stem diameter, plant height, petiole length, and the diameter of the flowering branches were measured with vernier calipers and straightedges. The leaf thickness was measured using a thickness gauge (Qifeng Digital High Precision Thickness Gauge MIMZI, Hangzhou, China). The number of stolons and flowering branches were counted through visual observation.

2.3. Determination of the Endogenous Phytohormone Content

We accurately weighed 0.5 g of the sample in a test tube and added 10 mL acetonitrile solution and 8 μL phytohormone standard solution. The sample was extracted overnight at 4 °C and then centrifuged (−80 °C, 12,000× *g*) for 5 min. After extracting the supernatant, 5 mL acetonitrile solution was added to the precipitate to be extracted twice. The supernatant was combined and the impurities were purified by adding the appropriate amount of C18 and graphitized carbon black (GCB). Then, the liquid was centrifuged (4 °C, 12,000× *g*) for 5 min, and retain supernatant was retained. The sample was blown dry under nitrogen, re-dissolved with 400 μL of methanol, passed through a 0.22 μm organic filter membrane, and placed in in a −20 °C refrigerator for further analysis. The endogenous phytohormones IAA, ABA, tZR, and GA_3 were determined using an Agilent 1290 HPLC system in series with an AB's SCIEX-6500Qtrap mass spectrometer. The samples were detected using a linear gradient eluent program: A total of 0.1% formic acid in methanol as eluent A and 0.1% formic acid in water as eluent B. A 2 μL sample was injected, and the flow rate remained at 0.3 mL/min. From 0 to 1 min, the eluent A was at 20%, and it increased from 20% to 50% and it kept on increasing until reaching 80% from 3 to 9 min. After staying at 80% for 1.5 min, the eluent A decreased to 20% in 1 min and remained at 20% for

2.9 min. The concentration of endogenous phytohormones was quantified by comparing them with the corresponding external standards, all of which were HPLC-grade standards purchased from Sigma (St. Louis, MI, USA). The experiments were independently repeated three times.

2.4. Determination of Sucrose, Fructose, and Glucose Contents

The contents of sucrose, glucose, and fructose were determined by high-performance liquid chromatography (Agilent HPLC 1260 Infinity II, Santa Clara, CA, USA) [31]. After weighing 0.3 g of the sample, it was extracted with 2 mL ultra-purified water and subsequently passed through an Agilent Athena NH2-RP column using an Agilent HPLC system with a differential refractive index detector. The elution program was: 75% acetonitrile for 10 min with a 10 μL injection and a flow rate of 1 mL/min.

2.5. Transcriptome Analysis

The total RNA was isolated using the cetyltrimethylammonium bromide (CTAB) method [32]. Libraries were then constructed and sequenced by Genepre in Chengdu, China. The quality of the libraries was assessed with an Agilent 2100 bioanalyzer. A total of 6 libraries were clustered and sequenced (150 bp, pair-end) on a Hiseq-2500 platform, with three replicates for both the control (CK) and the samples treated with 5 μmol/L of the GR24 treatment. To ensure data quality, low-quality reads with a quality score (Q) below 20 were screened using the FASTQ software (v0.11.9). The adaptors were trimmed using the Trim Galore software (v0.6.6). The resulting cleaned reads were then mapped onto the reference strawberry genome (v1.0.a2) and quantified using the Hisat 2 and stringtie (v) pipeline with the default parameters. The DESeq 2 (v3.34.1) R package was used to detect the differentially expressed genes (DEGs). The genes with $\log_2$-fold changes (FCs) >1 or <-1 and an adjusted $p \leq 0.05$ were considered as significant DEGs.

2.6. Real-Time Fluorescence Quantification (RT-qPCR)

A real-time quantitative PCR (RT-qPCR) assay was performed to verify the expression levels of the key genes (*FaAHP1-like, FaHK2-like, FaARR1-like, FaLAX2, FaMETTL5, FaAUX28-like, FaIAA8-like, FaIAA9, FaIAA27, FaARF5, FaPP2C10, FaPP2C12, FaPP2C13, FaPP2C27, FaABF1*, and *FaABF5*) involved in the CTK, IAA, and ABA signaling pathway. A 10 μL reaction mixture was prepared, consisting of 1 μL of cDNA template, 1 μL of gene specific primer pairs (Table S1), and 5 μL of TB Green Premix Ex Taq II (TaKaRa, Dalian, China). The qPCR assay involved a three-step PCR reaction, with a denaturation step at 94 °C for 30 s, an annealing step at 58 °C for 10 s, and an extension at 72 °C for 10 s. The total reaction circle was set to 40. The *FaActin2* gene (LOC101313255) was selected as an internal control. The relative expression levels of the detected genes were calculated by the $2^{-\Delta\Delta Ct}$ method. Three wells of each sample were conducted as three technological replicates, and three independent biological replicates were conducted for all the qPCR reactions.

2.7. Statistical Analysis

The software SPSS 22.0 was used for the statistical analysis of the data obtained from the experiments. A one-way ANOVA analysis using Duncan's test was performed to identify significant differences between the means ($p < 0.05$). The results are expressed as the mean ± standard deviation (SD). Three different biological replicates were obtained from each measurement.

3. Results

3.1. Effects of Different Concentrations of GR24 on Stem Diameter and Its Growth

The stem diameter of strawberry plants is a key index of their strength and resilience. In this study, the stem diameter of the plants treated with different concentrations of GR24 gradually increased over time. Starting from the day of treatment, the diameter was recorded as DAT 0, which represents day 0 after treatment. At DAT 18, the stem diameters of the plants treated with 5 μmol/L, 10 μmol/L, and 20 μmol/L GR24 were significantly increased by 1.43 mm, 1.47 mm, and 1.17 mm, respectively, compared to the control. However, there was no significant difference among the different GR24 treatments. At DAT 27, the stem diameters of the plants treated with 5 μmol/L and 10 μmol/L GR24 were significantly larger than that of CK by 1.67 mm and 0.94 mm, respectively. But there was no significant difference between the strawberry stem diameters treated with 20 μmol/L GR24 and the control (Figure 1A). To further understand the growth of the strawberry stem diameter after the treatments, the increase in the stem diameter was counted between each treatment. It was found that the increase in stem diameter under the 5 μmol/L GR24 treatment at DAT 9 was 1.11 mm, which was significantly smaller than that of the other three groups. Between DAT 9 and DAT 18, the increase in the stem diameter under the 5 μmol/L and 20 μmol/L GR24 treatments was 2.17 mm and 1.92 mm, respectively, which was significantly greater than that of CK at 1.07 mm. From DAT 18 to DAT 27, the increase in the plant stem diameter for all four treatments was 0.81 mm, 1.04 mm, 1.10 mm, and 0.93 mm, respectively, which were significantly smaller than the previous two increases, and there was no significant difference between each treatment and the control (Figure 1B). Based on these findings, it can be concluded that GR24 may have a promotion effect on the growth of strawberry stem diameter, especially at concentrations of 5 μmol/L and 10 μmol/L. The 20 μmol/L treatment also had a positive effect, although it was not significantly different from the control.

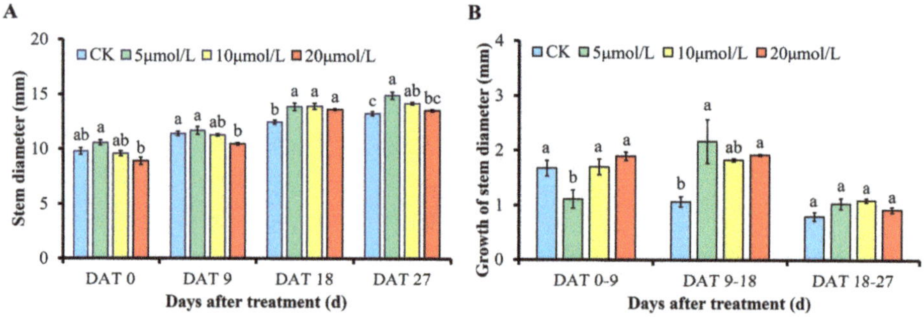

Figure 1. Effect of different concentrations of GR24 on the strawberry stem diameter (**A**) and its growth (**B**). DAT 0: the day when the 1st treatment was administered; DAT 9: the day when the 2nd treatment was administered and also the 9th day after the 1st treatment; DAT 18: the day when the 3rd treatment was administered and also the 9th day after the 2nd treatment; DAT 27: the 9th day after the 3rd treatment. The bars in the graph indicate the standard error (SE) (n = 3). Means followed by different letters are significantly different at $p \leq 0.05$ according to Tukey's multiple test (n = 3). Different letters indicate significant differences ($p < 0.05$). This legend also applies to the table below.

3.2. Effects of Different Concentrations of GR24 on Plant Height, Petiole Length, and Their Growth

Plant height and petiole length are two important indexes reflecting the nutritional growth of strawberry. The plant height and petiole length of strawberry under different concentrations of GR24 increased as the experiment progressed. It was found that the height of strawberry plants under the 5 μmol/L and 10 μmol/L GR24 treatments was significantly higher than that of CK at DAT 9, DAT 18, and DAT 27 (Figure 2A). The increase in the plant height showed that the overall growth rate of the strawberry plant height was the greatest in the first 9 days, with a maximum increase of 2.67 cm and 2.58 cm under the 5 μmol/L and 10 μmol/L GR24 treatments, respectively. However, from DAT 9 to DAT 18, there was no significant difference in the increase in the plant height between the strawberry plants under the three concentrations of GR24 treatments and the control. The obvious increase in the plant height was observed from DAT 18 to DAT 27 under the 10 μmol/L GR24 treatment, reaching 1.9 cm (Figure 2B).

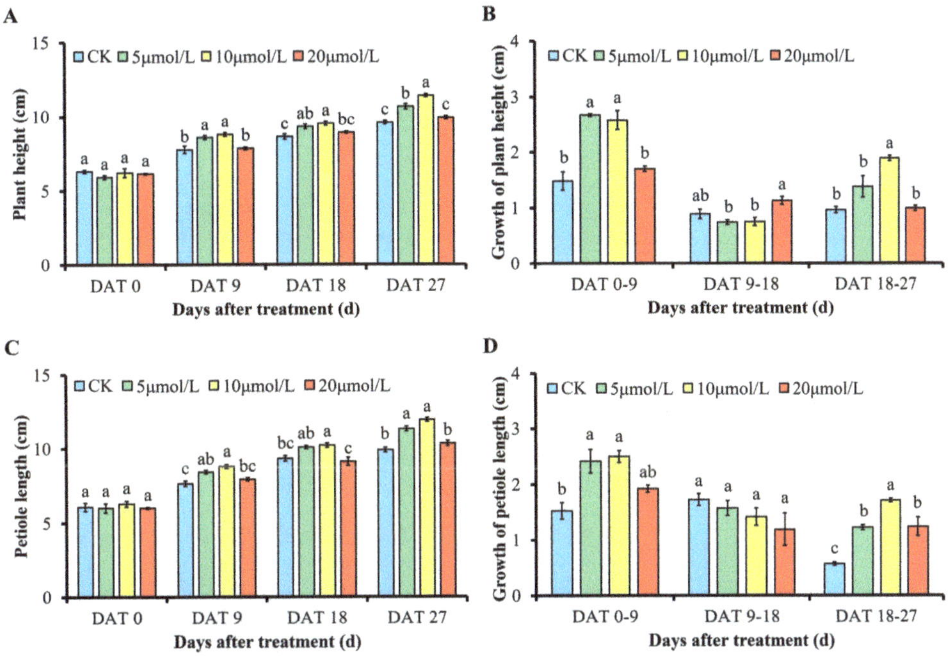

Figure 2. Effect of different concentrations of GR24 on the strawberry plant height (**A**) and its growth (**B**) and the petiole length (**C**) and its growth (**D**). Different letters indicate significant differences ($p < 0.05$).

The results of the petiole length measurement showed that a concentration of 10 μmol/L GR24 was the most effective in promoting the growth of the petiole length in strawberry, and the petiole length in this treatment was significantly higher than that of CK on DAT 9, DAT 18, and DAT 27, with increases of 1.13 cm, 0.88 cm, and 2.03 cm, respectively (Figure 2C). In addition, during the first 9 days, the maximum growth of the petiole length was found under the 5 μmol/L and 10 μmol/L GR24 treatments, with increases of 2.42 cm and 2.50 cm, respectively, which were significantly higher than the 1.53 cm increase in CK. From DAT 9 to DAT18, there was no significant difference in the growth rate of the petiole length between the three treated groups and the CK. However, from DAT 18 to DAT27, the increase in the strawberry petiole length was the highest (1.71 cm) under the treatment with 10 μmol/L GR24, which was significantly higher than that of CK (Figure 2D). Thus, GR24 may have a promotion effect on the growth of the strawberry plant height and petiole, and adding 10 μmol/L GR24 at the early growth stage promoted the increase in the plant height and petiole length more effectively.

3.3. Effect of Different Concentrations of GR24 on Leaf Thickness and Its Growth

We found no significant differences in the leaf thickness of plants in these four treatment groups at any of the four periods, and we subsequently measured the amount of growth in leaf thickness. But there was no significant difference between the treatments at any time (Figure 3). When studying the growth of leaf thickness, it was discovered that the increase in leaf thickness under the 10 μmol/L GR24 treatment was 0.09 mm from DAT 9 to DAT18, which was significantly higher than the 0.06 mm increase in CK (Figure 3). Therefore, the overall effect of GR24 in promoting leaf thickening was not significant, but 10 μmol/L GR24 could significantly promote strawberry leaf thickening in the middle term of the treatment.

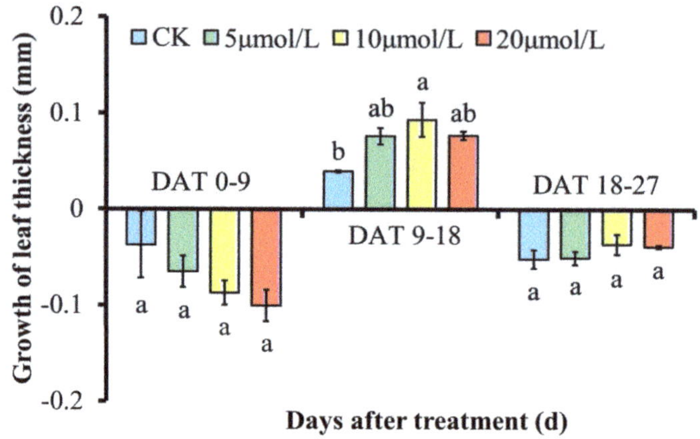

Figure 3. Effect of different concentrations of GR24 on the growth of strawberry leaf thickness. Different letters indicate significant differences ($p < 0.05$).

3.4. Effect of Different Concentrations of GR24 on Strawberry Stolons and Flowering Branches

The stolon is an important index of the vigorous growth of strawberry plants and is closely related to yield. This study found that the number of strawberry stolons gradually increased as the strawberry plants grew. However, the number of strawberry stolons in all three treatment groups was significantly lower than in CK at DAT 9 and DAT 18. By DAT 27, only the number of stolons under the 10 μmol/L GR24 treatment (1.91) was significantly lower than that in CK at 2.38 (Figure 4A). Based on the diameter and number of flowering branches of the plants after the last treatment, it was found that the plants treated with 5 μmol/L GR24 had the best promotion effect on these two indicators (Figure 4B,D). Since the purpose of reducing stolon growth is to increase fruit yield, based on the above results, we decided to use the strawberry plants under the 5 μmol/L GR24 treatment for the subsequent determination of the sugar and phytohormone contents as well as for the transcriptome analysis. Considering the above results, GR24 had a significant inhibitory effect on the stolon sprouting of the strawberry plants.

3.5. Effect of GR24 on the Sugar Content of the Strawberry Crowns

Since soluble sugars have been found to promote strawberry runner production in previous studies [10] 2020, this study also measured the sugar content of the crowns in strawberry plants under the different treatments. Sucrose was not detected in the strawberry crowns, but fructose and glucose were detected (Figure 4C). There was no significant difference in the fructose content between the CK and GR24-treated plants, but the glucose content in the 5 μmol/L GR24-treated plants was significantly lower than that in the plants of CK, accounting for 70% of the glucose content in CK.

3.6. Effect of GR24 on the Content of Endogenous Phytohormones

To systematically study the effect of GR24 on endogenous hormones in strawberry plants, we measured the content of four endogenous phytohormones in the crown. The results show that GR24 significantly promoted the production of tZR and GA_3 (Figure 5). Among them, GR24 most obviously promoted tZR, which was 4.5 times higher than that of CK after the treatment, while GA_3 levels increased by 39.1%. However, the contents of IAA and ABA were not significantly different compared to those of CK. In summary, the exogenous spraying of GR24 can influence the content of endogenous phytohormones in strawberry plants and thus regulate the growth of the plants.

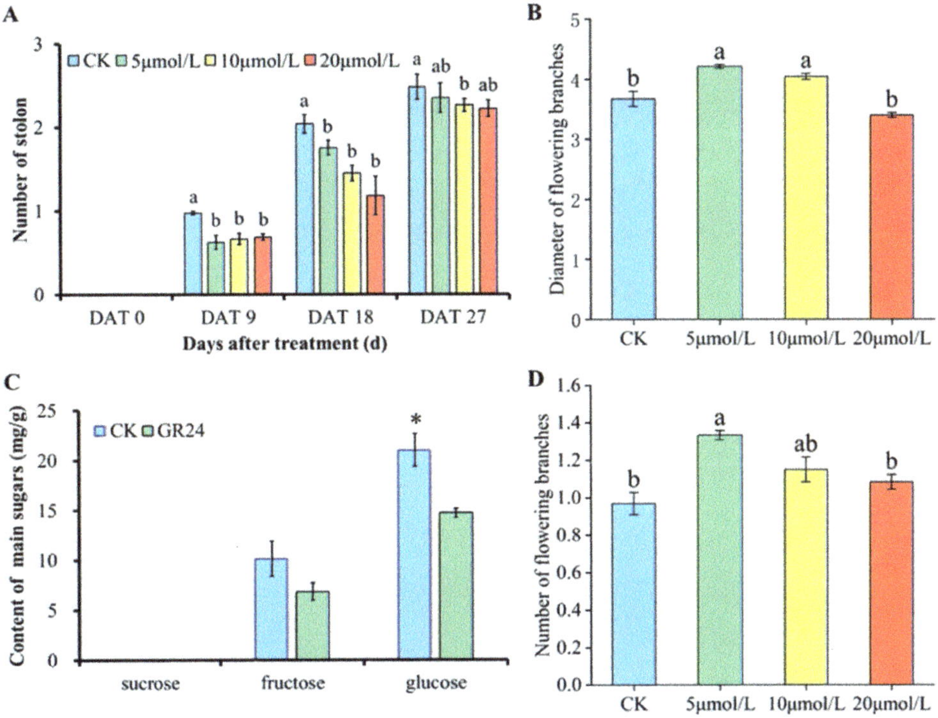

Figure 4. Effect of different concentrations of GR24 on the number of stolons (**A**), the diameter of the flowering branches (**B**), the content of the main sugars (**C**), and the number of the flowering branches (**D**). Different letters and * indicate significant differences ($p < 0.05$).

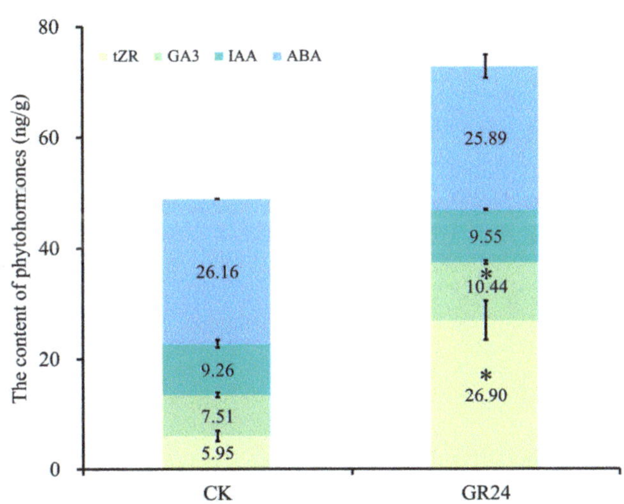

Figure 5. Effect of GR24 on the content of the endogenous phytohormones of strawberry plants. Comparison of the tZR content, GA_3 content, IAA content, and ABA content in CK and the strawberry plants treated with 5 μmol/L GR24 on the 9th day after the 3rd treatment. The * indicates statistically significant differences ($p \leq 0.05$, Student's *t*-test).

3.7. Transcriptome Analysis of the Cytokinin Anabolic and Signaling Pathways

Since higher levels of tZR and GA_3 were detected in the plants treated with 5 μmol/L GR24 (Figure 5), we sequenced and analyzed the transcriptome of crowns from both groups of materials to investigate the underlying molecular mechanisms. The results of the differential gene expression analysis of the CTK anabolic pathway as well as its signaling pathway (Figure 6) showed that *CKX* (FxaC_4g32960, FxaC_2g04210, FxaC_2g12570, FxaC_3g07760, FxaC_7g41590, FxaC_8g06620, and FxaC_6g42650), *ZOG* (FxaC_23g16160 and FxaC_24g30610), and *UGT76C2* (FxaC_26g11300), the key enzymes involved in CTK degradation, were significantly down-regulated in the GR24-treated plants. At the same time, the rate-limiting enzyme for cytokinin synthesis, *IPT* (FxaC_5g12240), and the key enzyme for CTK synthesis, *CYP735A* (FxaC_21g11720, FxaC_22g10010, FxaC_23g54500, and FxaC_24g54290), were significantly up-regulated. It is thus speculated that the significant increase in the tZR content in the GR24-treated plants may be attributed to the inhibition of the degradation pathway and the enhancement of CTK synthesis, leading to the accumulation of tZR in plants.

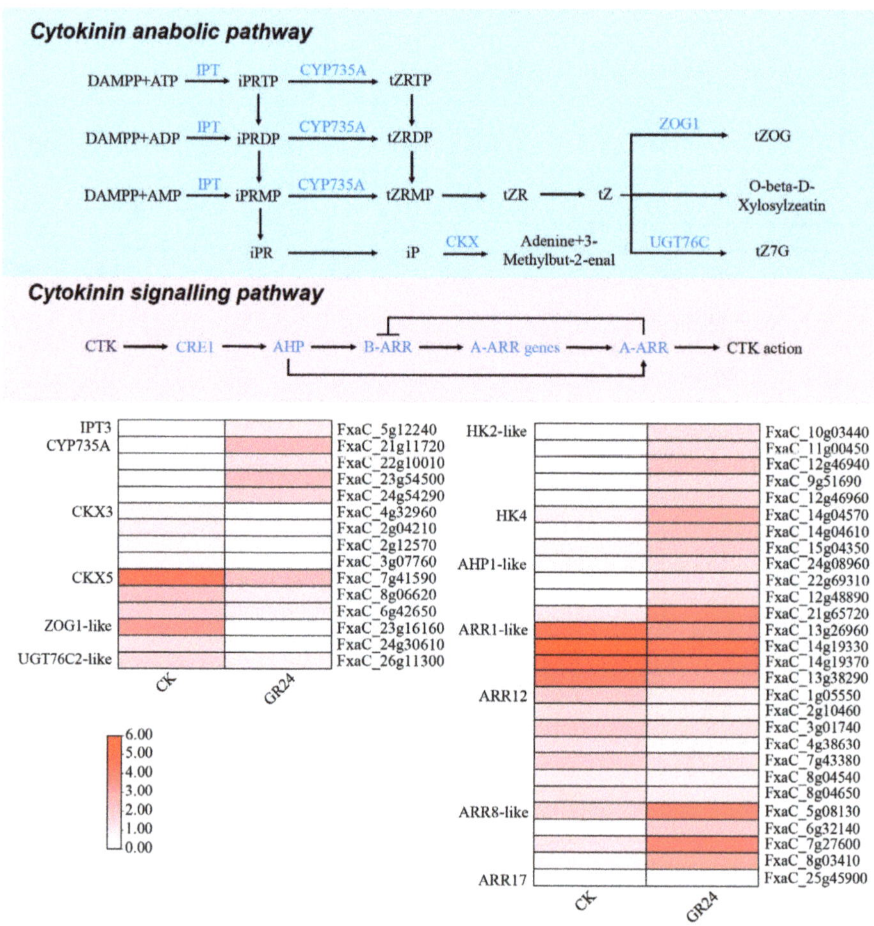

Figure 6. Transcriptome analysis of the cytokinin anabolic and signaling pathways. IPT, isopentenyl transferase; CYP735A, cytochrome P450 monooxygenase 735A; CKX, cytokinin oxidase; ZOG1,

zeatin O-glucosyltransferase; UGT76C, UDP-glycosyltransferase 76C2; CRE1, histidine kinase; AHP, histidine-containing phosphotransfer protein; B-ARR, B-type Arabidopsis response regulator; A-ARR, A-type Arabidopsis response regulator. For the gene expression comparison, the log_2-transformed CPM (counts per million) values were presented as a heatmap, with CK on the left and the material treated with 5 μmol/L GR24 on the right. Substances labeled in blue indicate that they were differentially expressed in CK and 5 μmol/L GR24-treated plants. The expression level gradually increases from white to red. The legend also applied to the figure below.

The CTK signaling pathway is a dual signaling system with the following core steps: the transmembrane protein CRE1 serves as a receptor for cytokinin from the extracellular environment, which are then transferred to AHP. The AHP proteins then transfer phosphate groups to response regulator (ARR) proteins and functions. B-ARR in the system is a v-myb avian myeloblastosis viral oncogene homolog (MYB) transcription factor that is activated upon phosphorylation and promotes the expression of downstream genes to regulate the growth and development of the plant. In addition, B-ARR can activate the transcription of A-ARR, while A-ARR can inhibit the activity of B-ARR, forming a negative regulatory network. In this study, the transcriptional results show that the expression of *CRE1* (FxaC_10g03440, FxaC_11g00450, FxaC_12g46940, FxaC_9g51690, FxaC_12g46960, FxaC_14g04570, FxaC_14g04610, and FxaC_15g04350) and *AHP* (FxaC_24g08960, FxaC_22g69310, FxaC_12g48890, and FxaC_21g65720) was significantly up-regulated in the GR24-treated plants. Conversely, the transcription factor *B-ARR* (FxaC_13g26960, FxaC_14g19330, FxaC_14g19370, FxaC_13g38290, FxaC_1g05550, FxaC_2g10460, FxaC_3g01740, 4g38630, FxaC_7g43380, FxaC_8g04540, and FxaC_8g04650) was significantly down-regulated and its repressor *A-ARR* (FxaC_5g08130, FxaC_6g32140, FxaC_7g27600, FxaC_8g03410, and FxaC_25g45900) was significantly up-regulated. This suggests that the signal transduction of CTK may be weakened in response to the suppression of B-ARR.

3.8. Transcriptome Analysis of the GA Anabolic and Signaling Pathways

A total of 25 differentially expressed transcripts directly related to the GA anabolic pathway were detected from the transcriptome data (Figure 7). *CPS* (FxaC_6g28150), *KS* (FxaC_9g40940, FxaC_11g13021, FxaC_9g41050, and FxaC_11g13020), and *KAO* (FxaC_28g-30380 and FxaC_26g31590) are genes involved in the synthesis of GA_{12}-aldehyde from geranylgeranylpyrophosphate (GGPP), whose expression levels were significantly up-regulated in the GR24-treated plants. The conversion of GA_{12}-aldehyde to differentially structured GAs is a complex oxidation process and the expression of the key gene *GA3ox* (FxaC_2g07540 and FxaC_22g32160) involved in this process was significantly up-regulated in the GR24-treated plants compared to the control. In contrast, GA2ox, which is involved in the inactivation of GA, was significantly down-regulated in the GR24-treated plants. Among the 16 differentially expressed transcripts, the expression of seven differentially coding genes in the GR24-treated plants was significantly down-regulated by more than 10 times compared to the control. To sum up, these finding suggest that the GR24 treatment enhanced the expression of genes encoding key enzymes in the GA biosynthesis pathway, while significantly inhibiting the expression of genes encoding key enzymes in the GA metabolism pathway, leading to an increase in the final GA_3 content in strawberry plants.

In addition, we analyzed the transcriptome of the GA signaling pathway. This pathway is mainly regulated by the GA receptor *GID1*, the GA transduction inhibitor DELLA, and the DELLA degradation complex $SCF^{GID2/SLY1}$. GA first binds to *GID1* and DELLA to form a GA-GID1-DELLA trimer, followed by the ubiquitination modification of the DELLA protein by $SCF^{GID2/SLY1}$, which initiates the degradation of DELLA by the 26s proteasome under the regulation of the transcription factor PIF3/4. This process relieves the repressive effect of DELLA on the downstream genes of the GA signaling pathway and exerts a bioregulatory effect on GA (Figure 7). In this study, the transcriptome data showed that the expression level of the GA receptor *GID1* (FxaC_10g12560, FxaC_11g09750, FxaC_22g63620, FxaC_22g69290, and FxaC_23g38030) was significantly up-regulated in the GR24-treated plants, and the inhibitory factor DELLA protein (FxaC_14g13060, FxaC_13g20790, and FxaC_15g13990) was significantly down-regulated. The expression levels of four DEGs encoding transcription factors *PIF3* and *PIF4* (FxaC_23g34990, FxaC_21g60640, FxaC_24g47710, and FxaC_22g16130) were all more than 3-fold up-regulated in the GR24-treated plants compared to the control. In summary, the analysis revealed an up-regulation in the expression of differential genes encoding both *GID1* and the transcription factors *PIF3/4*, indicating an increase in the binding of GA to GID1 and the subsequent ubiquitination and degradation of DELLA protein. This further suggests that the inhibitory effect of the DELLA protein on the GA signaling pathway was diminished, leading to an enhanced GA signal.

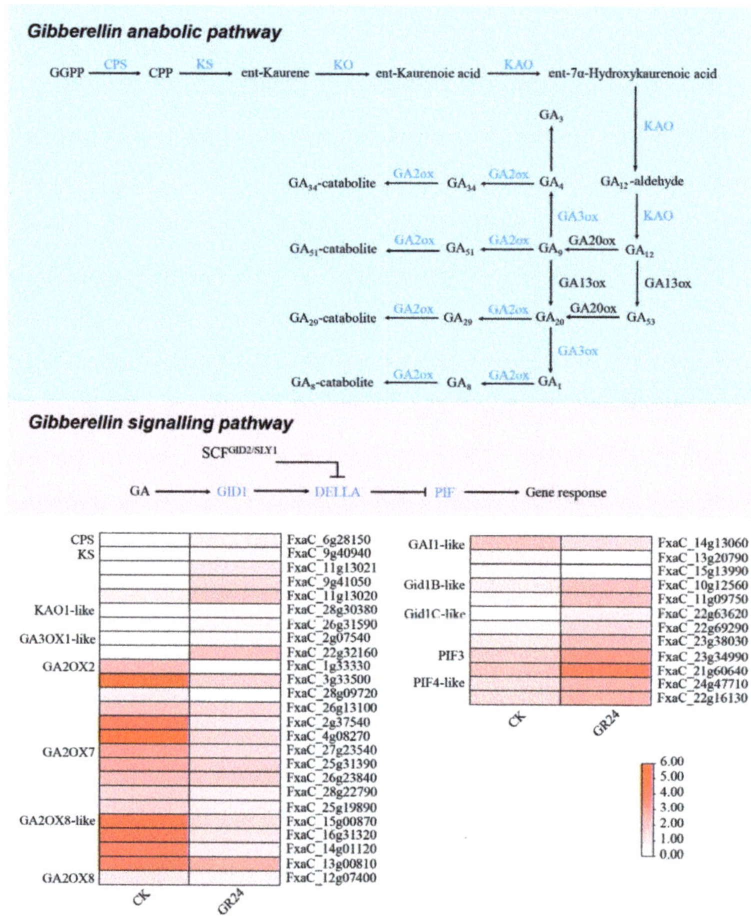

Figure 7. Transcriptome analysis of the GA anabolic and signaling pathways. The anabolic pathway of GA is divided into 3 steps: the synthesis of ent-kaurene by GGPP, the synthesis of GA_{12}-aldehyde from ent-kaurene, and the oxidation of GA_{12}-aldehyde to gibberellins with different structures. CPS, ent-copalyl diphosphate synthase; KS, ent-kaurene synthase; KAO, ent-kaurenoic acid oxidase; GA20ox, gibberellin 20 oxidase; GA3ox, gibberellin 3-beta-dioxygenase; GA2ox, gibberellin 2-beta-dioxygenase; GID1, gibberellin insensitive dwarf1; SCF, skp1-cullin-F-box protein; PIF, phytochrome-interacting factor.

3.9. Transcriptome Analysis of the Auxin Signaling Pathway

Since the contents of IAA and ABA were not significantly different between the GR24-treated plants and the control, and IAA and ABA were closely related to strawberry stolon sprouting [33,34], this experiment focused on the transcriptome analysis of the IAA and ABA signaling pathways. In the auxin signaling pathway (Figure 8), IAA binds to the auxin input carrier AUX1 and then the complex SCF^{TIR1}. This complex then modifies the inhibitory factor AUX/IAA through ubiquitination, leading to the release of the transcription factor ARF that it inhibits, and subsequently activating downstream response elements. The transcriptome results show that, after the GR24 treatment, the expression levels of auxin 1 (*AUX1*) (FxaC_21g55080, FxaC_21g56310, 22g56080, and FxaC_23g31230) and *ARF* (FxaC_7g41100, FxaC_17g12650, FxaC_17g16080, FxaC_19g11690, FxaC_27g35540, FxaC_27g31570, FxaC_28g16990, and FxaC_26g18090) were up-regulated. Additionally, the inhibitor *AUX/IAA* was also significantly up-regulated, thereby inhibiting the transmission of the auxin signal.

Figure 8. Transcriptome analysis of the auxin signaling pathway and the ABA signaling pathway. AUX/IAA, auxin/indole-3-acetic acid; ARF, auxin response factor; PYR/PYL, pyrabactin resistance 1/pyrabactin resistance 1-like; PP2C, protein phosphatase 2C; SnRK2, sucrose non-fermenting 1-related protein kinase 2; ABF, ABRE-binding factors.

3.10. Transcriptome Analysis of the ABA Signaling Pathway

In the ABA signaling pathway (Figure 8), PYL acts as an ABA receptor and regulates a series of downstream reactions. PP2C is also a key protein in the ABA signaling pathway, and PYR/PYL controls the ABA signal by inhibiting PP2C. The results of this study show that the GR24 treatment significantly up-regulated *PYL* (FxaC_10g08770 and FxaC_11g05280) expression while down-regulating *PP2C* expression. Additionally, the expression levels of *SnRK2* (FxaC_13g23250 and FxaC_8g32310) and the transcription factor *ABF* (FxaC_9g27080, FxaC_6g39990, and FxaC_7g36140) were also significantly up-regulated in the GR24-treated plants. In summary, the GR24 treatment may suppress ABA signal transduction by up-regulating the inhibitory factor *PYL*, which in turn inhibits *PP2C*.

3.11. Transcriptome Analysis of the Sugar Metabolism Pathway

The sugar metabolism pathway is divided into two parts: soluble sugar transformation metabolism and starch synthesis and metabolism in plastids. The metabolism of soluble sugars mainly involves the interconversion of sucrose, fructose, glucose, and trehalose. These monosaccharides are then further metabolized through glycolysis. Glucose-6-phosphate (D-glucose-6P) is converted into glucose 1-phosphate (D-glucose-1P) under the catalysis of glucose phosphate mutase (PGM), and then is polymerized to form starch under the catalysis of a series of enzymes in plastids. Subsequently, starch is broken down to maltose catalyzed by BMY or to glucose catalyzed by amylase (AMY). Maltose can also be further broken down to glucose by α-glucosidase (GAA) (Figure 9). The transcriptome data showed that INV (FxaC_24g24690, FxaC_25g57790, FxaC_25g48110, FxaC_28g06710, FxaC_25g48920, and FxaC_27g43570), involved in the decomposition of sucrose into glucose and fructose, was significantly up-regulated in the GR24-treated plants. On the other hand, TREH (FxaC_19g11430, FxaC_17g12410, and FxaC_20g12180), involved in the conversion of trehalose to glucose, was significantly reduced in the GR24-treated plants. Additionally, HK (FxaC_3g06340, FxaC_22g23850, FxaC_8g07590, FxaC_6g43410, and FxaC_7g40800) and FRK (FxaC_7g21700, FxaC_24g17000, FxaC_15g07820, FxaC_14g07810, FxaC_13g09120, FxaC_16g24240, FxaC_15g28090, FxaC_1g38700, and FxaC_26g01820) were also significantly up-regulated in the GR24-treated plants. And in the plastid, BMY (FxaC_13g49830, FxaC_15g35160, FxaC_16g05110, FxaC_17g12490, FxaC_17g15860, FxaC_18g34720, FxaC_19g11530, FxaC_25g19860, FxaC_25g31350, FxaC_26g23910, FxaC_27g23620, and FxaC_28g22720) participated in the reaction of starch decomposition to maltose, which was significantly down-regulated in the GR24-treated plants. This suggests that the decrease in the glucose content in the crowns of the GR24-treated plants may be due to the up-regulation of HK and the down-regulation of TREH and BMY.

3.12. Real-Time Quantitative PCR of the Key Genes

To verify the accuracy of the transcriptome data, we selected 16 key DEGs (FaAHP1-like, FaHK2-like, FaARR1-like, FaLAX2, FaMETTL5, FaAUX28-like, FaIAA8-like, FaIAA9, FaIAA27, FaARF5, FaPP2C10, FaPP2C12, FaPP2C13, FaPP2C27, FaABF1, and FaABF5) involved in the CTK, IAA, and ABA signaling pathways to perform a real-time quantitative PCR (Figure 10). The results show that the expression levels of FaAHP1-like and FaHK2-like in the CTK signaling pathway were significantly increased in the 5 μmol/L GR24-treated plants, while the expression of FaARR1-like was significantly down-regulated (Figure 10A–C). Additionally, the expression of FaMETTL5, FaAUX28-like, FaIAA8-like, FaIAA9, and FaIAA27 in the IAA signaling pathway was significantly up-regulated compared to the control, whereas the expression levels of FaLAX2 and FaARF5 were significantly down-regulated (Figure 10D–J). Among the key genes involved in the ABA signaling pathway, the expression levels of FaABF1 and FaABF5 were significantly higher in the 5 μmol/L GR24-treated plants than that in the control, while the expression levels of FaPP2C10, FaPP2C12, FaPP2C13, and FaPP2C27 were significantly lower in the 5 μmol/L GR24-treated plants than in the control (Figure 10K–P). In summary, the real-time quantitative PCR results of these 16 key genes are consistent with transcriptome data, indicating the reliability of the transcriptome data.

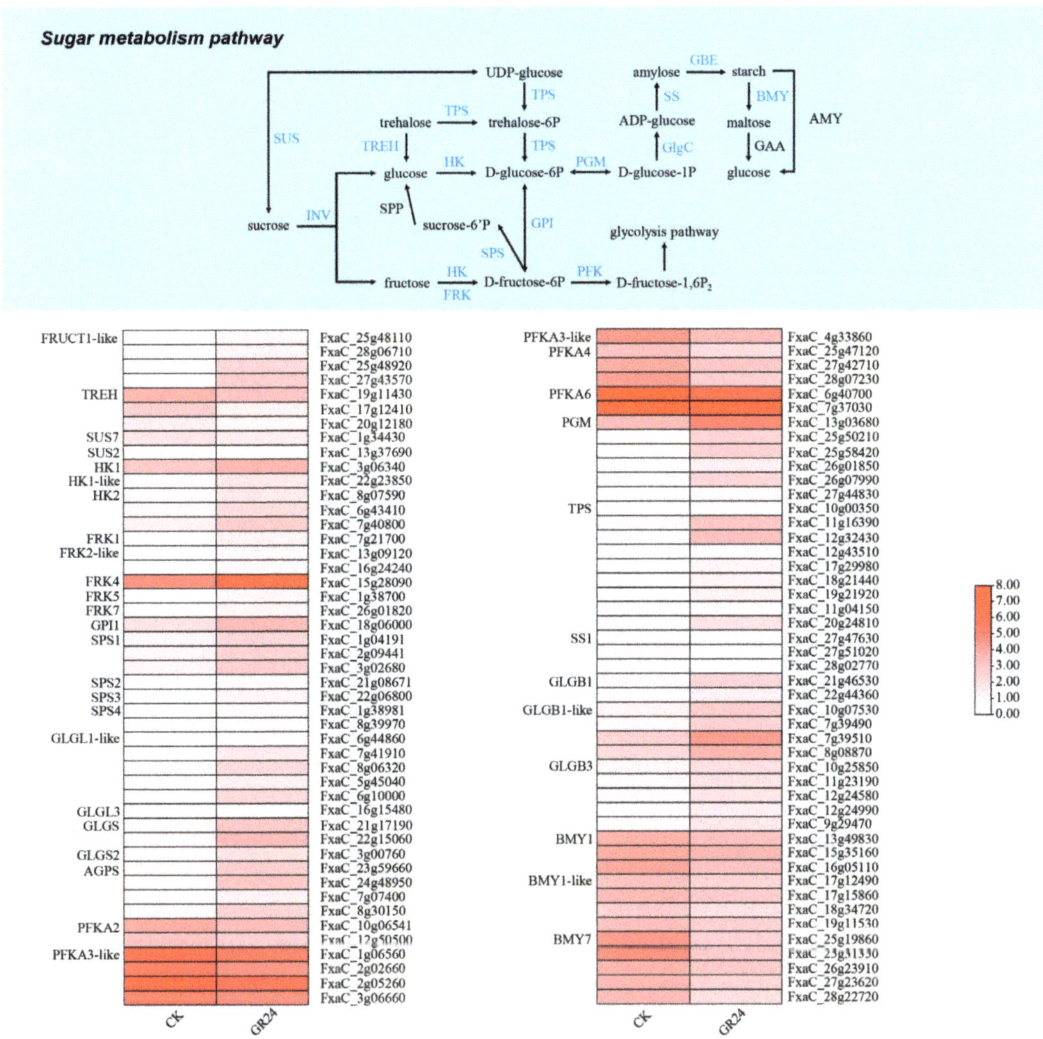

Figure 9. Transcriptome analysis of the sugar metabolism pathway. SUS, sucrose synthase; INV, invertase; TREH, trehalase; TPS, terpene synthase; HK, hexokinase; FRK, fructokinase; SPS, sucrose phosphate synthase; GPI, glucose phosphate isomerase; PGM, phosphoglucomutase; PFK, phosphofructokinase; SS, starch synthase; GlgC, ADP-glucose pyrophosphorylase; GBE, glycogen-branching enzyme; BMY, beta-amylase.

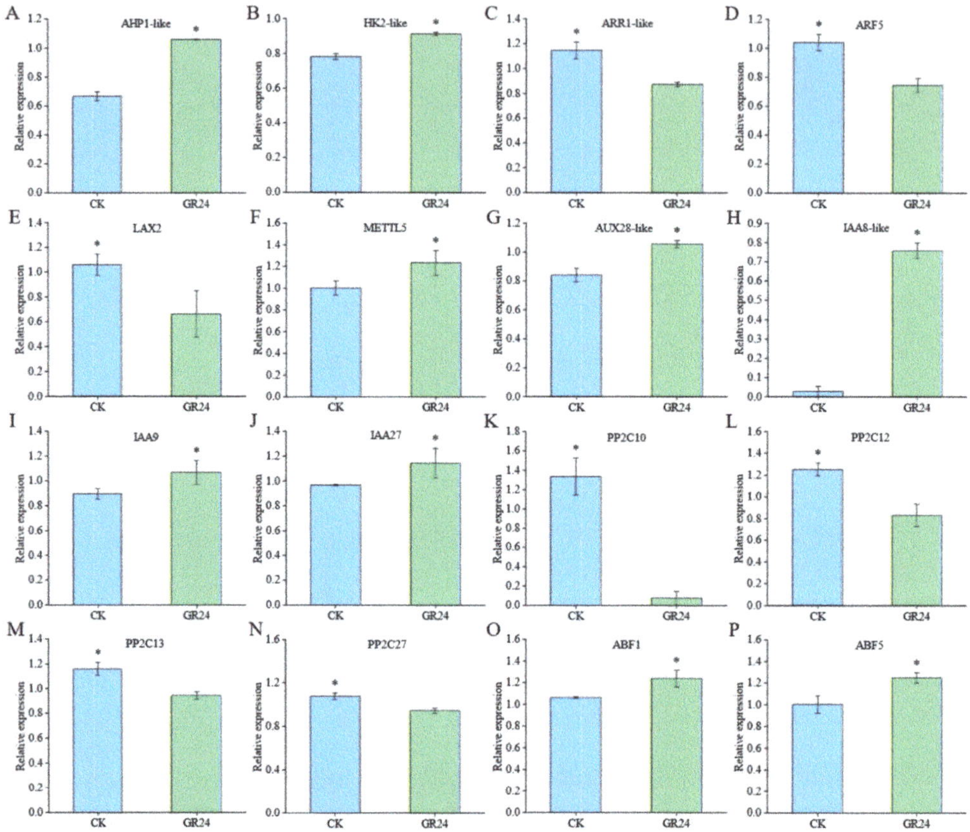

Figure 10. Real-time quantitative PCR analysis of the DEGs of CK and the 5 µmol/L GR24-treated plants. The relative expression levels of *FaAHP1-like* (**A**), *FaHK2-like* (**B**), *FaARR1-like* (**C**), *FaLAX2* (**D**), *FaMETTL5* (**E**), *FaAUX28-like* (**F**), *FaIAA8-like* (**G**), *FaIAA9* (**H**), *FaIAA27* (**I**), *FaARF5* (**J**), *FaPP2C10* (**K**), *FaPP2C12* (**L**), *FaPP2C13* (**M**), *FaPP2C27* (**N**), *FaABF1* (**O**), and *FaABF5* (**P**). The * indicates statistically significant differences ($p \leq 0.05$, Student's *t*-test).

4. Discussion

The occurrence of stolons in strawberries before fruit formation can hinder the allocation of nutrients for reproductive growth, resulting in a decline in both the yield and quality of strawberry fruits. In this study, we investigated the effects of spraying different concentrations of the SL analogue GR24 on the strawberry crown. We found that GR24 not only promoted the growth of strawberry plants but also effectively inhibited the sprouting of stolons. A further study revealed that GR24 inhibited the growth of the stolons by affecting the accumulation of soluble sugars and the signaling pathways of phytohormones in the strawberry crown.

SL participates in plant morphogenesis. Previous studies have shown that SL signaling in the vascular cortex can stimulate cortex growth and promote stem thickening in *Arabidopsis* [35]. SL can also promote the thickening of stems during maize (*Zea mays* L.) development [36]. In the present study, we found that the exogenous GR24 treatment at different concentrations (5 µmol/L, 10 µmol/L, and 20 µmol/L) promoted stem thickening in strawberry plants (Figure 1), which was consistent with previous research. In addition, after knocking out the SL receptor gene *dwarf14* in *Arabidopsis*, the plants showed dwarfing and shortened petioles [37]. In our experiment, both the 5 µmol/L and 10 µmol/L GR24 treatments promoted plant height (Figure 2A,B) and petiole length in strawberry plants

(Figure 2C,D). However, GR24 had no significant effect on the strawberry leaf thickness. In summary, exogenous GR24 promotes stem thickening, plant height, and petiole growth in strawberry plants, leading to an overall growth promotion.

The branches of plants develop from axillary buds (AXBs) in the leaf axils, which in turn are formed by axillary meristems (AMs) under suitable conditions [38,39]. The most prominent effect of SL in plants is to regulate the growth of the above-ground parts and inhibit plant branching. The levels of SL were found to be significantly reduced in the branching mutants of certain plants, but the application of SL inhibited branching in these mutants [40]. The mutants *d14* and *d10* in *Arabidopsis* exhibited increased branching and reduced plant height, while exogenous SL effectively inhibited tillering in the *d10* mutant [41]. Exogenous GR24 has been found to effectively inhibit the occurrence of secondary branches in rapeseed (*Brassica napus* L.) and improve the quality and yield of siliques [42]. In this study, exogenous GR24 effectively reduced the number of stolons in the strawberry plants (Figure 4), which is consistent with the results of previous studies.

Many studies have shown that SL can interact with various endogenous phytohormones in plants to jointly regulate plant growth and development. These phytohormones include CTK, GA, auxin, jasmonic acid, and ABA [43–46]. SL and CTK have opposite effects on axillary bud growth. For example, one study reported that exogenous GR24 inhibited the expression of CTK-synthesis-related genes in non-heading Chinese cabbage (*Brassica campestris ssp.* chinensis Makino). It also inhibited the promotion of axillary shoot growth by the CTK content in peas and increased the expression of the CTK-synthesis-related gene *PsIPT1* in the SL mutants [21,47]. Other studies have also shown that, after 12 h of the exogenous GR24 treatment, the tZR content in rice tillering buds significantly decreased, while the tZR content in the tillering nodes significantly increased [48]. SL can also inhibit the expression of *A-ARR* in rice buds [49]. In addition, the activation of *CRE1* and *ARR5* in the CTK signaling pathway was detected in plants with increased strawberry stolons [38], while the mutation of the B-type *ARR1* in *Arabidopsis* led to increased branching [50]. In this study, we found that the exogenous GR24 treatment significantly increased the tZR content and the expression of *IPT* and *CYP735A* in the CTK synthesis pathway in strawberry crowns (Figure 5). However, it significantly inhibited the expression of *CKX*, *ZOG*, and *UGT73C1* in the CTK metabolic pathways, resulting in an enhanced CTK synthesis and lowered metabolism (Figure 6). Additionally, the exogenous GR24 treatment activated the expression of *CRE1* and *AHP*, leading to the activation of *A-ARR* and the suppression of *B-ARR* in the CTK signaling pathway, thereby inhibiting this pathway (Figure 6). Therefore, we believe that, although exogenous GR24 promoted the accumulation of CTK in the strawberry crowns, it significantly inhibited the CTK signaling pathway. As a result, the development of strawberry axillary buds was hindered, and the sprouting of stolons was reduced.

GA has different effects on branching in different species. It can negatively regulate branching in pea and *Arabidopsis*, but can positively regulate the growth of axillary buds in perennial strawberry and jatropha (*Jatropha curcas* L.) [43,51–53]. GA20ox4 is responsible for the production of GA. The mutation of GA20ox4 results in fewer stolons in *F.vesca*, but this GA20ox4 can be rescued by the application of exogenous GA_3 [54]. In the present experiment, exogenous GR24 significantly increased the content of GA_3 in the strawberry crown (Figure 5), but the numbers of stolons were significantly inhibited (Figure 4). The further analysis of the transcriptome data revealed that exogenous GR24 induced the expression of *CPS*, *KS*, and *GA3ox* in the gibberellin synthesis pathway, while inhibiting the expression of *GA2ox* in the gibberellin metabolic pathway (Figure 7). In addition, the decreased expression of the repressor DELLA protein led to an enhanced GA signaling pathway (Figure 7). However, contrary to previous studies, the enhanced metabolism and signaling pathways did not promote strawberry tillering in this experiment. Thus, we suggest that GA_3 may not be the main factor that caused a decrease in the number of stolons in the GR24-treated plants.

IAA signal transduction is an important segment of IAA affecting plant growth and development. Studies have shown that it is the main signaling pathway for auxin in plants, that the signal passes through $SCF^{TIR1/AFB}$ receptors, and the ubiquitination degradation of AUX/IAA proteins; ultimately, it regulates the downstream gene expression through the ARF family of transcription factors [55]. Multiple studies have shown the significance of AUX/IAA in the lateral branches' growth in plants. For instance, the down-regulation of *SlIAA2*, *SlIAA4*, *SlIAA7*, and *SlIAA9* in transgenic tomatoes (*Solanum lycopersicum*) with multi-branching has been observed, while the overexpression of *ARF* transcription factors in eggplants (*Solanum melongena* L.) has been found to promote branching [56,57]. In the present study, although exogenous GR24 did not cause a significant difference in the content of IAA in the strawberry crowns, it did lead to a down-regulation of the *LAX2* gene, which encodes the IAA receptors in the IAA signaling pathway. Additionally, it promoted the expression of the inhibitory factor AUX/IAA, while inhibiting the transcription of the transcription factor *ARF* at the same time (Figure 8). Therefore, exogenous GR24 may inhibit the IAA signaling pathway by up-regulating *AUX/IAA* and down-regulating *ARF*, inhibiting strawberry tillering as a result.

ABA is a significant phytohormone regulating axillary bud growth and development. Previous studies have shown that ABA inhibits the growth of axillary buds in plants such as *Arabidopsis* and sugarcane (*Saccharum officinarum* L.) [58,59]. The ABA receptor PYR/PYL senses ABA and undergoes a conformational change. After interacting with PP2C, it can completely cover up the active site of PP2C phosphatase, thus activating SnRK2s and transmitting signals to the downstream transcription factors ABFs [60]. The overexpression of *SlTCP26* in tomatoes led to an increase in branching and the strong inhibition of the genes *SlSAPK2*, *SlSAPK3*, *SlSnRK2C*, and *SlSnRK2I* encoding the core component SnRK2 in the ABA signaling pathway. Meanwhile, the key gene *SlABF4* in the ABA signaling response was also strongly inhibited [61]. The down-regulation of *PYR/PYL* and the up-regulation of *PP2C* were detected in the apple multi-branch mutant, resulting in the inhibition of ABA signal transduction [62]. The exogenous cyclanilide treatment promoted apple branching, and the ABA signal responsive protein PP2C was down-regulated 168 h after the treatment [63]. In this experiment, although exogenous GR24 did not affect the ABA content in the strawberry crowns, it up-regulated the expression of *PYR/PYL*, *SnRK2*, and *ABF* and down-regulated the expression of *PP2C* (Figure 8). Therefore, we suggest that exogenous GR24 may promote ABA signal transduction by promoting the expression of *SnRK2* and *ABF*, which are key factors in the ABA signaling pathway, ultimately leading to the inhibition of strawberry plant branching.

In plants, sugar can serve as both a source of energy and a signal factor for regulating plant branching. Increasing the supply of sugar to the entire plant or explant buds can stimulate bud growth. However, dormant buds exhibit transcriptional characteristics similar to those observed during carbon starvation [64,65]. In rice, sucrose is hydrolyzed into glucose and fructose, promoting the elongation of rhizomes and the germination of basal axillary buds [66]. In this study, after being treated with exogenous GR24, a decrease in the fructose and glucose contents was detected in the strawberry crowns (Figure 4). The transcriptomic data showed that the expression of *INV*, which regulates the conversion of sucrose into fructose and glucose, was up-regulated, while the expression levels of *HK* and *FRK*, which regulate the degradation of these two monosaccharides, were also significantly up-regulated. In addition, the expression of *TREH*, which is involved in the conversion of trehalose into glucose, was significantly down-regulated. Moreover, the expression of *BMY*, which regulates the breakdown of starch into maltose in plastids, was also significantly down-regulated (Figure 9). And it is universally acknowledged that maltose is a disaccharide composed of two glucoses molecules. To sum up, we propose that exogenous GR24 may inhibit the transformation of trehalose into glucose and starch into maltose by promoting the decomposition of fructose and glucose, thus reducing the content of soluble sugar in the strawberry crowns and inhibiting strawberry tillering.

5. Conclusions

Our study suggests that exogenous GR24 can promote the robust growth of strawberry plants and inhibit tillering by affecting phytohormone signaling and the sugar metabolism pathway in the crowns. Therefore, we proposed a model in which exogenous GR24 may inhibit the germination of axillary buds and reduce tillering in strawberry plants by inhibiting the CTK and IAA signaling pathways, affecting sugar metabolism, and promoting the ABA signal transduction pathway (Figure 11). And the signaling pathways of auxin and cytokinin may play a synergistic role in affecting strawberry plant tillering. Their weakened signal led to a weakened nutritional growth and reduced tillering, while the abscisic acid signaling pathway antagonized both. These findings not only provide new technical strategies for properly controlling stolon formation during strawberry cultivation, but also provide valuable insights into the mechanism of tiller sprouting.

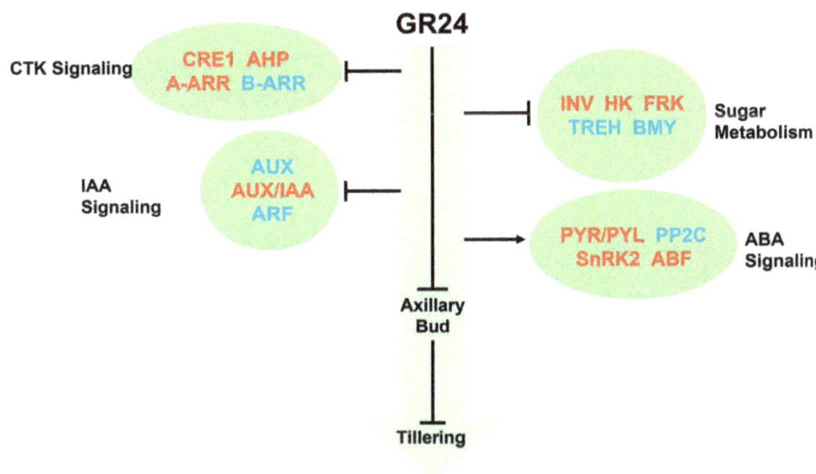

Figure 11. The model of exogenous GR24 influencing the tillering of strawberry plants. The up-regulated substances are reddened, and the down-regulated substances are marked in blue. GR24 up-regulated CRE1, AHP, and A-ARR and suppressed the expression of B-ARR, thereby inhibiting the CTK signaling transduction pathway. The IAA signaling pathway was also significantly inhibited, with AUX and ARF being repressed and AUX/IAA being up-regulated. In addition, GR24 also up-regulated INV, HK, and FRK; down-regulated TREH and BMY; and reduced the content of soluble sugars in the crown of strawberries. ABA inhibited plant tillering, and in this experiment, and its signaling transduction pathway was promoted. Among them, PYR/PYL, SnRK2, and ABF were up-regulated and PP2C was down-regulated. By acting on these four pathways, exogenous GR24 regulated the growth and development of axillary buds and reduced tillering in strawberries.

Supplementary Materials: The following supporting information can be downloaded at: https://www.mdpi.com/article/10.3390/agronomy13123078/s1. Table S1: Gene quantitative primers.

Author Contributions: Conceptualization, Y.L. (Ya Luo) and Y.P.; methodology, Y.J.; software, M.S.; validation, Y.J. and C.H.; formal analysis, M.L. and Q.C.; investigation, Y.P. and Y.J.; resources, Y.Z. (Yong Zhang), Y.L. (Yuanxiu Lin), and Y.Z. (Yunting Zhang); data curation, Y.W.; writing—original draft preparation, Y.P. and Y.J.; writing—review and editing, Y.L. (Ya Luo); visualization, W.H.; supervision, H.T., X.W., and Y.L. (Ya Luo). All authors have read and agreed to the published version of the manuscript.

Funding: This research was financially supported by the National Natural Science Foundation of China (No. 32272688); the Natural Science Foundation of Sichuan Province (No. 23NSFSC0777); the Key R&D Project of Science and Technology Department of Sichuan Province (No. 2023YFN0005); and Sichuan Tianfu New Area Rural Revitalization Research Institute "the open competition project to select the best candidates" (No. XZY108).

Data Availability Statement: The genome sequence data are available from the GDR database (accessed on 10 September 2022). The data presented in the study have been deposited in the NCBI SRA database under the accession number CNP0004398.

Acknowledgments: The authors would express their appreciation to Xiaoyang Liu for his help during the experiments.

Conflicts of Interest: The authors declare no conflict of interest.

References

1. Gasparrini, M.; Giampieri, F.; Forbes-Hernandez, T.Y.; Afrin, S.; Cianciosi, D.; Reboredo-Rodriguez, P.; Varela-Lopez, A.; Zhang, J.J.; Quiles, J.L.; Mezzetti, B.; et al. Strawberry extracts efficiently counteract inflammatory stress induced by the endotoxin lipopolysaccharide in Human Dermal Fibroblast. *Food Chem. Toxicol.* **2018**, *114*, 128–140. [CrossRef] [PubMed]
2. Guan, L.; Zhao, M.Z.; Qian, Y.M.; Yu, H.M.; Xia, J.; Wu, E.J. Phenotypic analysis combined with tandem mass tags (TMT) labeling reveal the heterogeneity of strawberry stolon buds. *BMC Plant Biol.* **2020**, *19*, 505. [CrossRef] [PubMed]
3. Tenreira, T.; Lange, M.J.P.; Lange, T.; Bres, C.; Labadie, M.; Monfort, A.; Hernould, M.; Rothan, C.; Denoyes, B. A specific gibberellin 20-oxidase dictates the flowering-runnering decision in diploid strawberry. *Plant Cell* **2017**, *29*, 2168–2182. [CrossRef] [PubMed]
4. Hytönen, T.; Elomaa, P.; Moritz, T.; Junttila, O. Gibberellin mediates daylength-controlled differentiation of vegetative meristems in strawberry (*Fragaria* × *ananassa* Duch). *BMC Plant Biol.* **2009**, *9*, 18. [CrossRef] [PubMed]
5. Konsin, M.; Voipio, I.; Palonen, P. Influence of photoperiod and duration of short-day treatment on vegetative growth and flowering of strawberry (*Fragaria* × *ananassa* Duch.). *J. Hort. Sci. Biotech.* **2001**, *76*, 77–82. [CrossRef]
6. Hytönen, T.; Palonen, P.; Mouhu, K.; Junttila, O. Crown branching and cropping potential in strawberry (*Fragaria* × *ananassa* Duch.) can be enhanced by daylength treatments. *J. Hort. Sci. Biotech.* **2004**, *79*, 466–471. [CrossRef]
7. Hytönent, T.; Kurokura, T. Control of flowering and runnering in strawberry. *Hortic. J.* **2020**, *89*, 96–107. [CrossRef]
8. Guo, L.; Plunkert, M.; Luo, X.; Liu, Z.C. Developmental regulation of stolon and rhizome. *Curr. Opin. Plant Biol.* **2021**, *59*, 101970. [CrossRef]
9. Andres, J.; Caruana, J.; Liang, J.H.; Samad, S.; Monfort, A.; Liu, Z.C.; Hytonen, T.; Koskela, E.A. Woodland strawberry axillary bud fate is dictated by a crosstalk of environmental and endogenous factors. *Plant Physiol.* **2021**, *187*, 1221–1234. [CrossRef]
10. Li, Y.L.; Hu, J.T.; Wei, H.; Jeong, B.R. A long-day photoperiod and 6-benzyladenine promote runner formation through upregulation of soluble sugar content in strawberry. *Int. J. Mol. Sci.* **2020**, *21*, 4917. [CrossRef]
11. Yu, J.J.; Li, M.; Li, Q.G.; Wang, R.Y.; Li, R.N.; Yang, Z.M. Reallocation of soluble sugars and IAA regulation in association with enhanced stolon growth by elevated CO_2 in creeping bentgrass. *Plants* **2022**, *11*, 1500. [CrossRef] [PubMed]
12. Ran, F.; Yuan, Y.J.; Bai, X.M.; Li, C.N.; Li, J.X.; Chen, H. Carbon and nitrogen metabolism affects kentucky bluegrass rhizome expansion. *BMC Plant Biol.* **2023**, *23*, 221. [CrossRef] [PubMed]
13. Qiu, Y.T.; Guan, S.C.; Wen, C.J.; Li, P.; Gao, Z.; Chen, X. Auxin and cytokinin coordinate the dormancy and outgrowth of axillary bud in strawberry runner. *BMC Plant Biol.* **2019**, *19*, 528. [CrossRef] [PubMed]
14. Lei, K.Q.; Tan, Q.W.; Zhu, L.Q.; Xu, L.B.; Yang, S.K.; Hu, J.L.; Gao, L.J.; Hou, P.; Shao, Y.H.; Jiang, D.; et al. Low red/far-red ratio can induce cytokinin degradation resulting in the inhibition of tillering in wheat (*Triticum aestivum* L.). *Front. Plant Sci.* **2022**, *13*, 971003. [CrossRef] [PubMed]
15. Djennane, S.; Hibrand-Saint, O.L.; Kawamura, K.; Lalanne, D.; Laffaire, M.; Thouroude, T.; Chalain, S.; Sakr, S.; Boumaza, R.; Foucher, F.; et al. Impacts of light and temperature on shoot branching gradient and expression of strigolactone synthesis and signalling genes in rose. *Plant Cell Environ.* **2014**, *37*, 742–757. [CrossRef] [PubMed]
16. Evers, J.B.; van der Krol, A.R.; Vos, J.; Struik, P.C. Understanding shoot branching by modelling form and function. *Trends Plant Sci.* **2011**, *16*, 464–467. [CrossRef]
17. Bhoi, A.; Yadu, B.; Chandra, J.; Keshavkant, S. Contribution of strigolactone in plant physiology, hormonal interaction and abiotic stresses. *Planta* **2021**, *254*, 28. [CrossRef]
18. Cook, C.E.; Whichard, L.P.; Turner, B.; Wall, M.E.; Egley, G.H. Germination of witchweed (*Striga lutea* Lour.): Isolation and properties of a potent stimulant. *Science* **1966**, *154*, 1189–1190. [CrossRef]
19. Gomez-Roldan, V.; Fermas, S.; Brewer, P.B.; Puech-Pages, V.; Dun, E.A.; Pillot, J.P.; Letisse, F.; Matusova, R.; Danoun, S.; Portais, J.C.; et al. Strigolactone inhibition of shoot branching. *Nature* **2008**, *455*, 189–1894. [CrossRef]
20. Ferguson, B.J.; Beveridge, C.A. Roles for auxin, cytokinin, and strigolactone in regulating shoot branching. *Plant Physiol.* **2009**, *149*, 1929–1944. [CrossRef]

21. Dun, E.A.; de Saint Germain, A.; Rameau, C.; Beveridge, C.A. Antagonistic action of strigolactone and cytokinin in bud outgrowth control. *Plant Physiol.* **2012**, *158*, 487–498. [CrossRef] [PubMed]
22. Ito, S.; Yamagami, D.; Umehara, M.; Hanada, A.; Yoshida, S.; Sasaki, Y.; Yajima, S.; Kyozuka, J.; Ueguchi-Tanaka, M.; Matsuoka, M.; et al. Regulation of strigolactone biosynthesis by gibberellin signaling. *Plant Physiol.* **2017**, *174*, 1250–1259. [CrossRef] [PubMed]
23. Koltai, H.; Prandi, C. Strigolactones: Biosynthesis, synthesis and functions in plant growth and stress responses. In *Phytohormones: A Window to Metabolism, Signaling and Biotechnological Applications*; Springer: Berlin/Heidelberg, Germany, 2014; pp. 265–288.
24. Zwanenburg, B.; Mwakaboko, A.S.; Reizelman, A.; Anilkumar, G.; Sethumadhavan, D. Structure and function of natural and synthetic signalling molecules in parasitic weed germination. *Pest. Manag. Sci.* **2009**, *65*, 478–491. [CrossRef]
25. Wang, L.; Wang, B.; Yu, H.; Guo, H.Y.; Lin, T.; Kou, L.Q.; Wang, A.Q.; Shao, N.; Ma, H.Y.; Xiong, G.S.; et al. Transcriptional regulation of strigolactone signalling in *Arabidopsis*. *Nature* **2021**, *158*, 277–281. [CrossRef] [PubMed]
26. Liu, X.J.; Xu, Y.; Sun, W.X.; Wang, J.Y.; Gao, Y.X.; Wang, L.; Xu, W.P.; Wang, S.P.; Jiu, S.; Zhang, C.X. Strigolactones modulate stem length and diameter of cherry rootstocks through interaction with other hormone signaling pathways. *Front. Plant Sci.* **2023**, *14*, 1092654. [CrossRef] [PubMed]
27. Vogel, D.; Hills, P.; Moore, J.P. Strigolactones GR-24 and Nijmegen applications result in reduced susceptibility of tobacco and grapevine plantlets to Botrytis cinerea infection. *Plants* **2023**, *12*, 3202. [CrossRef]
28. Huang, D.D.; Wang, Y.Y.; Zhang, D.C.; Dong, Y.F.; Meng, Q.X.; Zhu, S.H.; Zhang, L.L. Strigolactone maintains strawberry quality by regulating phenylpropanoid, NO, and H_2S metabolism during storage. *Postharvest Biol. Tech.* **2021**, *178*, 111546. [CrossRef]
29. Wu, H.; Li, H.H.; Chen, H.; Qi, Q.; Ding, Q.Q.; Xue, J.; Ding, J.; Jiang, X.N.; Hou, X.L.; Li, Y. Identification and expression analysis of strigolactone biosynthetic and signaling genes reveal strigolactones are involved in fruit development of the woodland strawberry (*Fragaria vesca*). *BMC Plant Biol.* **2019**, *19*, 73. [CrossRef]
30. Hu, P.P.; Zhang, X.F.; Zhao, X.; Li, G.; Zhao, F.L.; Li, L.J.; Zhou, H.C. Relationship between strigolactones and branching in strawberry. *J. Fruit. Sci.* **2019**, *36*, 578–589.
31. Aksic, M.F.; Tosti, T.; Sredojevic, M.; Milivojevic, J.; Meland, M.; Natic, M. Comparison of sugar profile between leaves and fruits of blueberry and strawberry cultivars grown in organic and integrated production system. *Plants* **2019**, *8*, 205. [CrossRef]
32. Chen, Q.; Yu, H.W.; Wang, X.R.; Xie, X.L.; Yue, X.Y.; Tang, H.R. An alternative cetyltrimethylammonium bromide-based protocol for RNA isolation from blackberry (*Rubus* L.). *Genet. Mol. Res.* **2012**, *11*, 1773–1782. [CrossRef] [PubMed]
33. Jung, H.; Lee, D.K.; Choi, Y.D.; Kim, J.K. *OsIAA6*, a member of the rice Aux/IAA gene family, is involved in drought tolerance and tiller outgrowth. *Plant Sci.* **2015**, *236*, 304–312. [CrossRef] [PubMed]
34. Kishor, P.B.K.; Tiozon, R.N.; Fernie, A.R..; Sreenivasulu, N. Abscisic acid and its role in the modulation of plant growth, development, and yield stability. *Trends Plant Sci.* **2022**, *27*, 1283–1295. [CrossRef] [PubMed]
35. Agusti, J.; Herold, S.; Schwarz, M.; Sanchez, P.; Ljung, K.; Dun, E.A.; Brewer, P.B.; Beveridge, C.A.; Sieberer, T.; Sehr, E.M.; et al. Strigolactone signaling is required for auxin-dependent stimulation of secondary growth in plants. *Proc. Natl. Acad. Sci. USA* **2011**, *108*, 20242–20247. [CrossRef] [PubMed]
36. Screpanti, C.; Fonne-Pfister, R.; Lumbroso, A.; Rendine, S.; Lachia, M.; De Mesmaeker, A. Strigolactone derivatives for potential crop enhancement applications. *Bioorg. Med. Chem. Lett.* **2016**, *26*, 2392–2400. [CrossRef] [PubMed]
37. White, A.R.F.; Mendez, J.A.; Khosla, A.; Nelson, D.C. Rapid analysis of strigolactone receptor activity in a Nicotiana benthamiana *dwarf14* mutant. *Plant Direct.* **2022**, *6*, e389. [CrossRef] [PubMed]
38. Liang, J.H.; Wu, Z.; Zheng, J.; Koskela, E.A.; Fan, L.J.; Fan, G.X.; Gao, D.H.; Dong, Z.F.; Hou, S.F.; Feng, Z.K.; et al. The GATA factor *HANABA TARANU* promotes runner formation by regulating axillary bud initiation and outgrowth in cultivated strawberry. *Plant J.* **2022**, *110*, 1237–1254. [CrossRef]
39. Liu, Y.H.; Yu, L.; Ding, J.H.; Wuang, R.Z.; Huang, Z.G.; Xiao, L.T. Research progress in synergistic regulatory roles of phytohormones in shoot branching. *Plant Physiol. J.* **2012**, *48*, 941–948.
40. Umehara, M.; Hanada, A.; Yoshida, S.; Akiyama, K.; Arite, T.; Takeda-Kamiya, N.; Magome, H.; Kamiya, Y.; Shirasu, K.; Yoneyama, K.; et al. Inhibition of shoot branching by new terpenoid plant hormones. *Nature* **2008**, *455*, 195–200. [CrossRef]
41. Arite, T.; Umehara, M.; Ishikawa, S.; Hanada, A.; Maekawa, M.; Yamaguchi, S.; Kyozuka, J. *d14*, a strigolactone-insensitive mutant of rice, shows an accelerated outgrowth of tillers. *Plant Cell Physiol.* **2008**, *50*, 1416–1424. [CrossRef]
42. Su, S.Y.; Luo, W.G.; Xu, M.; Yuan, F.; Li, L.X.; Xu, F.; Su, Y.; Xiao, L.T. Effects of strigolactone on the branching, growth and development in *Brassica napus*. *Mol. Plant Breed.* **2020**, *18*, 6822–6827.
43. Ni, J.; Gao, C.C.; Chen, M.S.; Pan, B.Z.; Ye, K.Q.; Xu, Z.F. Gibberellin promotes shoot branching in the perennial woody plant *Jatropha curcas*. *Plant Cell Physiol.* **2015**, *56*, 1655–1666. [CrossRef] [PubMed]
44. Torres-Vera, R.; Garcia, J.M.; Pozo, M.J.; Lopez-Raez, J.A. Do strigolactones contribute to plant defence? *Mol. Plant Pathol.* **2014**, *15*, 211–216. [CrossRef] [PubMed]
45. Brewer, P.B.; Dun, E.A.; Ferguson, B.J.; Rameau, C.; Beveridge, C.A. Strigolactone acts downstream of auxin to regulate bud outgrowth in pea and *Arabidopsis*. *Plant Physiol.* **2009**, *150*, 482–493. [CrossRef] [PubMed]
46. Liu, B.C.; Zhang, Y.; Wang, S.; Wang, W.N.; Xu, X.L.; Wu, J.R.; Fang, Y.L.; Ju, Y.L. Effects of strigolactone and abscisic acid on the quality and antioxidant activity of grapes (*Vitis vinifera* L.) and wines. *Food Chem. X* **2022**, *16*, 100496. [CrossRef] [PubMed]
47. Cui, H.M.; Cao, X.W.; Wang, J.J.; Xiong, A.S.; Hou, X.L.; Li, Y. Effects of exogenous GR24 on the growth of axillary bud of non-heading Chinese cabbage. *J. Nanjing Univ.* **2016**, *39*, 366–372.

48. Zha, M.; Wang, Y.; Chen, B.X.; Tan, Z.C. Strigolactones and cytokinin interaction in buds in the control of rice tillering. *Front. Plant Sci.* **2022**, *13*, 837136. [CrossRef]
49. Zha, M.; Imran, M.; Wang, Y.; Xu, J.; Ding, Y.; Wang, S.H. Transcriptome analysis revealed the interaction among strigolactones, auxin, and cytokinin in controlling the shoot branching of rice. *Plant Cell Rep.* **2019**, *38*, 279–293. [CrossRef]
50. Waldie, T.; Leyser, O. Cytokinin targets auxin transport to promote shoot branching. *Plant Physiol.* **2018**, *177*, 803–818. [CrossRef]
51. Mauriat, M.; Sandberg, L.G.; Moritz, T. Proper gibberellin localization in vascular tissue is required to control auxin-dependent leaf development and bud outgrowth in hybrid aspen. *Plant J.* **2011**, *67*, 805–816. [CrossRef]
52. Rameau, C.; Bertheloot, J.; Leduc, N.; Andrieu, B.; Foucher, F.; Sakr, S. Multiple pathways regulate shoot branching. *Front. Plant Sci.* **2015**, *5*, 741. [CrossRef] [PubMed]
53. Rinne, P.L.H.; Welling, A.; Vahala, J.; Ripel, L.; Ruonala, R.; Kangasjärvi, J.; van der Schoot, C. Chilling of dormant buds hyperinduces FLOWERING LOCUS T and recruits GA-inducible 1,3-β-Glucanases to reopen signal conduits and release dormancy in Populus. *Plant Cell.* **2011**, *23*, 130–146. [CrossRef]
54. Li, W.J.; Zhang, J.X.; Sun, H.Y.; Wang, S.M.; Chen, K.Q.; Liu, Y.X.; Li, H.; Ma, Y.; Zhang, Z.H. *FveRGA1*, encoding a DELLA protein, negatively regulates runner production in *Fragaria vesca. Planta* **2018**, *247*, 941–951. [CrossRef]
55. Salehin, M.; Bagchi, R.; Estelle, M. SCFTIR1/AFB-based auxin perception: Mechanism and role in plant growth and development. *Plant Cell.* **2015**, *27*, 9–19. [CrossRef] [PubMed]
56. Pattison, R.J.; Catala, C. Evaluating auxin distribution in tomato (*Solanum lycopersicum*) through an analysis of the PIN and AUX/LAX gene families. *Plant J.* **2012**, *70*, 585–598. [CrossRef] [PubMed]
57. Hu, R.L.; Wang, J.L.; Yang, H.Q.; Yuan, C.; Niu, Y.; Tang, Q.L.; Wei, D.Y.; Tian, S.B.; Yang, Y.; Wang, Z.M. Cloning and functional analysis of auxin response factor gene *SmARF5* in *Solanum melongena. Hortic. Plant J.* **2021**, *49*, 1895–1906.
58. Yao, C.; Finlayson, S.A. Abscisic acid is a general negative regulator of *Arabidopsis* axillary bud growth. *Plant Physiol.* **2015**, *169*, 611–626. [CrossRef]
59. Ortiz-Morea, F.A.; Vicentini, R.; Silva, G.F.F.; Silva, E.M.; Carrer, H.; Rodrigues, A.P.; Nogueira, F.T.S. Global analysis of the sugarcane microtranscriptome reveals a unique composition of small RNAs associated with axillary bud outgrowth. *J. Exp. Bot.* **2013**, *64*, 2307–2320. [CrossRef]
60. Guo, J.J.; Yang, X.H.; Weston, D.J.; Chen, J.G. Abscisic acid receptors: Past, present and future. *J. Integr. Plant Biol.* **2011**, *53*, 469–479. [CrossRef]
61. Wei, X.Y.; Yang, J.; Lei, D.; Feng, H.; Yang, Z.N.; Wen, G.Q.; He, Z.Y.; Zeng, W.J.; Zou, J. The SlTCP26 promoting lateral branches development in tomato. *Plant Cell Rep.* **2021**, *40*, 1115–1126. [CrossRef]
62. Ge, H.J.; Li, G.F.; Wan, S.W.; Zhao, A.H.; Huang, Y.; Ma, R.Q.; Zhang, R.F.; Song, Y.J.; Sha, G.L. Whole genome re-sequencing and transcriptome reveal an alteration in hormone signal transduction in a more-branching mutant of apple. *Gene* **2022**, *818*, 146214. [CrossRef] [PubMed]
63. Ma, J.J.; Xie, L.L.; Zhao, Q.; Sun, Y.T.; Zhang, D. Cyclanilide induces lateral bud outgrowth by modulating cytokinin biosynthesis and signalling pathways in apple identified via transcriptome analysis. *Int. J. Mol. Sci.* **2023**, *23*, 581. [CrossRef] [PubMed]
64. Wang, M.; Perez-Garcia, M.D.; Daviere, J.M.; Barbier, F.; Oge, L.; Gentilhomme, J.; Voisine, L.; Peron, T.; Launay-Avon, A.; Clement, G.; et al. Outgrowth of the axillary bud in rose is controlled by sugar metabolism and signalling. *J. Exp. Bot.* **2021**, *72*, 3044–3060. [CrossRef] [PubMed]
65. Tarancón, C.; González-Grandío, E.; Oliveros, J.C.; Nicolas, M.; Cubas, P. A conserved carbon starvation response underlies bud dormancy in woody and herbaceous species. *Front. Plant Sci.* **2017**, *8*, 788. [CrossRef]
66. Fan, Z.Q.; Huang, G.W.; Fan, Y.R.; Yang, J.Y. Sucrose facilitates rhizome development of perennial rice (*Oryza longistaminata*). *Int. J. Mol. Sci.* **2022**, *23*, 13396. [CrossRef]

agronomy

Article

Transcriptomic Analysis of *Cucumis hystrix* and the Functional Identification of *ChTrxh* under NaCl Stress

Zhaolai Guo †, Senlin Zeng †, Kunzhi Li and Huini Xu *

Faculty of Life Science and Technology, Kunming University of Science and Technology, Jingming South Street, Kunming 650224, China; slzeng0708@163.com (S.Z.); likz@kust.edu.cn (K.L.)
* Correspondence: xuhn@kust.edu.cn
† These authors contributed equally to this work.

Abstract: Salinity is a prominent environmental stressor that significantly impacts plant growth and development. Here, we conducted research on the physiological and transcriptomic mechanism of a wild cucumber, *Cucumis hystrix* Chakr, under NaCl stress. Physiological data showed that contents of malondialdehyde, peroxide (H_2O_2), proline, soluble sugar, and activities of antioxidant enzymes of superoxide dismutase, peroxidase, ascorbate peroxidase, and glutathione reductase in wild cucumber plants were increased significantly after NaCl treatment. Transcriptomic analysis revealed that 3509 transcripts were differentially expressed in leaves and 5516 transcripts in roots after NaCl treatment. Numerous genes were related to the signal transduction, transcription factor, ion transport, osmotic metabolism, and reactive oxygen species scavenging. Moreover, the thioredoxin H type gene of *Cucumis hystrix* Chakr (*ChTrxh*) was isolated and characterized. Our study demonstrated that the transgenic tobacco plants overexpressing *ChTrxh* exhibited enhanced tolerance to NaCl stress compared to wild-type plants. These findings contribute valuable insights into the functional characteristics of important genes in wild cucumber under NaCl stress.

Keywords: *Cucumis hystrix*; salt stress; ROS; differentially expressed genes; transgenic tobacco; antioxidant enzymes activities

Citation: Guo, Z.; Zeng, S.; Li, K.; Xu, H. Transcriptomic Analysis of *Cucumis hystrix* and the Functional Identification of *ChTrxh* under NaCl Stress. *Agronomy* **2023**, *13*, 2931. https://doi.org/10.3390/agronomy13122931

Academic Editor: Ainong Shi

Received: 27 October 2023
Revised: 21 November 2023
Accepted: 25 November 2023
Published: 28 November 2023

1. Introduction

Salinity stress is a worldwide concern that severely affects the growth, development, and yield of crops. The area of soil salinization is gradually increasing due to environmental factors and human-induced soil degradation. This stress leads to osmotic stress and ion toxicity, resulting in the excessive accumulation of ions within plants. Consequently, the primary effects of salinity stress often lead to secondary stresses, such as oxidative damage [1]. Plants employ various biochemical and molecular mechanisms to deal with salt stress [2].

The responses of plants to salt stress have been extensively investigated, particularly with regard to transcriptomic analysis in various plants species. Many stress-responsive genes have been identified by RNA sequencing (RNA-seq) technology under salt stress [3–6]. Some wild species of plants, having high levels of salt tolerance, are important resources for investigating the adaptive mechanisms that form the basis of salt tolerance. RNA-Seq analysis was employed to investigate the transcriptome of wild barley (*H. spontaneum*) [7], wild cotton (*Gossypium klotzschianum*) [8], wild rice [9], and wild recretohalophyte *Reaumuria trigyna* [10] under salt stress. These plants offer the potential to provide new insights into stress-tolerant mechanisms and theoretical guidance for the cultivation of resilient agronomic crops.

Cucumis hystrix Chakr. (HH, 2n = 24) serves as the wild counterpart to cultivated cucumber [11]. Previous research highlighted the presence of several potentially advantageous disease-resistant and abiotic stress-tolerant traits in *Cucumis hystrix* Chakr., which could be utilized for enhancing cucumber (*C. sativus* L., CC, 2n = 14) traits [12]. Additionally, new resistance against downy mildew has also been discovered in the wild species, *C. hystrix Chakr* [13]. Nonetheless, the response of wild cucumber to salt stress remains uninvestigated.

Therefore, this study aimed to investigate the physiological and transcriptomic responses of wild cucumber under NaCl stress. The research will contribute new insights into the salt tolerance mechanisms exhibited by wild cucumber and will serve as valuable research for identifying novel salt resistance genes to enhance the salt tolerance of various crops.

2. Materials and Methods

2.1. Plant Materials and Salt Treatment

The seeds of *C. hystrix* were collected in PuEr, Yunnan province, China (Figure S1). The seeds were surface-sterilized, washed with sterile water, and germinated in a dark incubator at 28 °C on moisture filter paper. After germination, the seeds were transferred to vermiculite pots for 15 days. Subsequently, plantlets of uniform size were transplanted into plastic containers filled with a nutrient solution, following our previously established hydroponic growth protocol [14]. Experiments were conducted with two treatment groups: the control group (CK), receiving 0 mM NaCl treatment), and the NaCl treatment group, receiving 100 mM NaCl treatment for a duration for 24 h. Each treatment was replicated three times using individual samples.

2.2. Malondialdehyde, Peroxide Contents, Antioxidant Enzyme Activities Analysis

The malondialdehyde (MDA) contents measurement was carried out with the thiobarbituric acid reaction method [15]. Peroxide (H_2O_2) contents were measured according to a previous method [16].

Superoxide dismutase (SOD) activity was assessed by measuring the inhibition of photochemical reduction of nitroblue tetrazolium (NBT) spectrophotometrically at 560 nm [15]. One unit (U) of SOD activity is defined as the enzyme amount needed to inhibit 50% of NBT reduction under the given experimental conditions. Peroxidase (POD) activities were assessed by monitoring the increase in absorbance at 470 nm, resulting from the oxidation of guaiacol by H_2O_2 [17]. Ascorbate peroxidase (APX) activities were assessed by Cakmak and Marschner [17]. Glutathione reductase (GR) activities were assessed by analyzing the NADPH oxidation rate [18].

2.3. Proline and Soluble Sugar Contents Analysis

The free proline and soluble sugar contents in the leaves and roots of wild cucumber were analyzed following the methods described by Bates (1973) [19] and Yemm (1954) [20], respectively.

2.4. Library Construction and Sequencings

When leaves and roots of the control and NaCl-treated wild cucumber were collected, a total of 12 samples were sent to the Gene Denovo Biotechnology Co., Ltd., located in Guangzhou, China, for RNA extraction services. Subsequently, library construction and RNA sequencing were performed using an Illumina HiSeq™ 2000 Sequence system (San Diego, CA, USA).

2.5. De Novo Transcriptome Assembly and Functional Annotation

The raw sequence data underwent filtering to eliminate the adaptor sequence as well as low-quality reads. This resulted in the acquisition of clean reads, which were used for subsequent analyses due to their high quality. The clean reads were de novo assembled using the "Trinity" program. Gene expression levels were determined using the fragments per kilobase of transcript per million mapped reads (FPKM) method. To provide gene functional descriptions, all assembled transcripts were compared with the NCBI non-redundant (nr) protein database, Gene Ontology terms (GO), and the Kyoto Encyclopedia of Genes and Genomes (KEGG), using the BLASTX algorithm with an E value threshold of 10^{-5}. In our study, we assessed differentially expressed genes (DEGs) by implementing a default screening condition of a false discovery rate (FDR) <0.05 and an absolute fold change ≥2. Subsequently, the DEGs underwent enrichment analysis of GO functions and KEGG pathways. The creation of heatmaps was performed in R using the pheatmap package (R3.6.2 version).

2.6. Total RNA Extraction and qRT-PCR Validation

The total RNA was extracted using an EastepTM Universal RNA Extraction Kit (Promega, Shanghai, China). Gene-specific primers were designed based on the unique sequences of the reference using Primer Premier 5.0. The actin gene was selected as the internal control (Table S1). The qRT-PCR assays were carried out using an SYBR Premix Ex Taq (Takara, Dalian, China) and an iCycler iQ real-time PCR detection system (BIORAD), following the instructions provided by the manufacturers. Three biological replicates were performed, and the relative expression levels were calculated using the $2^{-\Delta\Delta CT}$ method.

2.7. The Isolation of ChTrxh and Its Overexpression Transgenic Tobacco Plants Characterization

The complete coding sequence of *ChTrxh* was identified in Unigene0040254 and amplified using PCR. To construct the overexpression vector, the full-length *ChTrxh* sequence was amplified with the gene-specific primers, *ChTrxh-F* (5′- CTGTTGATACATATGATGGCAGAAATGGGGAAAG-3′), and *ChTrxh-R* (5′- ATCATCGATGAATTCATAGGCTTGAATGTTTGG-3′) utilizing a Pfu DNA polymerase (Vazyme, Nanjing, China). The resulting PCR fragment was then ligated into the binary plant vector pRI101 using a ClonExpress II one-step cloning kit (Vazyme, Nanjing, China). The binary plasmid was subsequently transferred into Agrobacterium LBA4404 (*Agrobacterium tumefaciens*). Tobacco seedlings (*Nicotiana tabacum* cv. NC89) were cultivated on a sterile MS medium and were utilized for leaf-disc transformation. Transgenic tobacco plants were obtained using the leaf plate method [21]. To confirm the presence of the transgene, genomic PCR and Western blot analysis were performed. The T_2-generation transgenic plants were selected for subsequent stress treatment.

2.8. Western Blot Analysis

Western blot was performed following the protocol described by Bai et al. (2023) [22]. Proteins derived from both wild-type (WT) and transgenic plants were separated on a 12% (*v*/*w*) SDS-polyacrylamide gel. Subsequently, the proteins were transferred onto a polyvinylidene difluoride (PVDF) membrane and incubated with a mouse polyclonal anti-GFP antibody from Beyotime (Nanjing, China). To detect the presence of the target protein, a goat peroxidase-conjugated anti-mouse antibody at a dilution of 1:4000 from Sigma was used. The visual representation of the detected proteins was achieved through chemiluminescence using an ECL system from Bio-Rad (Hercules, CA, USA).

2.9. Analysis of Transgenic Plants under Salt Stress

Seeds were germinated on petri dishes containing Murashige and Skoog (MS) medium with either 0 mM or 100 mM NaCl. The daily germination rate was documented. The phenotype of the seedlings was subsequently photographed after an 11-day period.

2.10. Statistical Analysis

The experiments were conducted with a minimum of three repetitions, each consisting of three or more biological replicates. The data were subjected to a one-way analysis of variance (ANOVA) or a *t*-test. Error bars were included on the graphs, where different letters denoted significant differences ($p < 0.05$).

3. Results

3.1. Physiological Responses of Wild Cucumber to NaCl Stress

Hydroponically grown wild cucumber seedlings were subjected to 24-h with 100 mM NaCl. The MDA contents increased by 94.44% and 116.67%, respectively, in the root and leaves of the wild cucumber after NaCl treatment ($p < 0.05$). The H_2O_2 levels in the root and leaves of the wild cucumber increased by 81.81% and 30.91%, respectively, following treatment with NaCl, with the most significant increase observed in the leaves. Furthermore, the levels of proline and soluble sugar, two well-known osmoprotectants, were analyzed. The proline contents were approximately 2.10-fold levels of the control in the root after NaCl treatment ($p < 0.05$). The soluble sugar contents increased by 29.51% and 13.00%, respectively, after NaCl treatment in the root and leaves ($p < 0.05$) (Figure 1).

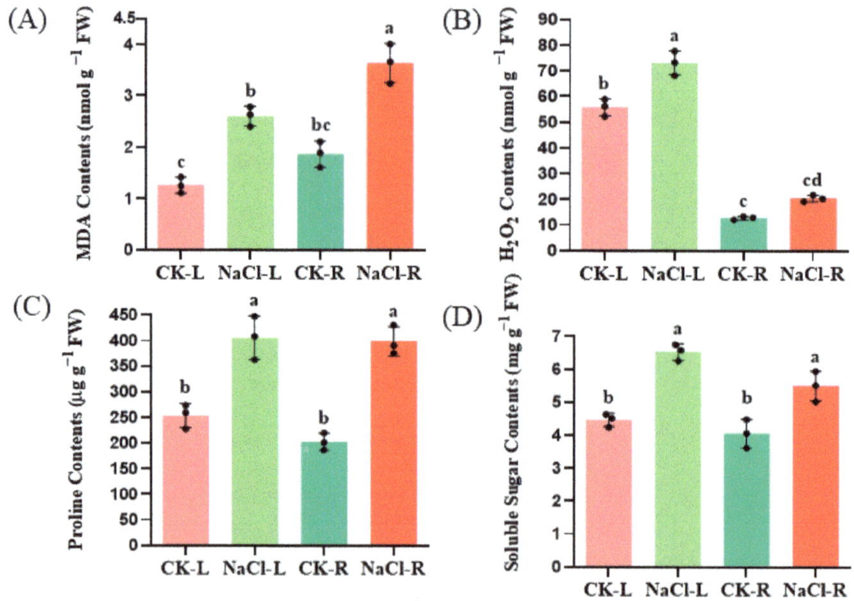

Figure 1. Effect of NaCl stress on the level of lipid peroxidation (**A**), H_2O_2 contents (**B**), proline (**C**) and soluble sugar (**D**) contents in the roots and leaves of wild cucumber. CK means control. The values presented are the means of three replicates ± standard error (SE). Different letters indicate statistical significance at a level of $p < 0.05$.

Furthermore, the activities of superoxide dismutase (SOD), peroxidase (POD), ascorbate peroxidase (APX), and glutathione reductase (GR) in the roots increased by 1.31-, 1.38-, 1.26-, and 1.72-fold, respectively, following NaCl treatment ($p < 0.05$). These findings suggest that these ROS-scavenging enzymes participate in the detoxification of NaCl-induced ROS induced in wild cucumber plants (Figure 2).

3.2. RNA-Seq and De Novo Assembly

To investigate the underlying molecular mechanisms of salt response in the wild cucumber, RNA-Seq analysis was conducted on leaf and root samples from both NaCl-treated and control seedlings. Each condition was represented by three biological replicates, resulting in a total of 12 libraries for analysis. The sequencing process generated an average of 20.00–33.95 million reads per sample. A high-quality dataset was obtained, with over 95% of the reads having Phred-like quality scores at the Q20, ensuring accurate sequencing results. Following the removal of adaptors, reads with unknown nucleotides, and low-quality reads, a set of 19.99–33.35 million clean reads with a GC content ranging from 44.4 to 46.27% were obtained (Table S2).

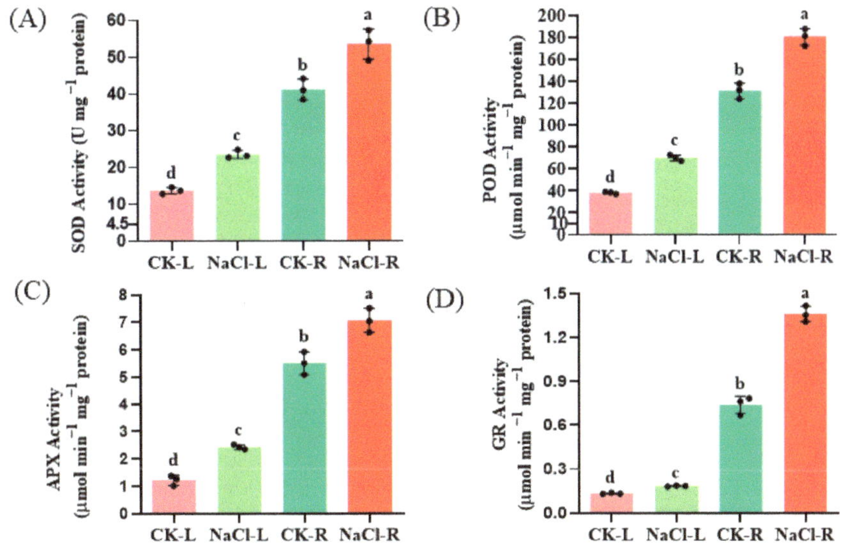

Figure 2. Effect of NaCl stress on the antioxidant enzyme activities of SOD (**A**), POD (**B**), APX (**C**) and GR (**D**) in the roots and leaves of wild cucumber. CK means control. The presented values represent the means of three replicates ± standard error (SE), with different letters indicating statistical significance at a level of $p < 0.05$.

Using the Trinity assembling program, a de novo assembly was performed, resulting in 49,005 unigenes. The N_{50} value, which represents the length at which 50% of the total assembly is contained in contigs of that length or longer, was found to be 1896 bp. In total, the assembled bases amounted to 54,803,657 bp. The length distribution of the unigenes varied, with the maximum length being 24,241 bp and the minimum length being 201 bp. On average, the length of the unigenes was 1118 bp. The most frequent length range for the unigenes was between 100 and 299 bp, accounting for 19.80% of the total unigenes. The length distribution of the unigenes and the distribution of unique-mapped reads of the assembled unigenes are presented in Figure S2. The RNA-seq sequence data have been deposited in the NCBI. The Biosample accessions for the data range from SAMN117789 to SAMN1177800.

3.3. Unigene Function Annotation

The unigenes were annotated using a Blastx program, with an E-value threshold of 10^{-5}, against protein databases in the following priority order: Nr, Swiss-Prot, KEGG and KOG. Out of the 49,005 unigenes, a total of 32,141 (65.59%) were successfully annotated using the Nr database. Similarly, 24,323 (49.63%) unigenes were annotated using the Swiss-Prot database. The KEGG database annotated 13,912 (28.39%) unigenes, while the KOG database annotated 19,995 (40.80%) unigenes (Figure S3A).

In the analysis, it was found that 43.21% of the sequences were successfully mapped to known genes in the Nr database for plants, indicating that a significant proportion of the unigenes had a match with existing genes. Additionally, around 17.11% of the unigenes displayed a similarity of over 80% with previously deposited sequences (Figure S3B). Further investigation revealed that approximately 72.86% of the annotated unigenes could be identified to species from the top six hits: *Cucumis sativus* (50.13%), *Cucumis melo* (15.01%), *Theobroma cacao* (3.14%), *Medicago truncatula* (2.35%), *Brassica napus* (2.21%), *Gossypium arboretum* (1.21%) (Figure S3C).

The functional categories of the 49,005 unigenes were analyzed and categorized into 44 Gene Ontology (GO) terms. In the three main categories, the assignment to the "metabolic processes" (10,103), accounted for the majority, followed by "cellular processes" (9088), and "single organisms" (7006). Within the cellular components category, the majority of the unigenes were classified into three highly represented GO terms: "cell part" (7271), "cell" (7273) and "organelle" (5339). For the molecular functions category, the two largest proportions of genes were found in the "binding" (9144) and "catalytic activity" (8654) categories. Together, these two categories accounted for 84.72% of the total molecular functions identified in the dataset (Figure S4).

Out of the 34,987 unigenes analyzed, they were assigned to KOG functional classification and divided into 25 specific categories based on sequence homology (Figure 3). The most common category was "general functional prediction only", with 6282 unigenes falling into this category. Following that, the categories of "posttranslational modification, protein turnover, chaperones" and "signal transduction mechanisms" were well represented with 4137 and 3227 unigenes, respectively.

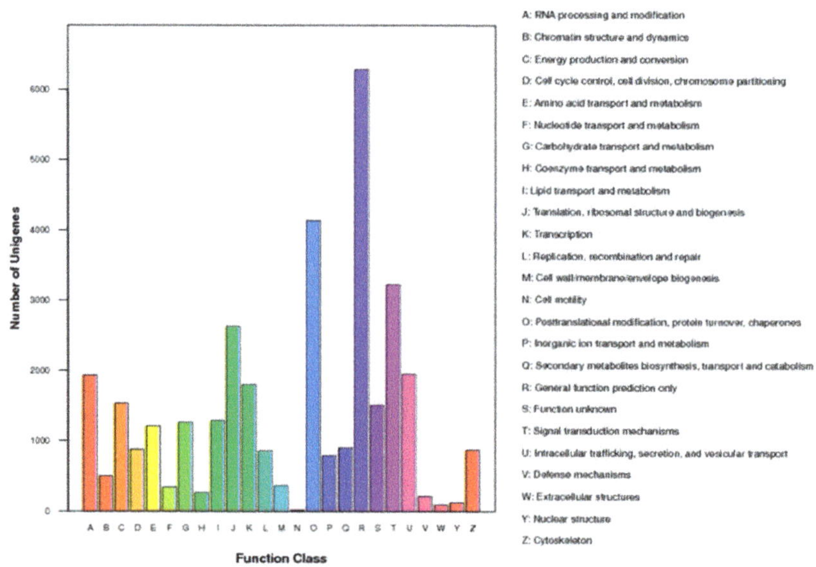

Figure 3. Histogram presentation of KOG classification. The *y*-axis indicates the number of unigenes in each specific functional category.

3.4. Analysis and Functional Classifications of DEGs in Leaves and Roots of Wild Cucumber

To determine the genes that exhibited differential expression under NaCl-induced stress in wild cucumber, we conducted a comparative analysis of the DEG profiles to assess the variations in gene expression. The comparison between CK-L and NaCl-L revealed a significant up-regulation of 1932 genes and a down-regulation of 1577 genes. Similarly, the comparison between CK-R and NaCl-R showed an up-regulation of 4245 genes and a down-regulation of 1271 genes (Figure 4A).

The distribution of DEGs among the comparisons of CK-L vs. NaCl-L and CK-R vs. NaCl-R is illustrated in Figure 4B using a Venn diagram. Remarkably, there were 701 DEGs that exhibited similar expression patterns in both CK-L vs. NaCl-L and CK-R vs. NaCl-R.

The significant changes in DEGs in leaves and roots of the CK vs. the NaCl treatment were analyzed for GO classification and enrichment. Within the up- and down-regulated unigenes of CK-R vs. NaCl-R and CK-L vs. NaCl-L, the most prominent genes in the biological process category of GO annotation, were related to the "metabolic process", "cellular process" and "single-organism process". In the cellular component category, the major genes were associated with the "cell", "cell part" and "organelle". Regarding the molecular function category, the dominant genes exhibited "catalytic activity", "binding", and "structural molecule activity" or "nucleic acid transcription factor activity" (the up-regulated genes in CK-L vs. NaCl-L) (Figure S5).

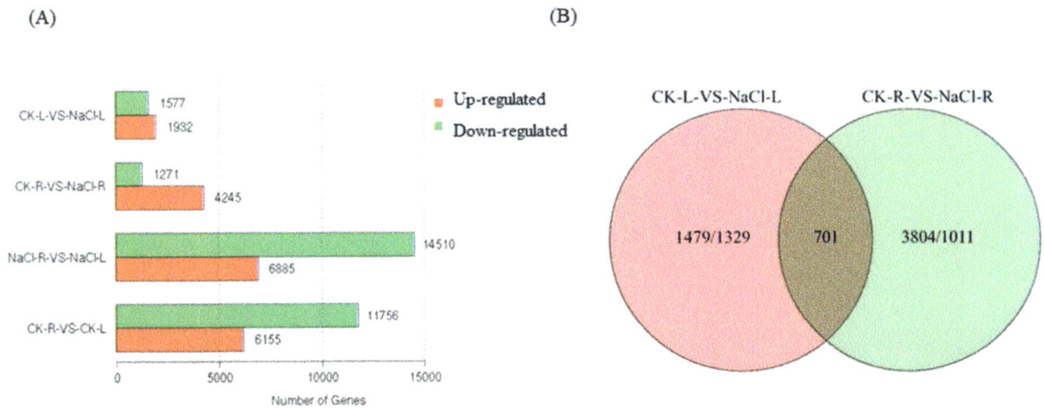

Figure 4. The analysis of DEGs in the leaves and roots of wild cucumber under NaCl stress. (**A**) The number of DEGs in each comparison, with up-regulated genes indicated in red and down-regulated genes in green. (**B**) Venn diagrams illustrate the distribution of DEGs among the comparisons of CK-L vs. NaCl-L and CK-R vs. NaCl-R.

Pathway enrichment analysis was conducted to identify the metabolic pathways and signal transduction pathways that showed significant enrichment in DEGs. In CK-R vs. NaCl-R, a total of 1526 genes were assigned a KEGG ID and categorized into 122 pathways (Table S3). Similarly, in CK-L vs. NaCl-L, 646 genes were assigned a KEGG ID and categorized into 112 pathways (Table S4). Subsequently, pathways with Q values indicating significant enrichment among the DEGs were selected after multiple testing corrections. The 20-pathway enrichment is shown in Figure S6. In CK-L vs. NaCl-L, the top five pathways are "ribosome", "plant hormone signal transduction", "phenylpropanoid biosysnthesis", "starch and sucrose metabolism", and "plant-pathogen metabolism". In CK-R vs. NaCl-R, the top five pathways are "ribosome", "protein processing in endoplasmic reticulum", "endocytosis", "oxidative phosphorylation", and "phagosome".

3.5. Identification of Salt Responsive Genes Expressed during the Salt Stress Treatment

3.5.1. DEGs Encoding Signaling Transduction Proteins

In the leaves, DEGs encoding signaling-related genes were identified, including six calmodulin or calmodulin-like proteins, two calcium-dependent protein kinase, two CBL-interacting serine/threonine-protein kinases, and six mitogen-activated protein kinase kinase kinase. In the roots, eight calmodulin or calmodulin-like proteins, one calcium-dependent protein kinase, five DEGs of mitogen-activated protein kinase, one mitogen-activated protein kinase kinase, and four mitogen-activated protein kinase kinase kinase were found, and all these DEGs were up-regulated. Additionally, four calcineurin genes were found and up-regulated in the root, while they were not found in the leaves (Table S5, Figure 5A).

3.5.2. DEGs Encoding Transcription Factors

In leaves, we identified 159 DEGs encoding a transcription factor; among them, 45 were down-regulated and 114 were up-regulated. The predominantly expressed transcription factors were the ethylene-responsive transcription factor (ERF) (39), the basic helix-loop-helix (bHLH) (18), MYB (16), WRKY(23), the heat stress transcription factor (10), and GATA (4) families. In roots, we identified 138 DEGs encoding a transcription factor; among these, 55 were down-regulated and 83 were up-regulated. The transcription factors predominantly expressed in response to salt stress include members of the ERF (40), bHLH (11), MYB (13),

WRKY (18), heat stress transcription factor (7), and GATA (6, all down-regulated) families. This suggests their significant role in response to salt stress (Table S5).

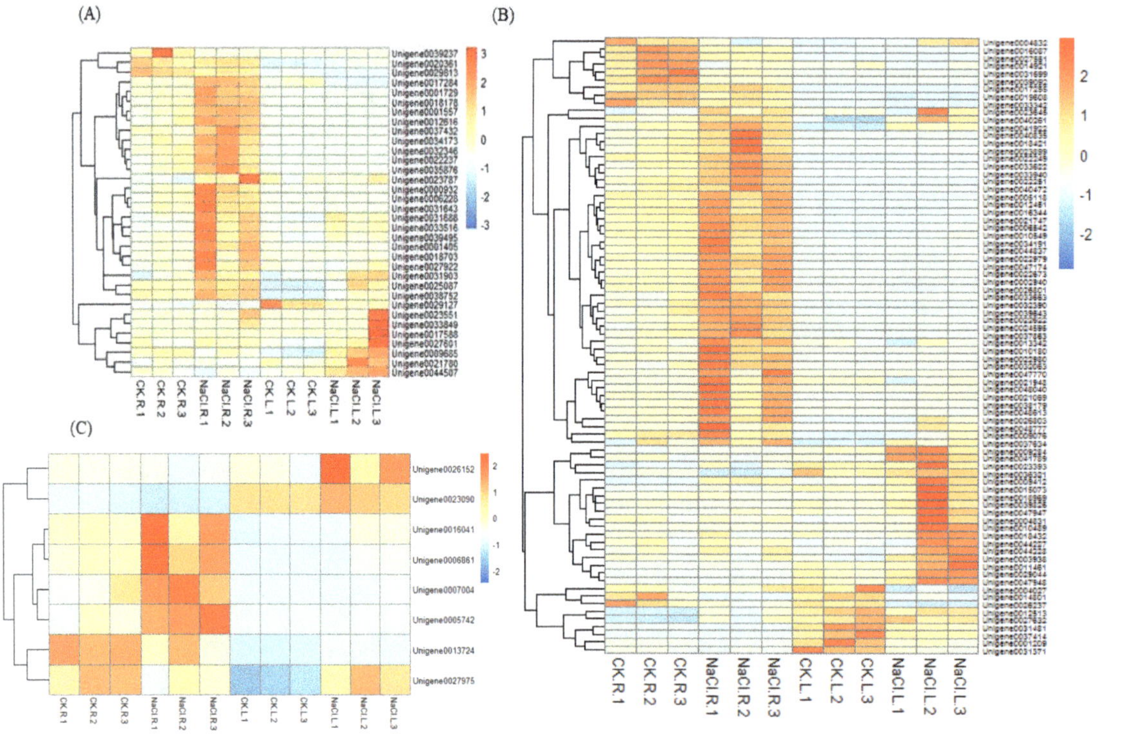

Figure 5. Heatmap analysis of the expression levels of signal transduction DEGs (**A**), ROS scavenging DEGs (**B**) and osmotic stress-related DEGs (**C**).

3.5.3. DGEs Encoding Transporters

Additionally, 50 DEGs in leaves were predicted to encode diverse transporters, such as the ABC transporter (8), auxin transporter or auxin transporter-like protein (3), amino acid transporter (5), vacuolar type H^+-ATPase (3), metal transporter (3, all up-regulated), polyol transporter (4), phosphate transporter (4), choline transporter-like protein (1), potassium transporter (1, up-regulated), etc. In roots, 102 DEGs encoding transporters were identified, such as the ABC transporter (24), amino acid transporter (10), choline transporter or choline-like transporter (4), the potassium transporter (2, down-regulated), vacuolar type H^+-ATPase (15, all up-regulated), H^+-PPase family transporter (7, all up-regulated), sodium transporter HKT1 (1, down-regulated), zinc transporter (4), phosphate transporter (5), sugar transporter (4), polyol transporter (5), etc. (Table S5).

3.5.4. DEGs Encoding ROS Scavenging Enzymes

A total of 41 and 59 DEGs were identified as putative enzyme-coding genes involved in the ROS scavenging system in the leaves and root of wild cucumber under salt stress, respectively. These DEGs were further categorized into glutathione S-transferase (GST) family proteins (12), superoxide dismutase (SOD, 2), peroxidase (POD, 15), glutaredoxin (7), thioredoxin superfamily protein isoform and thioredoxin-like protein (4), and 1-Cys peroxiredoxin (1) in leaves. In the root, they were categorized into GST family proteins (25), SOD (3, up-regulated), POD (23), glutathione peroxidase (4, up-regulated), L-ascorbate peroxidase (1, up-regulated), glutathione reductase (1, up-regulated), and glutaredoxin (2, up-regulated) (Table S5, Figure 5B).

3.5.5. DEGs Encoding Osmotic Substances

In the present study, one unigene encoding pyrroline-5-carboxylate reductase was found up-regulated in the root. Three genes encoding trehalose-phosphate phosphatase were found up-regulated in the leaves after NaCl treatment. Three genes encoding alpha, alpha-trehalose-phosphate synthase were found up-regulated only in the root (Table S5, Figure 5C).

3.6. Verification of the Expression Level of Some Transcripts Using qRT-PCR

To validate the gene expression levels obtained from RNA-seq, 12 DEGs with distinct expression patterns were randomly selected for qRT-PCR analysis. The findings demonstrate that although there were variations in the fold-change values between the qRT-PCR and RNA-seq, the expression patterns of the unigenes were consistent with the RNA-Seq data. The correlation coefficients between qPCR- and FPKM-derived expressions in the leaves and roots of the wild cucumber were 0.8462 and 0.9253, respectively. Collectively, these results demonstrate the high reliability of the RNA-seq data (Figure 6).

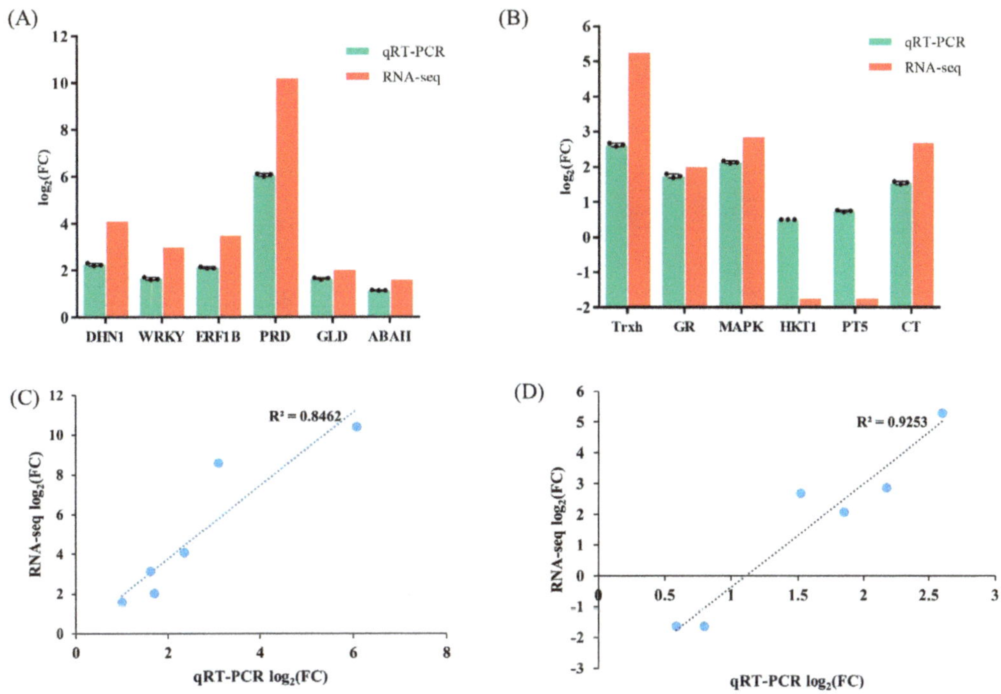

Figure 6. The differential gene expression values of selected genes obtained through total RNA sequencing (RNA-seq) and qPCR in leaves (**A**) and roots (**B**) of wild cucumber. Additionally, the correlations between the RNA-Seq and qRT-PCR of selected DEGs were analyzed in leaves (**C**) and roots (**D**) of wild cucumber. R^2 was the Pearson correlation coefficient. The bars plot represents the mean $\log_2$ fold change ± SE with three replicates (n = 3).

3.7. The Characterization and Functional Analysis of Overexpressed ChTrxh Transgenic Tobacco under NaCl Stress

According to the sequence of Unigene0040254, the ORF of *ChTrxh* was isolated. The ORF amplified from *Cucumis hystrix* was 348 bp and the sequence is shown in Figure S7. Trxs play important roles in defending against oxidative stress [23]. To further investigate the function of *ChTrxh* under salt stress, 10 transgenic tobacco lines overexpressing *ChTrxh* were obtained through kanamycin screening. The genomic PCR analysis confirmed the presence of the target band in the expected size, which was absent in WT (Figure S8A).

Additionally, the Western blot results demonstrated a significant increase in ChTrxh protein expression in OE1, OE2, and OE3 transgenic plants compared to the WT (Figure S8B). The findings collectively indicate the successful overexpression of *ChTrxh* in tobacco.

To assess the salt tolerance of *ChTrxh* overexpressing tobacco plants, the seeds of the transgenic tobacco and WT plants were subjected to MS medium with 0 or 100 mM NaCl for a period of 12 days. As depicted in Figure 7A,C, there was no difference in the seed germination rate between the WT and transgenic tobacco grown on the MS medium. However, under NaCl treatment, the germination rate of *ChTrxh* transgenic tobacco plants exhibited a significant increase compared to the WT plants. This observation suggested that the overexpression of *ChTrxh* enhanced the tolerance to NaCl-induced stress (Figure 7B,D).

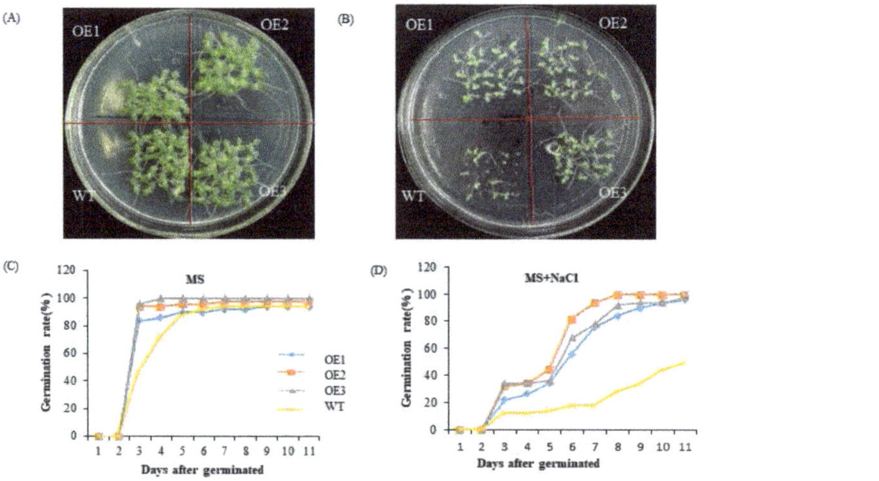

Figure 7. The functional analysis of *ChTrxh* overexpression transgenic tobacco seeds under NaCl stress. (**A**,**B**) The phenotype of tobacco seeds treated with 0 or 100 mM NaCl on MS medium for 11 days. (**C**,**D**) The germination rate of overexpression *ChTrxh* transgenic tobacco and WT seeds under 0 or 100 mM NaCl stress.

4. Discussion

Wild cucumber holds the potential for plant improvement due to its genes associated with resistance to various pathogens, including powdery mildew, downy mildew, anthracnose, and fusarium wilt [24]. However, the physiological and transcriptome analysis of wild cucumber in response to salt stress is still largely unknown. In this study, we sequenced the transcriptomes of the leaves and roots of wild cucumber under NaCl treatment. A total of 49,005 unigenes were assembled, with 66.7% of them being annotated. Our transcriptome results shed light on the molecular mechanisms underlying the response of wild cucumber to salinity stress.

4.1. Signaling Network of Wild Cucumber in Response to NaCl Stress

The tolerance of plants to NaCl stress is activated through complex signaling pathways that aim to maintain cellular homeostasis. Our data revealed significant differential expression of numerous genes involved in signaling pathways in both the leaves and roots of NaCl-treated wild cucumber plants compared to the control groups.

Calmodulin (CaM), calcineurin B-like proteins (CBLs), calcium-dependent protein kinases (CDPK), and CBL-interacting protein kinases are important components of Ca^{2+}-mediating signaling pathways. The overexpression of *GmCaM4* in soybean was found to enhance tolerance to salt stress [25]. Genes of CaM-like (CML) play a significant role in regulating plant growth, development, and responses to salinity and drought stress. In rice, a CML gene, *OsDSR-1*, improved drought tolerance through scavenging ROS [26]. Calcineurin, a protein phosphatase dependent on both Ca^{2+}- and CaM, is involved in the regulation of Na^+, K^+ and Ca^{2+} homeostasis, as well as hormone response in yeast [27]. CDPKs are versatile proteins that possess calcium-binding and signaling capabilities within a single gene product. The involvement of the CDPK gene *VaCPK21* from the wild grapevine in response to salt stress was observed [28]. Through overexpression of a CDPK gene, increased tolerance to both salt and drought was achieved in rice [29]. Our study identified several CaM, CBLs, and CDPK genes that showed a significant response to NaCl stress, indicating their crucial roles in salt stress signaling.

Osmotic stress triggers the activation of protein kinases, particularly mitogen-activated kinases (MAPK), which play a crucial role in mediating osmotic homeostasis and/or detoxification responses [30]. MAPKs cascade pathways, consisting of a MAPK kinase kinase (MAPKKK), a MAPK kinase (MAPKK), and a MAPK, have been identified in response to various abiotic stresses, including NaCl stress. Additionally, the isolation of a MAPK gene (*CsNMAPK*) from cucumber has demonstrated its capability to enhance the tolerance of transgenic tobacco plants in response to both salt and osmotic stress [31]. The overexpression of *Medicago* MAPKKKK in *Arabidopsis* was found to enhance salt sensitivity [32]. Our study identified a strong expression of transcripts encoding MAPK, MAPKK, and MAPKKKK, with a particular emphasis on MAPKKKK, indicating the significant involvement of the MAPK cascade t in the salt response of wild cucumber (Table S4).

4.2. Transcription Factors of Wild Cucumber Are Affected by Salt Stress

Several classes of transcription factors were expressed under salt stress in wild cucumber. In our study, the predominantly expressed transcription factors are ERF, bHLH, MYB, and WRKY (Table S4). In *G. hirsutum*, members of AP2/EREBP, MYB, NAC, and WRKY transcription factor families were identified as highly enriched in the regulation of salt stress responses [33]. Similarly, in the hexaploid hulless oat, the MYB- and MYB-related TF family, bHLH, WRKY, and NAC transcription factor families exhibited expression patterns in response to salinity stress [34].

R2R3-MYB transcription factors in plants play important roles in responding to various abiotic and biotic stresses. For instance, the R2R3-type MYB gene SlMYB102 from the tomato has been found to increase salt tolerance in transgenic tomato [35]. In soybean, the MYB gene GmMYB68 confers salt-alkali resistance [36]. Another example is the wild soybean R2R3-MYB transcription factor GsMYB15, which, when overexpressed, enhances salt stress resistance in transgenic Arabidopsis [37]. Additionally, ethylene response factors, belonging to AP2/ERF superfamily proteins, are part of the largest transcription factors and participate in multiple abiotic stress tolerances, including salt. A transcriptome analysis was conducted to investigate differentially expressed ERF transcription factors in cotton in response to salt stress [38]. Overexpression of the JcERF1, an AP2/ERF-type transcription factor from *Jatropha curcas*, has been shown to increase salt tolerance in transgenic tobacco [39]. However, potato StERF3 has a negative regulatory effect on salt tolerance in potato [40]. Further investigation is required to understand the function of these transcription factors under salt stress in the next context of our study.

4.3. Transporters Are Affected by Salt Stress in Wild Cucumber

In our study, 152 unigenes encoding transporter proteins were identified including ATPase, ABC transporters, the ion transporter, amino acid transporter, choline transporter, potassium transporter, H^+-PPase family transporter, sodium transporter HKT1, zinc transporter, phosphate transporter, sugar transporter, polyol transporter, etc., in wild cucumber roots and leaves (Table S4). Plasma membrane H^+-ATPases play a vital role in plant growth and development. The up-regulation of these ATPase genes indicates that wild cucumber possesses a strong ability to maintain an osmotic balance. ATP-binding cassette transporters from extensive gene superfamilies are responsible for the transportation of diverse molecules including heavy metal ions, secondary metabolites, and phytohormones [41]. The overexpression of AtABCG36 in *Arabidopsis* resulted in enhanced tolerance to salt stress [42].

Ionic stress plays a crucial role in salinity, primarily caused by the excessive accumulation of Na^+ in the aboveground portions of plants. Maintaining a balanced Na^+/K^+ ratio within the cytosolic has emerged as a crucial mechanism for salinity tolerance. To achieve this homeostatic balance, the activity of Na^+ and K^+ transporters and/or channels is necessary [43]. In our study, we identified three potassium transporters. The results support previous findings that genes involved in ion transport are significantly induced in responses to salinity stress, aiming to maintain or restore cytoplasmic homeostasis [44,45]. Additionally, other transporters, including magnesium transporters, sulfate transporters, and sugar transporters, were up-regulated in salt-stressed wild cucumber. The up-regulation of these genes potentially contributes to the maintenance of homoeostasis in wild cucumber.

4.4. ROS Scavenging Substances of Wild Cucumber Play Important Roles in Response to NaCl Stress

Salt stress triggers a rapid elevation in ROS. In this study, the MDA and H_2O_2 contents increased levels in the roots and leaves, indicating that NaCl stress induced lipid peroxidation and oxidative damage to wild cucumber. Antioxidant enzymes play a crucial role in protecting cells form oxidative damage [46,47]. Following exposure to NaCl stress, the wild cucumber exhibited heightened activities of antioxidant enzymes, including SOD, POD, APX, and GR. This study further identified several DEGs associated with SOD, POD, and APX. These findings strongly suggest that the ROS scavenging system enhances the antioxidative ability of the wild cucumber under NaCl stress.

The ubiquitous and multifunctional conjugating proteins, glutathione transferases (GSTs), have significant importance in stress responses due to their ability to prevent oxidative damage [48]. *AtGSTU19* has been identified as a key player in salt stress tolerance in *Arabidopsis* [49]. In our study, 37 unigenes of GSTs were found in wild cucumber root and leaves, suggesting that GST may play an important role under NaCl stress.

Thioredoxin (Trx) acts as an antioxidant by facilitating redox states changes in target proteins through cysteine thiol-disulfide exchanges, thus modulating ROS scavenging. In *Arabidopsis thaliana*, Trxs are categorized into six main groups, each of which is distributed across different subcellular compartments such as the chloroplast, mitochondria, and cytosol [50]. The genome of *Arabidopisis thaliana* contains nine H-type Trx proteins, which are presumed to be located in the cytosol. Trxs play a vital role in response to abiotic stresses. Previous research has shown that an H-Type Trx is involved in tobacco defense responses under oxidative stress [51]. In our study, we observed a significant increase in the expression of thioredoxin H type (Trxh). Transgenic tobacco lines, expressing *LmTrxh2* from the halophyte plant *Lobularia maritima*, exhibited enhanced tolerance to salt stress compared to wild type plants [52]. Trx CDSP32 has been found to mitigate photo-inhibition-induced cadmium in tobacco leaves [53]. The overexpression of *ThTrx5*, a thioredoxin protein, in *Arabidopsis* has been shown to enhance salt tolerance. In transgenic *ThTrx5* plants, the activities of SOD, POD, and CAT were significantly higher than WT plants under salt stress [54]. To further investigate the role of thioredoxin under NaCl stress, we isolated the full-length sequence of ChTrxh and subsequently overexpressed it in tobacco plants. The germination rate of *ChTrxh* overexpressed transgenic plants was higher than that of WT, indicating that overexpression of *ChTrxh* increased NaCl stress tolerance.

4.5. Osmotic Substance Metabolism of Wild Cucumber in Response to NaCl Stress

Organic osmolytes, including proline, soluble sugar, and trehalose, serve as effective osmoprotectants that alleviate osmotic damage induced by salt stress [1]. Physiological measurements demonstrated a significant increase in proline and soluble sugar content in response to salt stress, indicating that wild cucumber possesses a robust capacity for osmotic adjustment.

Pyrroline-5-carboxylate reductase (P5CR) serves as the final enzymatic step in the biosynthesis of proline. One DEG encoding P5CR was identified in the root. On the other hand, pyrroline-5-carboxylate synthase (P5CS) plays a crucial role in proline biosynthesis by facilitating the conversion of glutamate into P5C. In our study, no DEG encoding P5CS in the root and leaves of wild cucumber were found.

Trehalose (Tre) is a growth regulator extensively utilized for enhancing plant stress tolerance. Its synthesis is mediated by the enzymes trehalose phosphate synthase (TPS) and trehalose phosphate phosphatase (TPP) [55]. Here, we observed three TPPs up-regulated in the leaves of wild cucumber and three TPS up-regulated in the roots of wild cucumber after NaCl treatment, suggesting that trehalose accumulation was part of the wild cucumber's tolerance to NaCl stress.

5. Conclusions

This study aimed to investigate the physiological and molecular responses of wild cucumber to NaCl stress. The results demonstrated that under NaCl stress, there was a significant increase in the levels of MDA, H_2O_2, proline and soluble sugar. Additionally, the activities of antioxidant enzymes, including SOD, POD, APX, and GR, were found to be elevated. The transcript of differentially expressed genes analyzed by RNA-Seq showed that numerous genes were related to signal transduction, the transcription factor, ion transport, osmotic metabolism, and ROS scavenging in adaptation to salinity stress in wild cucumber. Furthermore, the overexpression of *ChTrxh* in transgenic tobacco plants resulted in increased NaCl stress tolerance. Some of these novel genes may subsequently be utilized to improve crops in saline soils.

Supplementary Materials: The following supporting information can be downloaded at: https://www.mdpi.com/article/10.3390/agronomy13122931/s1, Figure S1: The morphology of the wild cucumber in Puer, Yunnan Province, China; Figure S2 Random distribution of the assembled unigenes and assessment of assembly quality; Figure S3 Characteristics of similarity search of unigenes against Nr databases; Figure S4 Histogram of the gene ontology (GO) classification; Figure S5 Histogram presentation of DEGs' Gene Ontology classification; Figure S6 Top 20 pathway enrichment of the DEGs in the leaves and roots of wild cucumber; Figure S7 The cDNA sequence of thioredoxin H-type-like of Cucumis hystrix Chakr designated as ChTrxh (Unigene0040254); Figure S8 The characterization of ChTrxh overexpression transgenic tobacco; Table S1: Primer sequences for qRT-PCR analysis; Table S2 Summary of sequences analysis; Table S3 CK-R vs. NaCl-R pathway enrichment; Table S4 CK-L VS. NaCl-L pathway enrichment; Table S5 The main category of up and down regulated DEGs in roots and leaves of wild cucumber.

Author Contributions: H.X. designed and drafted the manuscript; Z.G. and S.Z. performed the experiments and analyzed the data; K.L. helped to analyze the data and revise the paper. All authors have read and agreed to the published version of the manuscript.

Funding: This work was supported by the National Natural Science Foundation of China (grant no. 32260753).

Data Availability Statement: The data presented in this study are available on request from the corresponding author.

Conflicts of Interest: The authors declare no conflict of interest.

References

1. Zhu, J.K. Plant salt tolerance. *Trends Plant Sci.* **2001**, *6*, 66–71. [CrossRef] [PubMed]
2. Parida, A.K.; Das, A.B. Salt tolerance and salinity effects on plants: A review. *Ecotoxicol. Environ. Saf.* **2005**, *60*, 324–349. [CrossRef]
3. Yu, Y.; Huang, W.G.; Chen, H.Y.; Wu, G.W.; Yuan, H.M.; Song, X.X.; Kang, Q.H.; Zhao, D.S.; Jiang, W.D.; Liu, Y.; et al. Identification of differentially expressed genes in flax (*Linum usitatissimum* L.) under saline-alkaline stress by digital gene expression. *Gene* **2014**, *549*, 113–122. [CrossRef] [PubMed]
4. Diray-Arce, J.; Clement, M.; Gul, B.; Khan, M.A.; Nielsen, B.L. Transcriptome assembly, profiling and differential gene expression analysis of the halophyte *Suaeda fruticosa* provides insights into salt tolerance. *BMC Genom.* **2015**, *16*, 353. [CrossRef] [PubMed]
5. Shen, X.; Wang, Z.; Song, X.; Xu, J.; Jiang, C.; Zhao, Y.; Ma, C.; Zhang, H. Transcriptomic profiling revealed an important role of cell wall remodeling and ethylene signaling pathway during salt acclimation in *Arabidopsis*. *Plant Mol. Biol.* **2014**, *86*, 303–317. [CrossRef] [PubMed]

6. Villarino, G.H.; Bombarely, A.; Giovannoni, J.J.; Scanlon, M.J.; Mattson, N.S. Transcriptomic Analysis of *Petunia hybrida* in Response to Salt Stress Using High Throughput RNA Sequencing. *PLoS ONE* **2014**, *9*, e94651. [CrossRef]
7. Ahmad, A.; Olah, G.; Szczesny, B.; Wood, M.E.; Whiteman, M.; Szabo, C. AP39, A Mitochondrially Targeted Hydrogen Sulfide Donor, Exerts Protective Effects in Renal Epithelial Cells Subjected to Oxidative Stress in Vitro and in Acute Renal Injury in Vivo. *Shock* **2016**, *45*, 88–97. [CrossRef]
8. Wei, Y.; Xu, Y.; Lu, P.; Wang, X.; Li, Z.; Cai, X.; Zhou, Z.; Wang, Y.; Zhang, Z.; Lin, Z.; et al. Salt stress responsiveness of a wild cotton species (*Gossypium klotzschianum*) based on transcriptomic analysis. *PLoS ONE* **2017**, *12*, e0178313. [CrossRef]
9. Zhou, Y.; Yang, P.; Cui, F.; Zhang, F.; Luo, X.; Xie, J. Transcriptome Analysis of Salt Stress Responsiveness in the Seedlings of Dongxiang Wild Rice (*Oryza rufipogon* Griff.). *PLoS ONE* **2016**, *11*, e0146242. [CrossRef]
10. Dang, Z.H.; Zheng, L.L.; Wang, J.; Gao, Z.; Wu, S.B.; Qi, Z.; Wang, Y.C. Transcriptomic profiling of the salt-stress response in the wild recretohalophyte *Reaumuria trigyna*. *BMC Genom.* **2013**, *14*, 29. [CrossRef]
11. Sebastian, P.; Schaefer, H.; Telford, I.R.; Renner, S.S. Cucumber (*Cucumis sativus*) and melon (*C. melo*) have numerous wild relatives in Asia and Australia, and the sister species of melon is from Australia. *Proc. Natl. Acad. Sci. USA* **2010**, *107*, 14269–14273. [CrossRef] [PubMed]
12. Chen, J.F.; Luo, X.D.; Qian, C.T.; Jahn, M.M.; Staub, J.E.; Zhuang, F.Y.; Lou, Q.F.; Ren, G. *Cucumis* monosomic alien addition lines: Morphological, cytological, and genotypic analyses. *Theor. Appl. Genet.* **2004**, *108*, 1343–1348. [CrossRef] [PubMed]
13. Wan, H.; Zhao, Z.; Malik, A.A.; Qian, C.; Chen, J. Identification and characterization of potential NBS-encoding resistance genes and induction kinetics of a putative candidate gene associated with downy mildew resistance in Cucumis. *BMC Plant Biol.* **2010**, *10*, 186. [CrossRef]
14. Xu, H.N.; Wang, X.F.; Sun, X.D.; Shi, Q.H.; Yang, F.J.; Du, D.L. Molecular cloning and characterization of a cucumber MAP kinase gene in response to excess NO^{3-} and other abiotic stresses. *Sci. Hortic.* **2008**, *117*, 1–8. [CrossRef]
15. Rao, K.V.M.; Sresty, T.V.S. Antioxidative parameters in the seedlings of pigeonpea (*Cajanus cajan* (L.) Millspaugh) in response to Zn and Ni stresses. *Plant Sci.* **2000**, *157*, 113–128.
16. Xu, H.N.; Sun, X.D.; Shi, Q.H.; Yang, F.J.; Yang, X.Y.; Wang, X.F. Physiological Responses of Two Cucumber Cultivars to Nitrate Stress. *J. Plant Nutr.* **2012**, *35*, 2167–2179. [CrossRef]
17. Cakmak, I.; Marschner, H. Magnesium deficiency and high light intensity enhance activities of superoxide dismutase, ascorbate peroxidase, and glutathione reductase in bean leaves. *Plant Physiol.* **1992**, *98*, 1222–1227. [CrossRef]
18. Foyer, C.H.; Halliwell, B. The presence of glutathione and glutathione reductase in chloroplasts: A proposed role in ascorbic acid metabolism. *Planta* **1976**, *133*, 21–25. [CrossRef]
19. Bates, L.S.; Waldren, R.P.; Teare, I.D. Rapid determination of free proline for water-stress studies. *Plant Soil* **1973**, *39*, 205–207. [CrossRef]
20. Yemm, E.W.; Willis, A.J. The estimation of carbohydrates in plant extracts by anthrone. *Biochem. J.* **1954**, *57*, 508–514. [CrossRef]
21. Horsch, R.B.; Fry, J.E.; Hoffmann, N.L.; Wallroth, M.; Eichholtz, D.; Rogers, S.G.; Fraley, R.T. A simple and general method for transferring genes into plants. *Science* **1985**, *227*, 1229–1231. [CrossRef] [PubMed]
22. Bai, X.G.; Long, J.; He, X.Z.; Yan, J.P.; Chen, X.Q.; Tan, Y.; Li, K.Z.; Chen, L.M.; Xu, H.N. Overexpression of spinach non-symbiotic hemoglobin in Arabidopsis resulted in decreased NO content and lowered nitrate and other abiotic stresses tolerance. *Sci. Rep.* **2016**, *6*, 26400. [CrossRef] [PubMed]
23. Duan, X.H.; Wang, Z.Y.; Zhang, Y.; Li, H.; Yang, M.; Yin, H.; Cui, J.; Chai, H.; Gao, Y.H.; Hu, G.F.; et al. Overexpression of a Thioredoxin-Protein-Encoding Gene, MsTRX, from Medicago sativa Enhances Salt Tolerance to Transgenic Tobacco. *Agronomy* **2022**, *12*, 1467. [CrossRef]
24. Chen, J.; Staub, J.; Qian, C.; Jiang, J.; Luo, X.; Zhuang, F. Reproduction and cytogenetic characterization of interspecific hybrids derived from *Cucumis hystrix* Chakr. × *Cucumis sativus* L. *Theor. Appl. Genet.* **2003**, *106*, 688–695. [CrossRef] [PubMed]
25. Rao, S.S.; El-Habbak, M.H.; Havens, W.M.; Singh, A.; Zheng, D.M.; Vaughn, L.; Haudenshield, J.S.; Hartman, G.L.; Korban, S.S.; Ghabrial, S.A. Overexpression of GmCaM4 in soybean enhances resistance to pathogens and tolerance to salt stress. *Mol. Plant Pathol.* **2014**, *15*, 145–160. [CrossRef]
26. Yin, X.M.; Huang, L.F.; Wang, M.L.; Cui, Y.C.; Xia, X.J. *OsDSR-1*, a calmodulin-like gene, improves drought tolerance through scavenging of reactive oxygen species in rice (*Oryza sativa* L.). *Mol. Breed.* **2017**, *37*, 75. [CrossRef]
27. Mendoza, I.; Rubio, F.; Rodriguez-Navarro, A.; Pardo, J.M. The protein phosphatase calcineurin is essential for NaCl tolerance of *Saccharomyces cerevisiae*. *J. Biol. Chem.* **1994**, *269*, 8792–8796. [CrossRef]
28. Dubrovina, A.S.; Kiselev, K.V.; Khristenko, V.S.; Aleynova, O.A. *VaCPK21*, a calcium-dependent protein kinase gene of wild grapevine Vitis amurensis Rupr., is involved in grape response to salt stress. *Plant Cell Tissue Organ Cult.* **2016**, *124*, 137–150. [CrossRef]
29. Campo, S.; Baldrich, P.; Messeguer, J.; Lalanne, E.; Coca, M.; San Segundo, B. Overexpression of a Calcium-Dependent Protein Kinase Confers Salt and Drought Tolerance in Rice by Preventing Membrane Lipid Peroxidation. *Plant Physiol.* **2014**, *165*, 688–704. [CrossRef]
30. Zhu, J.K. Regulation of ion homeostasis under salt stress. *Curr. Opin. Plant Biol.* **2003**, *6*, 441–445. [CrossRef]
31. Xu, H.N.; Li, K.Z.; Yang, F.J.; Shi, Q.H.; Wang, X.F. Overexpression of CsNMAPK in tobacco enhanced seed germination under salt and osmotic stresses. *Mol. Biol. Rep.* **2010**, *37*, 3157–3163. [CrossRef] [PubMed]

32. Ovecka, M.; Takac, T.; Vyplelova, P.; Komis, G.; Bekesova, S.; Luptovciak, I.; Vadovic, P.; Samajova, O.; Hirt, H.; Samaj, J. Overexpression of Medicago mitogen-activated protein kinase kinase SIMKK in *Arabidopsis causes* salt stress-induced subcellular relocation and enhanced salt sensitivity. *New Biotechnol.* **2016**, *33*, S159–S160. [CrossRef]
33. Peng, Z.; He, S.; Gong, W.; Sun, J.; Pan, Z.; Xu, F.; Lu, Y.; Du, X. Comprehensive analysis of differentially expressed genes and transcriptional regulation induced by salt stress in two contrasting cotton genotypes. *BMC Genom.* **2014**, *15*, 760. [CrossRef] [PubMed]
34. Wu, B.; Hu, Y.N.; Huo, P.J.; Zhang, Q.; Chen, X.; Zhang, Z.W. Transcriptome analysis of hexaploid hulless oat in response to salinity stress. *PLoS ONE* **2017**, *12*, e0171451. [CrossRef] [PubMed]
35. Zhang, X.; Chen, L.C.; Shi, Q.H.; Ren, Z.H. SlMYB102, an R2R3-type MYB gene, confers salt tolerance in transgenic tomato. *Plant Sci* **2020**, *291*, 110356. [CrossRef]
36. He, Y.X.; Dong, Y.S.; Yang, X.D.; Guo, D.Q.; Qian, X.Y.; Yan, F.; Wang, Y.; Li, J.W.; Wang, Q.Y. Functional activation of a novel R2R3-MYB protein gene, GmMYB68, confers salt-alkali resistance in soybean (*Glycine max* L.). *Genome* **2020**, *63*, 13–26. [CrossRef]
37. Shen, X.J.; Wang, Y.Y.; Zhang, Y.X.; Guo, W.; Jiao, Y.Q.; Zhou, X.A. Overexpression of the Wild Soybean R2R3-MYB Transcription Factor *GsMYB15* Enhances Resistance to Salt Stress and *Helicoverpa armigera* in Transgenic *Arabidopsis*. *Int. J. Mol. Sci.* **2018**, *19*, 3958. [CrossRef]
38. Long, L.; Yang, W.W.; Liao, P.; Guo, Y.W.; Kumar, A.; Gao, W. Transcriptome analysis reveals differentially expressed ERF transcription factors associated with salt response in cotton. *Plant Sci.* **2019**, *281*, 72–81. [CrossRef]
39. Debbarma, J.; Sarki, Y.N.; Saikia, B.; Boruah, H.P.D.; Singha, D.L.; Chikkaputtaiah, C. Ethylene Response Factor (ERF) Family Proteins in Abiotic Stresses and CRISPR-Cas9 Genome Editing of ERFs for Multiple Abiotic Stress Tolerance in Crop Plants: A Review. *Mol. Biotechnol.* **2019**, *61*, 153–172. [CrossRef]
40. Tian, Z.; He, Q.; Wang, H.; Liu, Y.; Zhang, Y.; Shao, F.; Xie, C. The Potato ERF Transcription Factor StERF3 Negatively Regulates Resistance to *Phytophthora infestans* and Salt Tolerance in Potato. *Plant Cell Physiol.* **2015**, *56*, 992–1005. [CrossRef]
41. Saha, J.; Sengupta, A.; Gupta, K.; Gupta, B. Molecular phylogenetic study and expression analysis of ATP-binding cassette transporter gene family in *Oryza sativa* in response to salt stress. *Comput. Biol. Chem.* **2015**, *54*, 18–32. [CrossRef] [PubMed]
42. Kim, D.Y.; Jin, J.Y.; Alejandro, S.; Martinoia, E.; Lee, Y. Overexpression of AtABCG36 improves drought and salt stress resistance in Arabidopsis. *Physiol. Plant.* **2010**, *139*, 170–180. [CrossRef] [PubMed]
43. Assaha, D.V.M.; Ueda, A.; Saneoka, H.; Al-Yahyai, R.; Yaish, M.W. The Role of Na^+ and K^+ Transporters in Salt Stress Adaptation in Glycophytes. *Front. Physiol.* **2017**, *8*, 509. [CrossRef] [PubMed]
44. Qiu, Q.; Ma, T.; Hu, Q.J.; Liu, B.B.; Wu, Y.X.; Zhou, H.H.; Wang, Q.; Wang, J.; Liu, J.Q. Genome-scale transcriptome analysis of the desert poplar, *Populus euphratica*. *Tree Physiol.* **2011**, *31*, 452–461. [CrossRef] [PubMed]
45. Zhang, J.; Jiang, D.C.; Liu, B.B.; Luo, W.C.; Lu, J.; Ma, T.; Wan, D.S. Transcriptome dynamics of a desert poplar (*Populus pruinosa*) in response to continuous salinity stress. *Plant Cell Rep.* **2014**, *33*, 1565–1579. [CrossRef] [PubMed]
46. Abdelhamid, E.; Zoulfa, R.; Nada, N.; Zakia, Z.; Bouchra, B.; Azzouz, K.; Anass, K.; Imad, K.; Mohamed, N. Chaste plant extract is a promising biostimulant for tomato plants' growth under salt stress. *Biomass Convers. Biorefin.* **2022**, 1–12. [CrossRef]
47. Ennoury, A.; BenMrid, R.; Nhhala, N.; Roussi, Z.; Latique, S.; Zouaoui, Z.; Nhiri, M. River's Ulva intestinalis L. extract protects common bean plants (*Phaseolus vulgaris* L.) against salt stress. *S. Afr. J. Bot.* **2022**, *150*, 334–341. [CrossRef]
48. Vijayakumar, H.; Thamilarasan, S.K.; Shanmugam, A.; Natarajan, S.; Jung, H.J.; Park, J.I.; Kim, H.; Chung, M.Y.; Nou, I.S. Glutathione Transferases Superfamily: Cold-Inducible Expression of Distinct GST Genes in *Brassica oleracea*. *Int. J. Mol. Sci.* **2016**, *17*, 1211. [CrossRef]
49. Xu, J.; Tian, Y.S.; Xing, X.J.; Peng, R.H.; Zhu, B.; Gao, J.J.; Yao, Q.H. Over-expression of AtGSTU19 provides tolerance to salt, drought and methyl viologen stresses in *Arabidopsis*. *Physiol. Plant.* **2016**, *156*, 164–175. [CrossRef]
50. Calderon, A.; Sanchez-Guerrero, A.; Ortiz-Espin, A.; Martinez-Alcala, I.; Camejo, D.; Jimenez, A.; Sevilla, F. Lack of mitochondrial thioredoxin o1 is compensated by antioxidant components under salinity in *Arabidopsis thaliana* plants. *Physiol. Plant.* **2018**, *164*, 251–267. [CrossRef]
51. Sun, L.J.; Ren, H.Y.; Liu, R.X.; Li, B.Y.; Wu, T.Q.; Sun, F.; Liu, H.M.; Wang, X.M.; Dong, H.S. An h-Type Thioredoxin Functions in Tobacco Defense Responses to Two Species of Viruses and an Abiotic Oxidative Stress. *Mol. Plant Microbe Interact.* **2010**, *23*, 1470–1485. [CrossRef] [PubMed]
52. Kaur, H.; Kaur, H.; Kaur, H.; Srivastava, S. The beneficial roles of trace and ultratrace elements in plants. *Plant Growth Regul.* **2022**, *100*, 219–236. [CrossRef]
53. Ali, F.; Li, Y.H.; Li, F.G.; Wang, Z. Genome-wide characterization and expression analysis of cystathionine beta-synthase genes in plant development and abiotic stresses of cotton (*Gossypium* spp.). *Int. J. Biol. Macromol.* **2021**, *193*, 823–837. [CrossRef] [PubMed]
54. Luan, J.; Dong, J.; Song, X.; Jiang, J.; Li, H. Overexpression of *Tamarix hispida* ThTrx5 Confers Salt Tolerance to Arabidopsis by Activating Stress Response Signals. *Int. J. Mol. Sci.* **2020**, *21*, 1165. [CrossRef]
55. Iordachescu, M.; Imai, R. Trehalose biosynthesis in response to abiotic stresses. *J. Integr. Plant Biol.* **2008**, *50*, 1223–1229. [CrossRef]

agronomy

MDPI

Article

Insights into the Coloring Mechanism of Dark-Red and Yellow Fruits in Sweet Cherry through Transcriptome and Metabolome Analysis

Chaoqun Chen †, Yao Zhang †, Wanjia Tang, Hongxu Chen and Ronggao Gong *

College of Horticulture, Sichuan Agricultural University, Chengdu 611130, China;
2020205005@stu.sicau.edu.cn (C.C.); 2021305052@stu.sicau.edu.cn (Y.Z.); 17843559322@163.com (W.T.);
s20166113@stu.sicau.edu.cn (H.C.)

* Correspondence: gongronggao@sicau.edu.cn

† These authors contributed equally to this work.

Abstract: The color of sweet cherry fruits is an important indicator of their appearance and quality. That influences the purchasing desires of consumers. We performed a multi-omics analysis of two different colors of sweet cherry fruits (yellow "Bing Hu" and dark-red "Hong Deng" fruits). A total of 12 flavonoid differential metabolites, including hesperetin, rutin, and quercetin, and 18 differential structural genes, including *PAL*, *CHS*, *FLS*, and *DFR*, were identified. Possible key regulatory genes for the second stage of color change (from green to yellow) of "Bing Hu" sweet cherry fruits were identified as SBP, bHLH, WD40, and bZIP, which regulated the accumulation of flavonoids, including hesperetin and naringenin. In addition, the possible important roles of transcription factors, which were mainly MYB, bHLH, AP2, and WRKY, in the third stage of color change in both fruits were also identified. This study offers new insights into the changes in fruit coloration between yellow and dark-red sweet cherries, while the analysis of key metabolites and differential genes lays a molecular foundation for future color improvement and breeding programs.

Keywords: sweet cherry; flavonoids; fruit color; transcriptome; metabolome

Citation: Chen, C.; Zhang, Y.; Tang, W.; Chen, H.; Gong, R. Insights into the Coloring Mechanism of Dark-Red and Yellow Fruits in Sweet Cherry through Transcriptome and Metabolome Analysis. *Agronomy* **2023**, *13*, 2397. https://doi.org/10.3390/agronomy13092397

Academic Editors: Guanglong Wang, Lijun Ou and Aisheng Xiong

Received: 13 August 2023
Revised: 13 September 2023
Accepted: 15 September 2023
Published: 17 September 2023

1. Introduction

Plants contain various pigments, which are mainly chlorophyll, carotenoids, and anthocyanins [1]. These pigments, in turn, give plants their different colors. Plant pigments are also essential in plant growth and development and play a significant role in biological processes such as endogenous hormone synthesis and enhancement of tolerance to oxidative and drought stress [2–4]. Anthocyanins are highly safe and free of toxic side effects, as well as having good health functions, including antioxidant, antiaging, and blood pressure-lowering effects [5]. Among them, anthocyanins belong to flavonoids, which are the most widely distributed in plants. They mainly include cyanidin, delphinidin, malvidin, peonidin, petunidin, and pelargonidin. The components and proportions of anthocyanins determine the color of plant tissues. As a result, plants form different colors, such as red, purple, blue, and yellow [6].

In plants, large amounts of flavonoid compounds, such as flavones, flavonoids, and flavonols, are produced through phenylpropanoid biosynthesis, flavonoid biosynthesis, and flavone and flavonol biosynthesis [7]. In Chinese cherries, through targeted metabolism studies, Wang et al. found that the content of flavanol was much higher in yellow cherries than in red cherries [8]. Flavonoids are mainly synthesized from substances such as p-coumaroyl-CoA and cinnamoyl-CoA through various enzymatic reactions, which include the structure genes *CHI*, *CHS*, *F3H*, and *FLS* [9]. In addition, various transcription factors hold a key position in the synthesis of fruit flavonoids. The flavonoid biosynthesis pathway is transcriptionally regulated by the MBW complex formed by MYB transcription factors, the bHLH structure, and WD40 proteins [10]. In apples, *MdMYB90-like* plays a key role in the regulation of anthocyanin biosynthesis [11]. In strawberry studies, yeast two-hybrid and BiFC assays also confirmed that the MBW complex can function in the flavonoid metabolic pathway in the form of FaMYB5/FaMYB10-FaEGL3 (bHLH)-FaLWD1/FaLWD1-like (WD40) [12]. Similar findings were also presented in *Arabidopsis thaliana* and *Norway spruce* [13,14].

The sweet cherry (*Prunus avium* L.) is a perennial fruit tree of the genus *Prunus* in the family Rosaceae that is native to the Mediterranean region of Europe and is currently one of the fastest-developing economic fruit trees [15]. The sweet cherry is a diploid (2n = 16) species composed of 16 chromosomes, and its fruit is rich in sugars, organic acids, vitamins, trace elements, and other nutrients. In addition, it contains large numbers of biologically active substances, such as anthocyanins, phenolic acids, and flavonoids, with antioxidant activity and health effects [16]. Sweet cherry fruits are colorful and glossy, and the skin and flesh show purple-black, red, yellow, and other rich fruit colors. Three stages are involved in the natural development of red sweet cherries: the green ripening stage, the color change stage, and the full red stage [17]. The fruit is green in the first stage; it changes from green to yellow in the second stage; and it changes from yellow to red in the third stage. A similar finding was found in a comparative analysis of the yellow apple mutant and its parent. In this analysis, only part of the fruit turned red while the whole remained yellow, which is probably due to the collective suppression of 34 genes related to coloration [18].

Consumers mainly prefer sweet cherries with a gaily color compared with other qualities [19]. The abundant bioactive substances in plants play an essential role in plants' growth and development but are also closely related to human health. For the last few years, research on sweet cherries has developed rapidly, and important progress has been made mainly in postharvest preservation, fruit softening, hormone treatment, and growth and developmental stages [20–23]. However, the main substances that affect the coloring of the fruit still lack certainty. Therefore, the key candidate genes and potential molecular mechanisms that regulate the accumulation of flavonoids in sweet cherry fruits of different colors need to be explored. In this experiment, we used physiological, metabolomic, and transcriptomic analyses to explore the differences in flavonoids between the yellow-colored variety of sweet cherries "Bing Hu" and the dark-red cultivar of sweet cherries "Hong Deng". The results elucidated the key differential metabolites that form fruit color during ripening of sweet cherry fruits of both colors, along with the genes associated with flavonoid formation. These results provide new insights into changes in fruit coloration in yellow and dark-red sweet cherries, while the key screened metabolites and differential genes provide a molecular basis for future color improvement and breeding programs.

2. Materials and Methods

2.1. Plant Materials

Two types of sweet cherries, including "Hong Deng" (dark red) and "Bing Hu" (yellow), whose rootstocks were *Prunus tomentosa*, were collected in the sweet cherry test base in Xixi Township and Jiuxiang Township, Hanyuan County, Ya'an City, Sichuan Province, China (longitude 102.547, latitude 29.588, altitude 1350 m). Three sweet cherry fruit trees with consistent growth conditions were selected for each variety. Fruits of uniform size, identical appearance, and without damage were selected from each orientation at the time of ripening (55 days after flowering) for sampling. The samples were returned to the laboratory immediately for photographing and then quickly cut into uniform pieces, quick-frozen with liquid nitrogen, wrapped in tin foil, and stored in the refrigerator at −80 °C. Three biological replicates were set up for each sample.

2.2. Determination of Fruit Color

2.2.1. Determination of Anthocyanin Content

Anthocyanin content was determined using the 1% hydrochloric acid–methanol method [24]. A total of 0.5 g of the sample was weighed and added to 10 mL of hydrochloric acid–methanol (1/99, v/v). Then, it was shaken well and placed at 4 °C to avoid light extraction overnight. Thereafter, it was ultrasonicated at 4 °C for 30 min. Next, it was centrifuged at 8000 rpm at 4 °C for 5 min. Subsequently, the absorbance was measured at 530, 620, and 650 nm. Finally, the anthocyanin content was calculated according to the formula.

$$\text{OD}\lambda = (\text{OD}530 - \text{OD}620) - 0.1\ (\text{OD}650 - \text{OD}620)$$

$$\text{Anthocyanin content (nmol/g)} = (\text{OD}\lambda/\varepsilon\lambda) \times (\text{V/M}) \times 10^6$$

where ODλ denotes the calculated optical density of anthocyanins at each wavelength; ελ denotes the molar extinction coefficient of anthocyanins, which is 4.62×10^4; V denotes the total volume of extract; and M denotes the weight of sampling.

2.2.2. Determination of Total Flavonoid Content

The total flavonoid content (TFC) was determined using the method of Wang et al. with slight modifications [25]. The extraction solution was 70% methanol containing 2% formic acid (v/v). Fruits were weighed to 0.2 g, ground, and added to 5 mL of solution and sonicated for 30 min, followed by shaking at 250 rpm for 2 h at 30 °C on a shaker to fully extract. Centrifugation was performed at 8000× g for 10 min, and the supernatant was removed. By using 200 µL of fruit extract, 1.3 mL of methanol, 100 µL of $NaNO_2$ (0.5 M), and 100 µL of $AlCl_3$ (0.3 M) were added sequentially and mixed in a centrifuge tube. Then, 500 µL of NaOH (1 M) was added after 5 min, and the absorbance was measured at 510 nm.

2.3. *Metabolomic Analysis*

Tissues (100 mg) were individually ground with liquid nitrogen, and the homogenate was resuspended in prechilled 80% methanol by vortexing. Samples were incubated on ice for 5 min and then centrifuged at 15,000× g and 4 °C for 20 min. Some of the supernatant was diluted to a final concentration containing 53% methanol by liquid chromatography-mass spectrometry (LC-MS). The samples were subsequently centrifuged at 15,000× g and 4 °C for 20 min. Finally, the supernatant was injected into the LC-MS/MS system for analysis.

Samples were injected onto a Hypesil Gold column (C18) (100 × 2.1 mm, 1.9 µm) using a 12 min linear gradient at a flow rate of 0.2 mL/min and a column temperature of 40 °C with an automatic injector set at 8 °C. The eluents for the positive polarity mode were eluent A (0.1% FA in water) and eluent B (methanol). The eluents for the negative polarity mode were eluent A (5 mM ammonium acetate, pH 9.0) and eluent B (methanol). The solvent gradient was set as follows: 2% B, 1.5 min; 2–8% B, 3 min; 85–100% B, 10 min; 100–2% B, 10.1 min; and 2% B, 12 min. The Q Exactive TM HF-X mass spectrometer was operated in positive–negative polarity mode with a spray voltage of 3.5 kV, a capillary temperature of 320 °C, a sheath gas flow rate of 35 pounds per second (psig), and positive/negative polarity mode. The aux gas flow rate was 10 L/min, the S-lens RF level was 60, and the aux gas heater temperature was 350 °C.

The raw data files generated by UHPLC-MS/MS were processed using Compound Discoverer 3.1 (CD3.1, Thermo Fisher, San Jose, CA, USA) to perform peak alignment, peak picking, and quantitation for each metabolite. Then, peaks were matched with the mzCloud (https://www.mzcloud.org/ (accessed on 16 March 2023)), mzVault, and MassList databases to obtain accurate qualitative and relative quantitative results.

These metabolites were annotated using the KEGG database (https://www.genome.jp/kegg/pathway.html (accessed on 20 March 2023)), the HMDB database (https://hmdb.ca/ (accessed on 25 March 2023)), and the LIPID MAPS database (http://www.lipidmaps.org/ (accessed on 26 March 2023)). We applied univariate analysis (t-test) to calculate the statistical significance (p-value). The metabolites with VIP > 1, p-value < 0.05, and fold change ≥ 2 or FC ≤ 0.5 were considered to be differential metabolites.

2.4. Transcriptome Sequencing

Total RNA was extracted using a total RNA kit (TIANGEN Biotech, Beijing, China), and the RNA integrity was assessed using the RNA Nano 6000 Assay Kit of the Bioanalyzer 2100 system (Agilent Technologies, Santa Clara, CA, USA). mRNA was purified from total RNA using poly-T oligo-attached magnetic beads. First-strand cDNA was synthesized using random hexamer primers and M-MuLV reverse transcriptase (RNase H−). Second-strand cDNA synthesis was subsequently performed using DNA Polymerase I and RNase H. The library fragments were purified with the AMPure XP system (Beckman Coulter, Beverly, MA, USA). Then, PCR was performed with Phusion High-Fidelity DNA Polymerase, Universal PCR Primers, and Index (X) Primer. Finally, PCR products were purified (using the AMPure XP system (Beckman Coulter, Inc., Brea, CA, USA)). The index-coded samples were clustered on a cBot Cluster Generation System using TruSeq PE Cluster Kit v3-cBot-HS (Illumina, San Diego, CA, USA). After cluster generation, the library preparations were sequenced on an Illumina NovaSeq platform, and 150 bp paired-end reads were generated.

Raw data (raw reads) in fastq format were first processed using fastp software v0.19.7. Reference genome and gene model annotation files were downloaded directly from the genome website. The index of the reference genome was built using Hisat2 v2.0.5, and paired-end clean reads were aligned to the reference genome using Hisat2 v2.0.5. featureCounts v1.5.0-p3 was used to count the read numbers mapped to each gene. Then, the FPKM (expected number of fragments per kilobase of transcript sequence per million base pairs sequenced) of each gene was calculated based on the length of the gene and the read count mapped to this gene.

Differential expression analysis of two groups was performed using the DESeq2 R package (1.20.0). The resulting p values were adjusted using Benjamini and Hochberg's approach for controlling the false discovery rate. Genes with an adjusted p value ≤ 0.05 found by DESeq2 were assigned as differentially expressed.

Gene ontology enrichment analysis of differentially expressed genes was implemented by the clusterProfiler R package v3.8.1. We used the clusterProfiler R package to test the statistical enrichment of differentially expressed genes in KEGG pathways.

2.5. qRT-PCR Analysis

qRT-PCR analysis was performed using a Bio-Rad CFX96TM real-time PCR system (Bio-Rad, Hercules, CA, USA) and 2× TSINGKE® Master qPCR Mix (SYBR Green I) (TSINGKE, Beijing, China). The amplification program was as follows: predenaturation at 95 °C for 30 s, denaturation at 95 °C for 5 s, and annealing at 60 °C for 30 s. The number of amplification cycles was 40, and each sample was repeated three times. The internal reference was ACTIN. Gene-specific primers were designed by Primer Premier 6, and the primer sequences are shown in Table S3. Gene expression was calculated using $2^{-\Delta\Delta Ct}$ [26].

2.6. Data Processing and Analysis

All indicators were determined after three repetitions. The data is sorted, counted, and plotted through Excel 16.54, SPSS 27, and GraphPad Prism 9, respectively.

3. Results

3.1. Analysis of Fruit Pigmentation of Two Sweet Cherry Cultivars

The sweet cherry fruits of "Hong Deng" (named H) and "Bing Hu" (named B) were selected separately to analyze the differences in fruit color. As observed from the phenotypes at ripening (Figure 1), the sweet cherry fruits of "Hong Deng" were dark red at ripening, whereas "Bing Hu" was yellow, as was the flesh coloration. We then determined the contents of total flavonoids (Figure 1A), anthocyanins (Figure 1B), and carotenoids (Figure 1C). The anthocyanin content of "Hong Deng" sweet cherry fruits in the ripening stage was high, amounting to 723.78 nmol/g FW. By contrast, the anthocyanin content of "Bing Hu" fruits was only 40.58 nmol/g FW. Similarly, the total flavonoid and carotenoid contents of "Hong Deng" sweet cherry fruits during ripening were higher than those of

"Bing Hu" sweet cherries, at 40.77 mg/mg FW and 0.46 μg/g FW, respectively. Although "Bing Hu" fruits contained flavonoids and carotenoids, the flavonoid content was much higher than the carotenoid content.

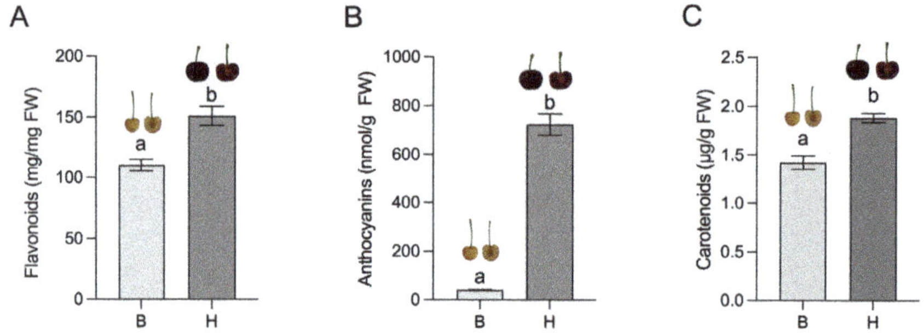

Figure 1. Differential analysis of the physiological properties of two sweet cherry fruits. (**A**) Total flavonoid content of two sweet cherry fruits at maturity; (**B**) anthocyanin content of two sweet cherry fruits at maturity; (**C**) carotenoid content of two sweet cherry fruits at maturity. Different lowercase letters indicate significant differences ($p \leq 0.05$). FW means fresh weight.

3.2. Analysis of Metabolomic Differences between Two Sweet Cherry Fruit Cultivars

A nontargeted metabolomics study was performed using LC-MS technology to investigate the metabolite variability in different color varieties of sweet cherries, and the data in positive and negative ion modes were analyzed using principal component analysis (PCA), as shown in Figure 2A. In the PCA model based on 3 quality control (QC) mix samples and 12 test samples, the first 2 principal components can be separated into 15 samples, which accounted for 57.1% and 8.6% of the total variability, respectively. In addition, two groups of test samples were separated and occupied the two ends of PC1, which indicated that sweet cherries of different colors varied greatly. A total of 532 metabolites were detected in two different varieties of sweet cherries, and the identified metabolites were annotated using the HMDB and LIPID MAPS databases. A total of 307 metabolites were categorized into 8 classes in the HMDB database (Figure 2B, Table S4), with the most abundant type being organic acids and derivatives (71 metabolites), followed by lipids and lipid-like molecules (50 metabolites), organoheterocyclic compounds (45 metabolites), and organic oxygen compounds (42 metabolites). In the LIPID MAPS database (Figure 2C, Table S5), the most abundant and diverse type was metabolism (10 classes and 305 metabolites), with amino acid metabolism and global and overview maps containing the most metabolites. Environmental Information Processing (2 classes and 14 metabolites) and Genetic Information Processing had a lesser extent. Three parameters, namely, VIP, FC, and p value (VIP > 1.0, FC > 1.5, or FC < 0.667 and p value < 0.05), were utilized to screen differential metabolites for the two varieties of sweet cherries. A total of 228 metabolites were identified as differential metabolites (150 upregulated and 78 downregulated), which indicated that most of the differential metabolites were highly abundant in the "Bing Hu" cultivars. Hierarchical cluster analysis was performed for all differential metabolites. As shown in Figure 2D, the metabolite cluster analysis of the 2 varieties showed a clear grouping pattern with good reproducibility, and all the differential metabolites were categorized into 2 groups. Group 2 metabolites were the most abundant and highly expressed in "Bing Hu" varieties, and group 1 metabolites were highly expressed in "Hong Deng" cultivars.

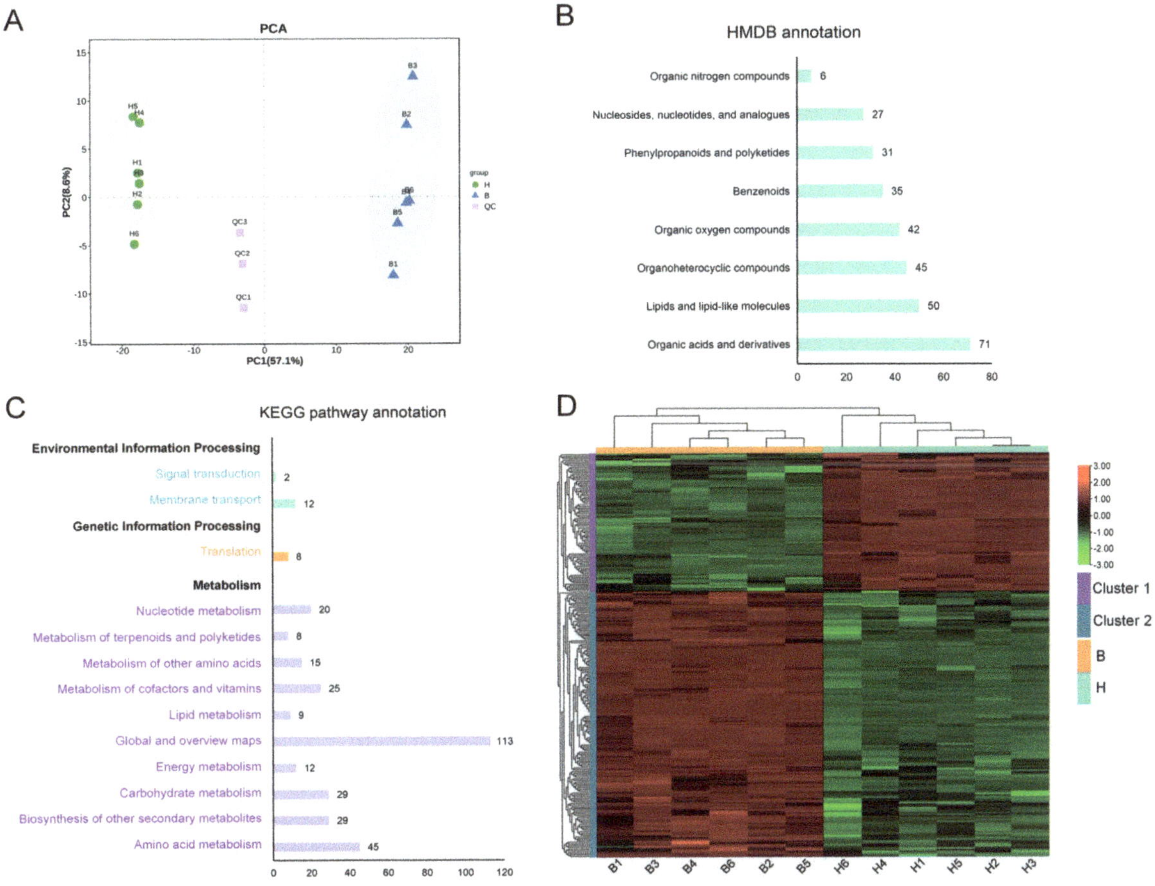

Figure 2. Metabolomic analysis of sweet cherry fruits of two cultivars. (**A**) PCA; (**B**) HMDB analysis; (**C**) KEGG pathway analysis; (**D**) Metabolite expression profile analysis.

3.3. Transcriptome Analysis of Two Cultivars of Sweet Cherry Fruits

RNA-seq technology was used to analyze the transcriptional sequencing of "Hong Deng" and "Bing Hu" sweet cherry fruits at the ripening stage. The results generated 47,709,534 raw data, and 47,119,642 high-quality clean reads were obtained after filtering out splice sequences, indeterminate reads, and low-quality reads; an average of 93.56% of the clean reads were localized to the sweet cherry genome (see Supplementary Table S1 for detailed results). A total of 44,819 transcripts were obtained, and the expression level of each gene was normalized to FPKM to get all genes' expression patterns.

The distribution of PCA among the samples is shown in Figure 3A. Among them, PC1 and PC2 accounted for 78.09% and 7.95%, respectively, and H and B were distributed at the positive and negative ends of PC1. Therefore, a large difference was observed between the two groups of previous transcripts at maturation. Meanwhile, clustering on PC1 among the three samples within each combination was high. The correlation between the samples was analyzed to further determine the repeatability between the samples, and the specific results are shown in Figure 3B. The higher correlation between the samples within each combination and the weaker correlation between the combinations further indicated that a large difference existed between the transcript levels of the sweet cherry fruit of the two cultivars.

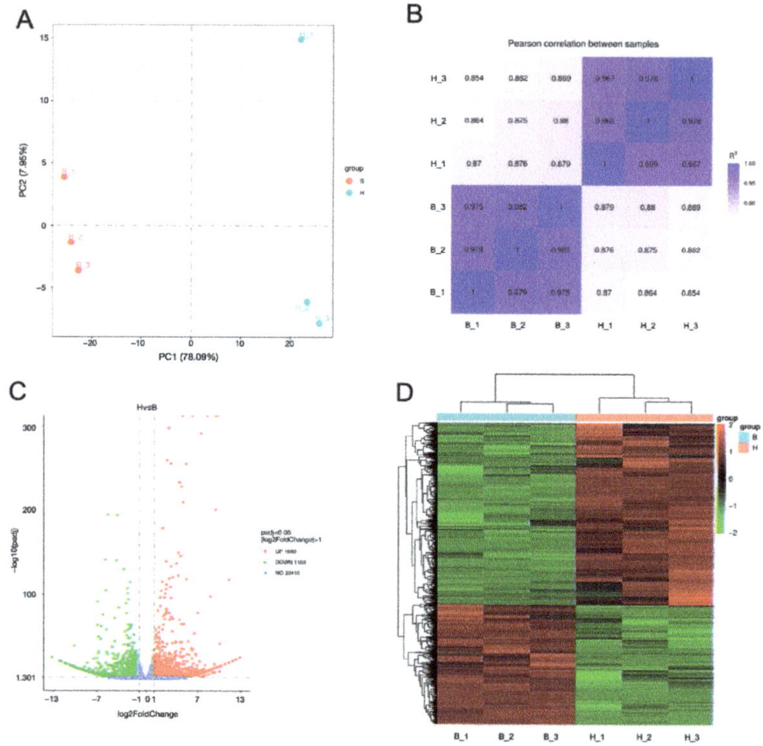

Figure 3. Transcriptome analysis of two cultivars of sweet cherry fruits. (**A**) PCA of two cultivars of sweet cherry fruits; (**B**) sample correlation analysis; (**C**) volcano plot analysis; (**D**) expression patterns of differential genes.

Using the criterion of |log2(FoldChange)| ≥ 1 and padj ≤ 0.05 for differential gene screening, 2783 differential genes were identified, of which 1680 genes were upregulated and 1103 genes were downregulated (Figure 3C). The expression patterns of all differential genes are shown in Figure 3D. As shown in the figure, large differences were observed in the expression of these differential genes, and most of them were mainly expressed at higher levels in "Hong Deng" sweet cherries. The differential expression of these genes might be the main reason for the large differences in the fruits of the two cultivars of sweet cherries.

3.4. Multi-Omics Analysis

3.4.1. KEGG Pathway Map Analysis

A KEGG pathway analysis was performed to further identify the pathways of differential metabolites and differential gene enrichment. As shown in Figure 4A, the flavone and flavonol biosynthesis pathway was the most significant and contained three differential metabolites as plotted against $-$log10(p-value). We also noted that the flavonoid biosynthesis, phenylpropanoid biosynthesis, and isoflavonoid biosynthesis pathways were all significantly enriched in the top 20, which suggests that flavonoids differed significantly between dark-red and yellow varieties. Similarly, the analysis of transcriptome-enriched pathways was conducted using $-$log10(padj) (Figure 4B). As observed from the figure, flavonoid biosynthesis was not only the most prominent pathway but also contained a large number of differential genes. Meanwhile, the phenylpropanoid biosynthesis pathway contained 33 differential genes and was the second most significantly enriched pathway. All these results indicate large differences in flavonoids between dark-red and yellow cultivars.

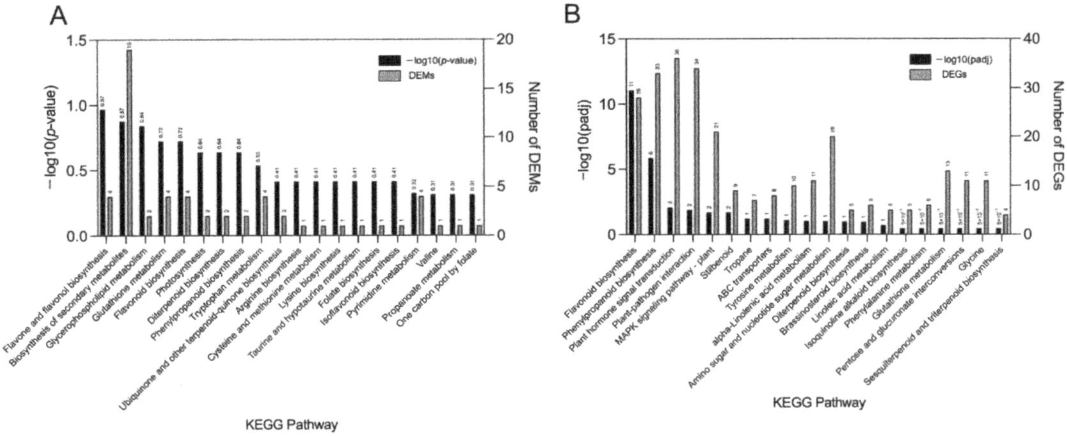

Figure 4. Annotation analysis of KEGG pathways in two sweet cherry fruit cultivars at the ripening stage. (**A**) Fruit differential metabolite KEGG pathway enrichment analysis of two cultivars as statistically mapped according to −log10(p-value). (**B**) Fruit differential gene KEGG pathway enrichment analysis of two cultivars as statistically mapped according to −log10(padj).

3.4.2. Flavonoid Anabolic Network Analysis

In a previous study, flavonoid-related pathways were found to be significantly enriched in the two cultivars of sweet cherry fruits; therefore, we further mapped the specific metabolic pathways (Figure 5 and Table S2). As shown in the figure, most of the flavonoids differed significantly between the two varieties of fruit. Among them, five substances, namely hesperetin, caffeic acid, ferulic acid, naringenin, and kaempferitrin, were higher in the yellow variety fruits than in the dark-red sweet cherry fruits. In addition, the remaining seven substances (luteolin, rutin, quercetin, eriodictyol, glycitein 7-O-glucoside, trifolin, and catechin) were all found at higher levels in dark-red fruit. We also analyzed the relevant differential genes in this pathway to further determine the regulatory mechanism of these substance changes. We identified a total of 18 differential genes, including E4.3.1.24 (PAL), EC: 2.3.1.74 (CHS), EC:1.14.20.6 (FLS), and EC: 1.1.1.219 (DFR). The expression of these genes was higher in dark-red fruits than in yellow varieties.

3.5. Transcription Factor Analysis

Their transcription factors were analyzed and identified, and the results are shown in Figure 6A. A total of 20 gene families with 780 genes, such as MYB, AP2, and bHLH, were identified, and the families with the most genes were MYB, AP2, FAR1, and bHLH. Among them, genes such as WRKY, which is related to plant resistance, were highly enriched. We further counted and analyzed the number of differential genes in each group of transcription factors and the expression pattern of differential transcription factors. The results are shown in Figure 6B,C. A total of 64 genes in 12 gene families were screened as differential transcription factors, and more differential genes and higher expression of the gene families of WRKY, AP2, bHLH, and MYB were related to flavonoid metabolism, among which the genes of the WRKY family were the most abundant (17, 26.56%). Most of the differential transcription factors scored higher expression in dark-red cultivars than in yellow cherries, and these differential transcription factors may be potential factors that regulate the existence of differences in fruit coloration between the two cultivars.

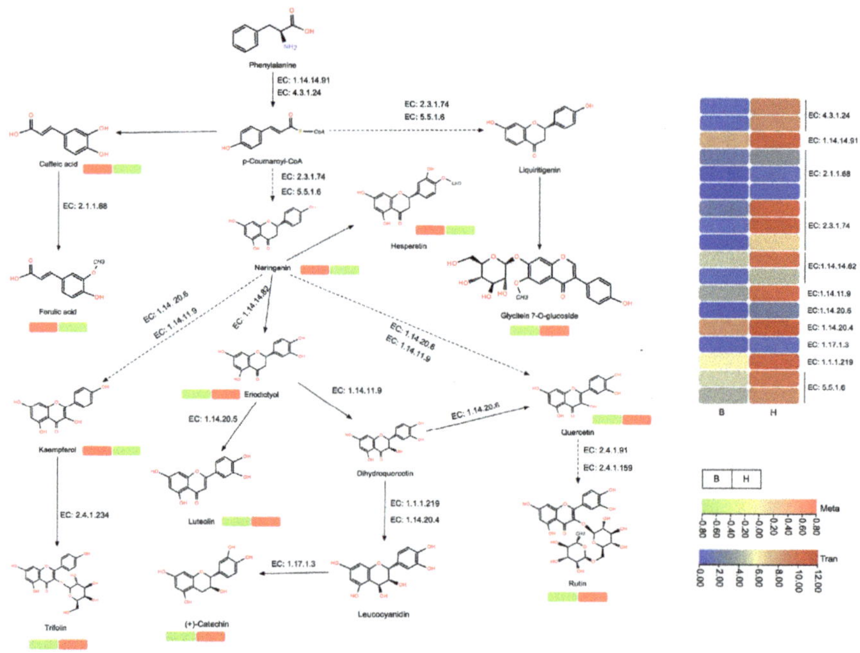

Figure 5. Flavonoid anabolic network analysis.

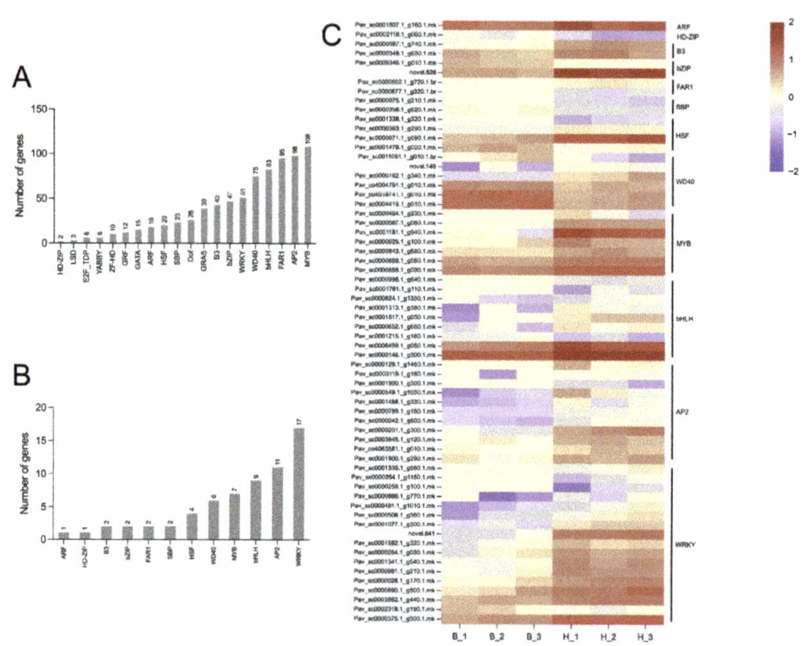

Figure 6. Transcription factor analysis of sweet cherry fruits from two cultivars. (**A**) Results of transcription factor analysis in the transcriptome; (**B**) statistics and analysis of differential transcription factors; (**C**) heatmap of the expression patterns of differential transcription factors as transformed to FPKM values of transcription factors according to log10.

3.6. Correlation Matrix Analysis

The key differential genes and transcription factors in the pathway were then analyzed in a correlation matrix. The results are shown in Figure 7. As shown in the figure, a positive correlation was observed between these transcription factors and the differential genes, except for hesperetin, caffeic acid, ferulic acid, naringenin, and kaempferitrin. In particular, some of the transcription factors boxed using black lines showed highly significant positive correlations between these differential genes and differential metabolites, which were mainly the four family genes MYB, bHLH, AP2, and WRKY. These transcription factors were more associated with fruit anthocyanin synthesis. Some transcription factors marked with red lines (mainly SBP, bHLH, WD40, and bZIP) showed high and highly significant correlations with several metabolites with higher content in yellow cultivars, which suggested that these transcription factors may be the key regulators in regulating the accumulation of flavonoids in the fruits of yellow cultivars of sweet cherries.

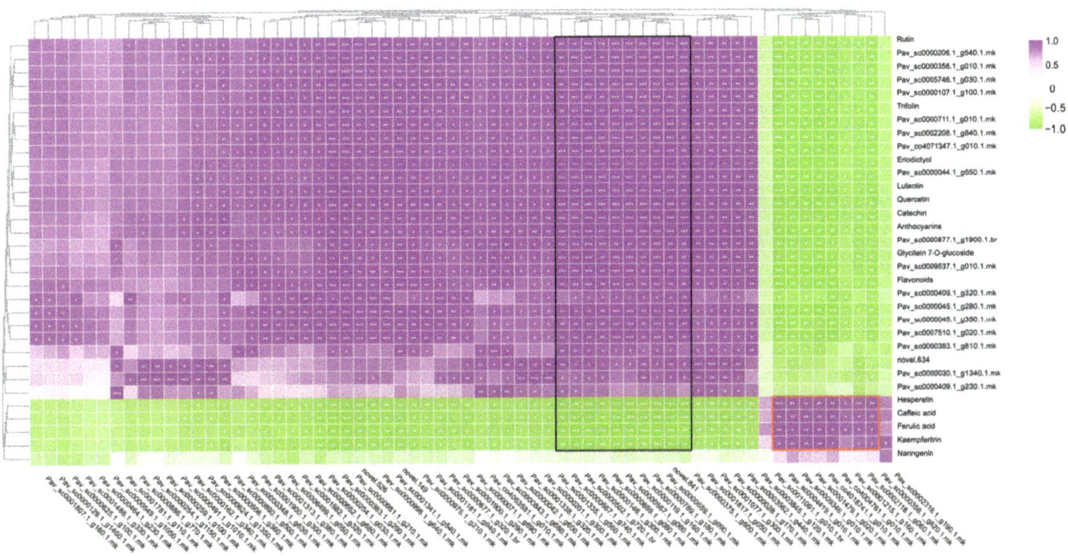

Figure 7. Correlation matrix analysis. * indicates a significant correlation at the 0.05 level; ** indicates a significant correlation at the 0.01 level; and *** indicates a highly significant correlation at the 0.001 level.

3.7. qPCR Validation Analysis

Key regulatory genes and transcription factors were randomly selected from the pathway and analyzed by qRT-PCR to verify the validity of the transcriptome, and the results are shown in Figure 8. Consistent with the transcriptome results (Table S3), most of these key structural genes were highly expressed in "Hong Deng" sweet cherries. Overall, the qRT-PCR results of most of the structural genes and transcription factors were consistent with the transcriptome data, which indicates that the transcriptome data had high confidence.

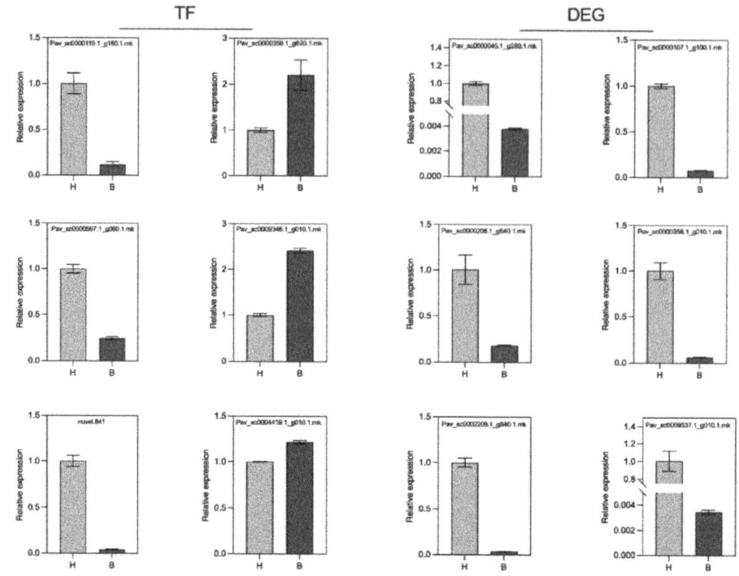

Figure 8. qRT-PCR analysis of genes and transcription factors related to flavonoid synthesis in different varieties of sweet cherry fruit.

4. Discussion

Flavonoids are a class of biologically active compounds that are beneficial to human health and are widely found in various fruits and vegetables [27–30]. Several epidemiological studies have shown that flavonoids play an important role in antioxidation, improvement of human immunity, prevention of urinary stones, treatment of hypertension, and prevention of cardiovascular diseases [31–35]. Color differences often exist between different cultivars of the same species, and flavonoids are an important source of yellow color in plants. In two differently colored cucumber cultivars, namely, L19 and L14, L14 with a yellow rind had a significantly higher flavonoid content than L19 with a green-white rind at late fruit ripening. Moreover, the carotenoid content was insignificantly different between the two cultivars [36]. Similar findings also occurred in paeonia, melon, and mulberry [37–39]. In this study, the plant source of yellow pigments (carotenoids) was much lower than the flavonoids in the "Bing Hu" fruit. Therefore, flavonoids may be the main source of substances responsible for the yellow coloration of "Bing Hu" sweet cherry fruits [40]. Therefore, flavonoids may be the main source of substances responsible for the yellow coloration of "Bing Hu" sweet cherry fruits.

The results of Metabolome showed that the pathways related to flavonoids were significantly enriched in the fruit of both sweet cherry cultivars. A total of 12 differential metabolites in the flavonoid pathway, including hesperetin, caffeic acid, luteolin, rutin, and quercetin, were identified (Table S3). Five substances (hesperetin, caffeic acid, ferulic acid, naringenin, and kaempferitrin) were higher in the yellow cultivar fruits than in the dark-red sweet cherry fruits. The remaining seven substances (luteolin, rutin, quercetin, eriodictyol, glycitein 7-O-glucoside, trifolin, and catechin) were more abundant in dark-red fruit.

Previous studies have shown that naringenin and hesperetin are two of the main substances responsible for the color differences in *Sorghum* bicolor seeds [41]. In a study of peach gum, high levels of hesperetin, naringenin, and eriodictyol were also significantly positively correlated with the deeper yellow coloration of *Sorghum* bicolor [42]. We hypothesized that higher levels of the five flavonoid metabolites, especially hesperetin and naringenin, in the yellow cultivar might be the primary reason for the yellow coloration of "Bing Hu" sweet cherry fruit. In another study on the coloration mechanism of *Cymbidium* sinense "Red Sun" leaves, the total phenol, total anthocyanin, and carotenoid contents of yellow leaves were significantly lower than those of red leaves, which is consistent with the results of the present study [43]. In our study, luteolin, rutin, quercetin, eriodictyol, glycitein 7-O-glucoside, trifolin, and catechin, which are seven differential flavonoid metabolites that are more abundant in the dark-red cultivar, may be the main reason for making the total flavonoid content higher in the dark-red sweet cherry cultivar. However, due to the extremely high anthocyanin content in the dark-red sweet cherry cultivar, the fruit color is still dark red.

In the analysis of the flavonoid pathway, we also identified 18 differential structural genes, including *PAL*, *CHS*, *CHI*, *FLS*, *ANS*, and *DFR*, and the expression of these genes was higher in the dark-red cultivar than in the yellow cultivar. These structural genes are extensively involved in the synthesis and metabolism of flavonoids and are responsible for the higher flavonoid content in dark red fruits [7]. In addition to structural genes, plant color formation is also regulated by transcription factors [44]. Further transcription factor analysis of the fruits of the two differently colored cultivars identified 64 genes in 12 gene families as differential transcription factors. The correlation matrix showed that these transcription factors were positively correlated with the differential genes. The seven flavonoid metabolites that were more abundant in the red sweet cherry varieties. In particular, the four families of genes, MYB, bHLH, AP2, and WRKY, were significantly positively correlated with some of the differential genes (PAL, C4H, F3′H, LDOX, DFR, CHI) and metabolites (Rutin, Trifolin, Luteolin, Quercetin, Catechin, and Anthocyanins). MYB and bHLH are two important transcriptional proteins in plants, and they have regulatory roles in anthocyanin biosynthesis in fruits such as apples [45,46], strawberries [12,47], and grapes [48,49]. They are usually involved in the regulation of color accumulation in plants through the MBW ternary complex with WD40 proteins [50]. AP2 and WRKY transcription factors primarily play a role in the response of plants to abiotic stresses [51,52]. In recent years, however, they have been found to also be involved in the anthocyanin biosynthesis of plants [53,54]. In the results of this experimental study, the four families of genes showed highly significant positive correlations with seven flavonoid metabolites and differentially structured genes that were more abundant in dark-red sweet cherry cultivars. The four classes of MYB, bHLH, AP2, and WRKY transcription factors identified in our study may therefore be the major regulatory genes that lead to the high anthocyanin and flavonoid contents in dark-red sweet cherry cultivars.

In addition, we found that several metabolites (hesperetin, caffeic acid, ferulic acid, naringenin, and kaempferitrin) that were more abundant in the yellow sweet cherry cultivar were correlated with transcription factors, including SBP, bHLH, WD40, and bZIP, at higher and highly significant levels. Previous studies have shown that SBP transcription factors can be mediated by MIR156a to inhibit isoflavone biosynthesis in *Pueraria thomsonii* Benth [55]. Given that the flavonoid pathway is the upstream pathway of anthocyanin biosynthesis, bHLH and WD40, which play important roles in anthocyanin biosynthesis, also regulate the biosynthesis of flavonoids. For instance, bHLH is involved in flavonoid biosynthesis in bryophytes [56]. In a study of *Oroxylum indicum*, transcription factor analysis showed that the bHLH structure and the WD40 transcription factor family regulated the abundance of flavonoid biosynthesis [57]. In another related study, after the authors knocked out one allele of *VvbZIP36*, the amount of naringenin in grapefruit increased [58]. All these studies indicate that SBP, bHLH, WD40, and bZIP have important roles in plant flavonoid biosynthesis. On the basis of the results of this experiment, we hypothesize that these transcription factors may also be key regulators of the yellow coloration of "Bing Hu" sweet cherry fruits and the accumulation of flavonoids such as hesperetin and naringenin.

Three stages are involved in the natural development of the "Hong Deng" sweet cherry, in which the fruit changes from green to yellow and finally to red [17], while the "Bing Hu" sweet cherry only changes color from green to yellow (Figure 1). Similarly, the "Bing Hu" sweet cherry changed from green to yellow during the second stage. However, only part of the fruit peel changed color during the process of changing from yellow to red, and the main peel and flesh of the fruit were still yellow. In the study of El-Sharkawy et al., it was shown that fruit will not turn color when coloring-related genes are repressed [18]. Activation of genes functioning in the later stages of the pathway leading to anthocyanin (pigmentation) production (*DFR, LDOX,* and *UFGT)* requires the MBW complex, and flavone/isoflavone and anthocyanins compete for the same precursor substances, which are the flavanones, to determine fruit coloration. We therefore deduced that SBP, bHLH, WD40, and bZIP might be the key regulatory genes for the change from green to yellow in the second stage of "Bing Hu" sweet cherry fruit. In the third stage, during the process of yellow sweet cherry fruit changing from yellow to red, transcription factors such as MYB, bHLH, AP2, and WRKY might be suppressed. We speculated that the structural genes failed to be activated, which led to the inhibition of anthocyanin accumulation and then led to the color change of only some of the peel of the yellow sweet cherry fruits. The remaining part of the peel and flesh failed to change color to red. In the "Hong Deng" sweet cherry, the aforementioned genes were expressed, and anthocyanins accumulated, which resulted in the dark-red color of the fruit.

5. Conclusions

We determined the pigmentation differences between two different cultivars of sweet cherry, namely, "Hong Deng" and "Bing Hu," through phenotypic observation and physiological index analysis. A total of 12 flavonoid differential metabolites, including hesperetin, rutin, and quercetin, and 18 differential structural genes, including *PAL, CHS, FLS,* and *DFR,* were identified in the two sweet cherry fruits by the combination of transcriptome and metabolome analyses. Transcription factor and correlation matrix analyses indicated that the transcription factors SBP, bHLH, WD40, and bZIP might be the key regulatory genes for the second stage of yellow sweet cherry fruits changing from green to yellow color, which regulates the accumulation of flavonoids such as hesperetin and naringenin. The transcription factors MYB, bHLH, AP2, and WRKY might function in the third stage of the fruit change from yellow to red. They were repressed in yellow sweet cherry fruit and normally expressed in the dark-red sweet cherry fruit cultivar, which in turn led to the final color difference between the two sweet cherry cultivars. These results provide new insights into the changes in fruit coloration in yellow and dark-red sweet cherries, while the key screened metabolites and differential genes provide a molecular basis for future color improvement and breeding programs.

Supplementary Materials: The following supporting information can be downloaded at: https://www.mdpi.com/article/10.3390/agronomy13092397/s1, Table S1: Transcriptome sequencing statistics; Table S2: Expression of DEMs and DEGs; Table S3: Primer sequences used for qPCR. Table S4: Classification results of the HMDB database. Table S5: Classification results from the KEGG database.

Author Contributions: C.C. and Y.Z.: designed the whole experiments. C.C., W.T. and Y.Z.: Data determination. Y.Z., C.C., W.T., H.C. and R.G.: analyzed the data. C.C. and Y.Z.: wrote the manuscript. R.G.: supervision. All authors have read and agreed to the published version of the manuscript.

Funding: This research was funded by the Sichuan Science and Technology Plan Project (Key R&D Project) (2021YFN0081, 2021YFN0082).

Institutional Review Board Statement: Not applicable.

Informed Consent Statement: Not applicable.

Data Availability Statement: The data presented in this study are available upon request from the corresponding author.

Conflicts of Interest: The authors declare no conflict of interest.

Sample Availability: The data is available from the authors.

References

1. Tanaka, Y.; Sasaki, N.; Ohmiya, A. Biosynthesis of plant pigments: Anthocyanins, betalains and carotenoids. *Plant J.* **2008**, *54*, 733–749. [CrossRef] [PubMed]
2. Kadomura-Ishikawa, Y.; Miyawaki, K.; Takahashi, A.; Masuda, T.; Noji, S. Light and abscisic acid independently regulated FaMYB10 in Fragaria × ananassa fruit. *Planta* **2015**, *241*, 953–965. [CrossRef] [PubMed]
3. Lee, S.G.; Vance, T.M.; Nam, T.G.; Kim, D.O.; Koo, S.I.; Chun, O.K. Contribution of Anthocyanin Composition to Total Antioxidant Capacity of Berries. *Plant Foods Hum. Nutr.* **2015**, *70*, 427–432. [CrossRef] [PubMed]
4. Cirillo, V.; D'Amelia, V.; Esposito, M.; Amitrano, C.; Carillo, P.; Carputo, D.; Maggio, A. Anthocyanins are Key Regulators of Drought Stress Tolerance in Tobacco. *Biology* **2021**, *10*, 139. [CrossRef]
5. Lu, W.; Shi, Y.; Wang, R.; Su, D.; Tang, M.; Liu, Y.; Li, Z. Antioxidant Activity and Healthy Benefits of Natural Pigments in Fruits: A Review. *Int. J. Mol. Sci.* **2021**, *22*, 4945. [CrossRef]
6. Saigo, T.; Wang, T.; Watanabe, M.; Tohge, T. Diversity of anthocyanin and proanthocyanin biosynthesis in land plants. *Curr. Opin. Plant Biol.* **2020**, *55*, 93–99. [CrossRef]
7. Falcone Ferreyra, M.L.; Rius, S.P.; Casati, P. Flavonoids: Biosynthesis, biological functions, and biotechnological applications. *Front. Plant Sci.* **2012**, *3*, 222. [CrossRef]
8. Wang, Y.; Wang, Z.; Zhang, J.; Liu, Z.; Wang, H.; Tu, H.; Zhou, J.; Luo, X.; Chen, Q.; He, W.; et al. Integrated Transcriptome and Metabolome Analyses Provide Insights into the Coloring Mechanism of Dark-red Yellow Fruits in Chinese Cherry [*Cerasus pseudocerasus* (Lindl), G. Don]. *Int. J. Mol. Sci.* **2023**, *24*, 3471. [CrossRef]
9. Liu, W.; Feng, Y.; Yu, S.; Fan, Z.; Li, X.; Li, J.; Yin, H. The Flavonoid Biosynthesis Network in Plants. *Int. J. Mol. Sci.* **2021**, *22*, 12824. [CrossRef]
10. Gonzali, S.; Perata, P. Fruit colour and novel mechanisms of genetic regulation of pigment production in tomato fruits. *Horticulturae* **2021**, *7*, 259. [CrossRef]
11. Sun, C.; Wang, C.; Zhang, W.; Liu, S.; Wang, W.; Yu, X.; Song, T.; Yu, M.; Yu, W.; Qu, S. The R2R3-type MYB transcription factor MdMYB90-like is responsible for the enhanced skin color of an apple bud sport mutant. *Hortic. Res.* **2021**, *8*, 156. [CrossRef] [PubMed]
12. Yue, M.; Jiang, L.; Zhang, N.; Zhang, L.; Liu, Y.; Lin, Y.; Zhang, Y.; Luo, Y.; Zhang, Y.; Wang, Y.; et al. Regulation of flavonoids in strawberry fruits by FaMYB5/FaMYB10 dominated MYB-bHLH-WD40 ternary complexes. *Front. Plant Sci.* **2023**, *14*, 1145670. [CrossRef] [PubMed]
13. Baudry, A.; Caboche, M.; Lepiniec, L. TT8 controls its own expression in a feedback regulation involving TTG1 and homologous MYB and bHLH factors, allowing a strong and cell-specific accumulation of flavonoids in Arabidopsis thaliana. *Plant J.* **2006**, *46*, 768–779. [CrossRef] [PubMed]
14. Nemesio-Gorriz, M.; Blair, P.B.; Dalman, K.; Hammerbacher, A.; Arnerup, J.; Stenlid, J.; Mukhtar, S.M.; Elfstrand, M. Identification of Norway Spruce MYB-bHLH-WDR Transcription Factor Complex Members Linked to Regulation of the Flavonoid Pathway. *Front. Plant Sci.* **2017**, *8*, 305. [CrossRef]
15. Yamamoto, T.; Terakami, S. Genomics of pear and other Rosaceae fruit trees. *Breed. Sci.* **2016**, *66*, 148–159. [CrossRef]
16. Chockchaisawasdee, S.; Golding, J.B.; Vuong, Q.V.; Papoutsis, K.; Stathopoulos, C.E. Sweet cherry: Composition, postharvest preservation, processing and trends for its future use. *Trends Food Sci. Technol.* **2016**, *55*, 72–83. [CrossRef]
17. Chen, C.; Chen, H.; Yang, W.; Li, J.; Tang, W.; Gong, R. Transcriptomic and Metabolomic Analysis of Quality Changes during Sweet Cherry Fruit Development and Mining of Related Genes. *Int. J. Mol. Sci.* **2022**, *23*, 7402. [CrossRef]
18. El-Sharkawy, I.; Liang, D.; Xu, K. Transcriptome analysis of an apple (Malus × domestica) yellow fruit somatic mutation identifies a gene network module highly associated with anthocyanin and epigenetic regulation. *J. Exp. Bot.* **2015**, *66*, 7359–7376. [CrossRef]
19. Correia, S.; Queirós, F.; Ribeiro, C.; Vilela, A.; Aires, A.; Barros, A.I.; Schouten, R.; Silva, A.P.; Gonçalves, B. Effects of calcium and growth regulators on sweet cherry (*Prunus avium* L.) quality and sensory attributes at harvest. *Sci. Hortic.* **2019**, *248*, 231–240. [CrossRef]
20. Karagiannis, E.; Michailidis, M.; Karamanoli, K.; Lazaridou, A.; Minas, I.S.; Molassiotis, A. Postharvest responses of sweet cherry fruit and stem tissues revealed by metabolomic profiling. *Plant Physiol. Biochem.* **2018**, *127*, 478–484. [CrossRef]
21. Zhai, Z.; Xiao, Y.; Wang, Y.; Sun, Y.; Peng, X.; Feng, C.; Zhang, X.; Du, B.; Zhou, X.; Wang, C.; et al. Abscisic acid-responsive transcription factors PavDof2/6/15 mediate fruit softening in sweet cherry. *Plant Physiol.* **2022**, *190*, 2501–2518. [CrossRef] [PubMed]
22. Chen, C.; Chen, H.; Chen, Y.; Yang, W.; Li, M.; Sun, B.; Song, H.; Tang, W.; Zhang, Y.; Gong, R. Joint metabolome and transcriptome analysis of the effects of exogenous GA3 on endogenous hormones in sweet cherry and mining of potential regulatory genes. *Front. Plant Sci.* **2022**, *13*, 1041068. [CrossRef]
23. Michailidis, M.; Karagiannis, E.; Tanou, G.; Samiotaki, M.; Sarrou, E.; Karamanoli, K.; Lazaridou, A.; Martens, S.; Molassiotis, A. Proteomic and metabolic analysis reveals novel sweet cherry fruit development regulatory points influenced by girdling. *Plant Physiol. Biochem.* **2020**, *149*, 233–244. [CrossRef] [PubMed]
24. Li, M.; Cheng, S.; Wang, Y.; Dong, Y. Improving Fruit Coloration, Quality Attributes, and Phenolics Content in 'Rainier' and 'Bing' Cherries by Gibberellic Acid Combined with Homobrassinolide. *J. Plant Growth Regul.* **2020**, *39*, 1130–1139. [CrossRef]
25. Wang, X.; Li, C.; Liang, D.; Zou, Y.; Li, P.; Ma, F. Phenolic compounds and antioxidant activity in red-fleshed apples. *J. Funct. Foods* **2015**, *18*, 1086–1094. [CrossRef]

26. Livak, K.J.; Schmittgen, T.D. Analysis of relative gene expression data using real-time quantitative PCR and the 2(-Delta Delta C(T)) Method. *Methods* **2001**, *25*, 402–408. [CrossRef]
27. Zhao, C.; Wang, F.; Lian, Y.; Xiao, H.; Zheng, J. Biosynthesis of citrus flavonoids and their health effects. *Crit. Rev. Food Sci. Nutr.* **2020**, *60*, 566–583. [CrossRef] [PubMed]
28. Nassiri-Asl, M.; Hosseinzadeh, H. Review of the Pharmacological Effects of *Vitis vinifera* (Grape) and its Bioactive Constituents: An Update. *Phytother. Res.* **2016**, *30*, 1392–1403. [CrossRef]
29. Jia, X.; Xie, H.; Jiang, Y.; Wei, X. Flavonoids isolated from the fresh sweet fruit of Averrhoa carambola, commonly known as star fruit. *Phytochemistry* **2018**, *153*, 156–162. [CrossRef]
30. Wu, M.; Xu, X.; Hu, X.; Liu, Y.; Cao, H.; Chan, H.; Gong, Z.; Yuan, Y.; Luo, Y.; Feng, B.; et al. SlMYB72 Regulates the Metabolism of Chlorophylls, Carotenoids, and Flavonoids in Tomato Fruit. *Plant Physiol.* **2020**, *183*, 854–868. [CrossRef]
31. Kozłowska, A.; Szostak-Wegierek, D. Flavonoids-food sources and health benefits. *Rocz. Panstw. Zakl. Hig.* **2014**, *65*, 79–85.
32. Peluso, I.; Miglio, C.; Morabito, G.; Ioannone, F.; Serafini, M. Flavonoids and immune function in human: A systematic review. *Crit. Rev. Food Sci. Nutr.* **2015**, *55*, 383–395. [CrossRef]
33. Zeng, X.; Xi, Y.; Jiang, W. Protective roles of flavonoids and flavonoid-rich plant extracts against urolithiasis: A review. *Crit. Rev. Food Sci. Nutr.* **2019**, *59*, 2125–2135. [CrossRef]
34. Mazurakova, A.; Koklesova, L.; Samec, M.; Kudela, E.; Sivakova, J.; Pribulova, T.; Pec, M.J.; Kello, M.; Büsselberg, D.; Golubnitschaja, O.; et al. Flavonoids exert potential in the management of hypertensive disorders in pregnancy. *Pregnancy Hypertens.* **2022**, *29*, 72–85. [CrossRef] [PubMed]
35. Luo, X.; Sun, D.; Wang, S.; Luo, S.; Fu, Y.; Niu, L.; Shi, Q.; Zhang, Y. Integrating full-length transcriptomics and metabolomics reveals the regulatory mechanisms underlying yellow pigmentation in tree peony (*Paeonia suffruticosa* Andr.) flowers. *Hortic. Res.* **2021**, *8*, 235. [CrossRef]
36. Chen, C.; Zhou, G.; Chen, J.; Liu, X.; Lu, X.; Chen, H.; Tian, Y. Integrated Metabolome and Transcriptome Analysis Unveils Novel Pathway Involved in the Formation of Yellow Peel in Cucumber. *Int. J. Mol. Sci.* **2021**, *22*, 1494. [CrossRef] [PubMed]
37. Zhao, D.; Jiang, Y.; Ning, C.; Meng, J.; Lin, S.; Ding, W.; Tao, J. Transcriptome sequencing of a chimaera reveals coordinated expression of anthocyanin biosynthetic genes mediating yellow formation in herbaceous peony (*Paeonia lactiflora* Pall.). *BMC Genom.* **2014**, *15*, 689. [CrossRef]
38. Zhang, A.; Zheng, J.; Chen, X.; Shi, X.; Wang, H.; Fu, Q. Comprehensive Analysis of Transcriptome and Metabolome Reveals the Flavonoid Metabolic Pathway Is Associated with Fruit Peel Coloration of Melon. *Molecules* **2021**, *26*, 2830. [CrossRef] [PubMed]
39. Li, H.; Yang, Z.; Zeng, Q.; Wang, S.; Luo, Y.; Huang, Y.; Xin, Y.; He, N. Abnormal expression of bHLH3 disrupts a flavonoid homeostasis network, causing differences in pigment composition among mulberry fruits. *Hortic. Res.* **2020**, *7*, 83. [CrossRef]
40. Khoo, H.E.; Prasad, K.N.; Kong, K.W.; Jiang, Y.; Ismail, A. Carotenoids and their isomers: Color pigments in fruits and vegetables. *Molecules* **2011**, *16*, 1710–1738. [CrossRef]
41. Zhou, Y.; Lv, J.; Yu, Z.; Wang, Z.; Li, Y.; Li, M.; Deng, Z.; Xu, Q.; Cui, F.; Zhou, W. Integrated metabolomics and transcriptomic analysis of the flavonoid regulatory networks in Sorghum bicolor seeds. *BMC Genom.* **2022**, *23*, 619. [CrossRef]
42. Liu, J.; Zhang, X.; Tian, J.; Li, Y.; Liu, Q.; Chen, X.; Feng, F.; Yu, X.; Yang, C. Multiomics analysis reveals that peach gum colouring reflects plant defense responses against pathogenic fungi. *Food Chem.* **2022**, *383*, 132424. [CrossRef]
43. Gao, J.; Ren, R.; Wei, Y.; Jin, J.; Ahmad, S.; Lu, C.; Wu, J.; Zheng, C.; Yang, F.; Zhu, G. Comparative Metabolomic Analysis Reveals Distinct Flavonoid Biosynthesis Regulation for Leaf Color Development of Cymbidium sinense 'Red Sun'. *Int. J. Mol. Sci.* **2020**, *21*, 1869. [CrossRef]
44. Yan, H.; Pei, X.; Zhang, H.; Li, X.; Zhang, X.; Zhao, M.; Chiang, V.L.; Sederoff, R.R.; Zhao, X. MYB-Mediated Regulation of Anthocyanin Biosynthesis. *Int. J. Mol. Sci.* **2021**, *22*, 3103. [CrossRef]
45. An, J.-P.; Wang, X.-F.; Zhang, X.-W.; Xu, H.-F.; Bi, S.-Q.; You, C.-X.; Hao, Y.-J. An apple MYB transcription factor regulates cold tolerance and anthocyanin accumulation and undergoes MIEL1-mediated degradation. *Plant Biotechnol. J.* **2020**, *18*, 337–353. [CrossRef]
46. An, J.-P.; Li, H.-H.; Song, L.-Q.; Su, L.; Liu, X.; You, C.-X.; Wang, X.-F.; Hao, Y.-J. The molecular cloning and functional characterization of MdMYC2, a bHLH transcription factor in apple. *Plant Physiol. Biochem.* **2016**, *108*, 24–31, Erratum in: *Plant Physiol. Biochem.* **2019**, *135*, 612. [CrossRef]
47. Li, Y.; Xu, P.; Chen, G.; Wu, J.; Liu, Z.; Lian, H. FvbHLH9 Functions as a Positive Regulator of Anthocyanin Biosynthesis by Forming a HY5-bHLH9 Transcription Complex in Strawberry Fruits. *Plant Cell Physiol.* **2020**, *61*, 826–837. [CrossRef]
48. Huang, Y.; Vialet, S.; Guiraud, J.; Torregrosa, L.; Bertrand, Y.; Cheynier, V.; This, P.; Terrier, N. A negative MYB regulator of proanthocyanidin accumulation, identified through expression quantitative locus mapping in the grape berry. *New Phytol.* **2014**, *201*, 795–809. [CrossRef]
49. Li, M.; Sun, L.; Gu, H.; Cheng, D.; Guo, X.; Chen, R.; Wu, Z.; Jiang, J.; Fan, X.; Chen, J. Genome-wide characterization and analysis of bHLH transcription factors related to anthocyanin biosynthesis in spine grapes (*Vitis davidii*). *Sci. Rep.* **2021**, *11*, 6863. [CrossRef]
50. Kim, J.; Kim, D.H.; Lee, J.Y.; Lim, S.H. The R3-Type MYB Transcription Factor BrMYBL2.1 Negatively Regulates Anthocyanin Biosynthesis in Chinese Cabbage (*Brassica rapa* L.) by Repressing MYB-bHLH-WD40 Complex Activity. *Int. J. Mol. Sci.* **2022**, *23*, 3382. [CrossRef] [PubMed]

51. Xie, Z.; Nolan, T.M.; Jiang, H.; Yin, Y. AP2/ERF Transcription Factor Regulatory Networks in Hormone and Abiotic Stress Responses in Arabidopsis. *Front. Plant Sci.* **2019**, *10*, 228. [CrossRef] [PubMed]
52. Jiang, J.; Ma, S.; Ye, N.; Jiang, M.; Cao, J.; Zhang, J. WRKY transcription factors in plant responses to stresses. *J. Integr. Plant Biol.* **2017**, *59*, 86–101. [CrossRef] [PubMed]
53. Alabd, A.; Ahmad, M.; Zhang, X.; Gao, Y.; Peng, L.; Zhang, L.; Ni, J.; Bai, S.; Teng, Y. Light-responsive transcription factor PpWRKY44 induces anthocyanin accumulation by regulating PpMYB10 expression in pear. *Hortic. Res.* **2022**, *9*, uhac199. [CrossRef] [PubMed]
54. Chen, Y.; Wu, P.; Zhao, Q.; Tang, Y.; Chen, Y.; Li, M.; Jiang, H.; Wu, G. Overexpression of a Phosphate Starvation Response AP2/ERF Gene From Physic Nut in Arabidopsis Alters Root Morphological Traits and Phosphate Starvation-Induced Anthocyanin Accumulation. *Front. Plant Sci.* **2018**, *9*, 1186. [CrossRef]
55. He, M.; Yao, Y.; Li, Y.; Yang, M.; Li, Y.; Wu, B.; Yu, D. Comprehensive transcriptome analysis reveals genes potentially involved in isoflavone biosynthesis in *Pueraria thomsonii* Benth. *PLoS ONE* **2019**, *14*, e0217593. [CrossRef] [PubMed]
56. Zhao, Y.; Zhang, Y.-Y.; Liu, H.; Zhang, X.-S.; Ni, R.; Wang, P.-Y.; Gao, S.; Lou, H.-X.; Cheng, A.-X. Functional characterization of a liverworts bHLH transcription factor involved in the regulation of bisbibenzyls and flavonoids biosynthesis. *BMC Plant Biol.* **2019**, *19*, 497. [CrossRef]
57. Deshmukh, A.B.; Datir, S.S.; Bhonde, Y.; Kelkar, N.; Samdani, P.; Tamhane, V.A. De novo root transcriptome of a medicinally important rare tree Oroxylum indicum for characterization of the flavonoid biosynthesis pathway. *Phytochemistry* **2018**, *156*, 201–213. [CrossRef]
58. Tu, M.; Fang, J.; Zhao, R.; Liu, X.; Yin, W.; Wang, Y.; Wang, X.; Wang, X.; Fang, Y. CRISPR/Cas9-mediated mutagenesis of VvbZIP36 promotes anthocyanin accumulation in grapevine (*Vitis vinifera*). *Hortic. Res.* **2022**, *9*, uhac022. [CrossRef] [PubMed]

agronomy

MDPI

Article

Ethylene Signaling Pathway Genes in Strawberry and Their Expression Patterns during Fruit Ripening

Yunting Zhang [1,†], Meiyi Deng [1,†], Xianjie Gu [2], Chenhui Guo [1], Yan Chen [1], Yuanxiu Lin [1], Qing Chen [1], Yan Wang [1], Yong Zhang [1], Ya Luo [1], Xiaorong Wang [1] and Haoru Tang [1,*]

1 College of Horticulture, Sichuan Agricultural University, Chengdu 611130, China
2 Institute of Pomology & Olericulture, Mianyang Academy of Agricultural Sciences, Mianyang 621000, China
* Correspondence: htang@sicau.edu.cn
† These authors contributed equally to this work.

Abstract: Ethylene at least partly regulates some aspects during non-climacteric ripening in strawberry. However, the ethylene signaling pathway genes in the strawberry fruit have not been comprehensively and systematically analyzed. In the present study, 15 *FaETRs* and 14 *FaEIN3/EINs* were identified in the octoploid strawberry genome. Subcellular localization analysis predicted that FaETRs and FaEIN3/EINs are respectively localized to the endoplasmic reticulum and the nucleus. The phylogenetic trees showed that FaETRs were classified into two subgroups, while FaEIN3/EINs were divided into three clades, which was supported by gene structure and conserved motif analysis. FaETRs and FaEIN3/EINs could interact with several components, such as CTR1, RTE1, EIN2 and ERF1B, in the ethylene signaling pathway by protein–protein interaction network analysis. Transcriptomic data showed that *FaETRs* were mainly expressed at the early stage of fruit development in three strawberry cultivars. Additionally, a couple of *FaETRs* (*FaETR2* and *FaETR13*) and *FaEINs* (*FaEIN2* and *FaEIN7*) could be induced by 1 μM ABA and inhibited by 100 μM nordihydroguaiaretic acid (NDGA, an ABA biosynthesis blocker). These findings suggested that the FaETR- and FaEIN3/EIN-mediated ethylene signaling pathway might play a role in strawberry fruit ripening.

Keywords: strawberry; FaETRs; FaEIN3/EINs; fruit ripening

Citation: Zhang, Y.; Deng, M.; Gu, X.; Guo, C.; Chen, Y.; Lin, Y.; Chen, Q.; Wang, Y.; Zhang, Y.; Luo, Y.; et al. Ethylene Signaling Pathway Genes in Strawberry and Their Expression Patterns during Fruit Ripening. *Agronomy* **2023**, *13*, 1930. https://doi.org/10.3390/agronomy13071930

Academic Editor: Chengdao Li

Received: 26 June 2023
Revised: 19 July 2023
Accepted: 19 July 2023
Published: 21 July 2023

1. Introduction

Ethylene, the simplest olefin gas, is the first gaseous molecule shown to function as a hormone. It is biosynthesized by plants and is well-known to control a variety of aspects of plant growth and development, such as cell division and expansion, seed germination, root hair formation, organ senescence, leaf flower abscission and fruit ripening, as well as responses to various abiotic and biotic stresses, such as wounding, salt, heat, chilling, heavy metals, drought, flooding, insect infestation and pathogen invasion [1–4].

Ethylene biosynthesis starts from methionine, which sequentially converts into S-adenosyl-l-methionine (SAM) and the ethylene precursor 1-aminocyclopropane-1-carboxylate (ACC) through two-step enzymatic reactions, catalyzed by SAM synthetase (SAMS) and ACC synthase (ACS). ACC is then metabolized to ethylene by ACC oxidase (ACO) [5]. During the past three decades, a combination of genetic and molecular analyses that relies on the triple-response phenotype as a morphological marker has allowed the characterization of a collection of ethylene-response mutants and pivotal components of the ethylene signaling pathway, which has proposed a primarily linear model of ethylene signal transduction that starts with hormone perception and ends in transcriptional regulation [6]. Briefly, once the ethylene molecule is perceived by ethylene receptors (ETR1, ERS1, ETR2, EIN4 and ERS2) at the endoplasmic reticulum (ER) membrane, the receptors signal to the negative regulator CTR1, a Raf-like serine/threonine (Ser/Thr) kinase, preventing the phosphorylation of the positive regulator EIN2 and causing the C-terminal end of EIN2 to translocate to the nucleus, where the EIN2 C-end leads to the stabilization of EIN3/EILs and the initiation of transcriptional responses to ethylene [7,8].

Fruit ripening, an irreversible phenomenon, is accompanied by genetically programmed and highly coordinated biochemical changes in color, aroma, flavor, texture and nutritional characteristics, and this process is regulated by an intricate network of different phytohormones. According to the respiration pattern and ethylene production during fruit ripening, fleshy fruits are broadly segmented into two groups: climacteric such as tomato, peach and banana and non-climacteric such as strawberry, litchi and citrus [9,10]. The phytohormone ethylene is well-known to be predominately involved in climacteric fruit ripening [11]. Therefore, the importance of the ethylene biosynthetic and signal transduction pathways in climacteric fruits has been deeply and extensively studied including the characterization of gene families involving these pathways [12–14]. It has been demonstrated that ABA also plays a synergistic role during climacteric fruit ripening [15,16]. The peak value of ABA content occurs prior to ethylene, and ABA was proposed to be an upstream regulator of the ethylene pathway [17]. Exogenous ABA treatment or manipulation of genes related to ABA signaling and the biosynthetic pathway regulates climacteric fruit ripening by affecting ethylene biosynthesis and signal transduction [16,18,19]. Moreover, auxins, jasmonic acid (JA), gibberellin (GA), brassinosteroid (BR), salicylic acid (SA) and melatonin are involved in climacteric fruit ripening through ethylene-related pathways [20]. In non-climacteric fruits, ABA is mainly regarded to participate in the regulation of fruit ripening [21]. However, an increasing body of evidence suggests ethylene contributes at least partly to the control of some aspects of non-climacteric ripening [22–25], but significant progress has not been achieved in revealing the role of ethylene in this process. It was demonstrated that ethylene played a potential role in grape development and ripening [22]. Li et al. (2016) [26] summarized the effect of 1-Methylcyclopropene (1-MCP), an ethylene antagonist, on the postharvest storage performance of non-climacteric fruits and proposed that 1-MCP application could be a method of inhibiting the color change, retarding senescence processes and reducing physiological disorders in certain non-climacteric fruits. Moreover, ethylene signaling has been found to participate in regulating non-climacteric fruit ripening as well by interacting with the ABA signaling, as elucidated by AREB/ABF-mediated ACS/ACO expression and ERF-mediated NCED expression [20,21]. In addition, ethylene response factors (ERFs) can affect different aspects (color, aroma and flavor) of fruit ripening and senescence in non-climacteric fruits [27–29].

Strawberry, a typical example of a non-climacteric fruit, has great economic and nutritional value because of its health-promoting compounds, pleasant flavor and attractive appearance. ABA is the core factor that promotes strawberry ripening [30]. It has been found that ethylene also accelerates some aspects of strawberry ripening such as anthocyanin accumulation, total sugar increment and cell wall degradation, while 1-MCP treatment has the opposite effect [31,32], and some ripening-related genes could be regulated by ethylene [31,33]. Downregulation of the *FaSAMS1* or *FaCTR1* expression level by the RNAi technique retarded strawberry red-coloring and softening. Accordingly, ethephon application could promote natural strawberry fruit red-coloring and softening and partially rescue anthocyanin accumulation in *FaSAMS1* and *FaCTR1*-RNAi fruits but could not significantly influence firmness [34]. Additionally, a high transcript level of ethylene receptors seems required for strawberry fruit ripening [35]. Our previous study showed that the expression level of *FaERFs* was associated with strawberry fruit development and ripening and was correspondingly regulated by exogenous ABA treatment [36]. These findings indicate that the ethylene signal transduction pathway plays a role in strawberry fruit ripening. In this study, the *FaETR* (starting) and *FaEIN3/EIN* (ending) genes in the ethylene signal transduction pathway were identified, and their transcript profiles were measured during fruit ripening, with the aim to obtain new information on various components of the pathway and gain insight into their effect on fruit ripening.

2. Materials and Methods

2.1. Plant Materials

The octoploid strawberries ('Xiaobai' cultivar) were harvested at three different developmental stages according to the fruit peel color and days after full bloom (DAFB) during ripening (big green, 21 DAFB; turn red, 28 DAFB; full red, 35 DAFB) from a local orchard located in Wenjiang County, Sichuan Province, China, on 12 March 2021. The samples were quickly frozen in liquid nitrogen, delivered to the laboratory and stored in −80 °C freezer for further analysis.

2.2. Screening of Ethylene Signaling Pathway Genes in Strawberry

The files of octoploid cultivated strawberry (*Fragaria* × *ananassa* genome v1.0.a1) were downloaded from the Genome Database for Rosaceae (GDR) (https://www.rosaceae.org, accessed on 12 August 2022). Two methods (local Blast and HMMER search) were used to screen ethylene signaling pathway genes in strawberry. The amino acid sequences in this pathway from tomato and *Arabidopsis thaliana* were used as queries for BLASTP analysis against *Fragaria* × *ananassa* Camarosa genome v1.0.a1 (https://www.rosaceae.org, accessed on 14 August 2022). The specific hidden Markov Model (HMM) profile (PF04873) of EIN3/EINs was downloaded from the Pfam database (https://pfam.xfam.org, accessed on 16 August 2022) and used as the query to search for FaEIN3/EINs genes. Finally, all proteins were delivered to the Conserved Domain Database (CDD) (https://www.ncbi.nlm.nih.gov/Structure/cdd/wrpsb.cgi, accessed on 20 August 2022) to confirm the reliability of candidates.

2.3. Physicochemical Property and Subcellular Localization Analysis

The basic physicochemical properties of protein size, grand average of hydropathicity (GRAVY), theoretical pI, aliphatic index, molecular weight (MW), instability index and subcellular localization of FaETR and FaEIN3/EINs were analyzed using ExPASy-ProtParam online servers (http://web.expasy.org/protparam/, accessed on 22 August 2022) and ProtComp v.9.0 (http://linux1.softberry.com/berry.phtml?topic=protcomppl&group=programs&subgroup=proloc, accessed on 22 August 2022), respectively.

2.4. Chromosomal Mapping and Phylogenetic Tree Analysis

The information on the chromosomal locations of FaETR and FaEIN3/EIN genes was retrieved from the annotated file of the GDR database and graphically visualized with Tbtools. A phylogenetic tree of FaETR and FaEIN3/EINs was built with MEGA X using neighbor-joining (NJ) algorithm with bootstrap support (1000 replicates), according to the multiple sequence alignment in MUSCLE.

2.5. Gene Structure and Conserved Motif Analysis

The exon–intron organization of FaETR and FaEIN3/EIN genes was analyzed with the online tool (Gene Structure Display Server v.2.0, http://gsds.cbi.pku.edu.cn, accessed on 25 August 2022) via the comparison of the full-length genome sequence and the corresponding coding sequences. Conserved motifs were characterized using the Multiple Em for Motif Elicitation (MEME) online tool (http://meme-suite.org/tools/meme, accessed on 27 August 2022) with default parameters.

2.6. Protein–Protein Interaction Network Prediction

Fifteen FaETR and fourteen FaEIN3/EIN protein sequences were used as queries; protein–protein interaction network was analyzed with the STRING online tool (https://string-db.org/, accessed on 5 September 2022). The orthologous genes of diploid strawberry (*Fragaria* × *vesca*) were selected as references. After mapping, the interaction network was constructed using the highest-score gene (bitscore).

2.7. Transcriptome Analysis during Fruit Development and Ripening

The expression patterns of FaETR and FaEIN3/EIN genes in ethylene signaling pathway were calculated using the transcriptomic data that had been published on the NCBI database (PRJNA552213; PRJNA338879) [37,38]. The fruit samples in the transcriptome (PRJNA552213) were harvested from three stages of middle green (MG), initial red (IR) and full red (FR) of three strawberry varieties ('Benihoppe', 'Xiaobai' and 'Snow Princess'). The samples in the transcriptome (PRJNA338879) were obtained from 'Toyonoka' cultivar that was subjected to 1 μM ABA and 100 μM nordihydroguaiaretic acid (NDGA, an ABA biosynthesis blocker) treatments.

2.8. RNA Extraction, cDNA Synthesis and qRT-PCR

The total RNA from strawberry varieties ('Xiaobai') was isolated using the improved CTAB (cetyltrimethylammonium bromide) method [39]. After evaluation of the quantity and quality with NanoDrop ND 2000 spectrophotometer and agarose gel electrophoresis, the total RNA was transcribed into cDNA using the PrimeScriptTM RT reagent Kit with gDNA Eraser (Takara, Japan) according to the manufacturing instruction. qRT-PCR was performed on the CFX96 real-time PCR system (Bio-Rad, Hercules, CA, USA) with SYBR Premix (Takara, Japan). The total 10 μL reaction mixture consisted of 1 μL template (1:10 diluted cDNA), 0.4 μL each primer for a final concentration of 0.4 μM, 5 μL SYBR Premix and 3.2 μL of RNase-free water. Reaction procedure was performed using two-step cycling conditions: 95 °C for 5 min, followed by 40 cycles of 95 °C for 10 s, 58 °C for 20 s and 72 °C for 20 s. Melting curve was inserted after the final cycle, ramping from 65 °C to 95 °C (increment 0.5 °C/5 s). The presence of a single and sharp peak demonstrated the specificity of primer amplification. Controls without template were included in each run to check the potential reagent contamination. Gene-specific primers used for qRT-PCR were designed using Primer 3 and are listed in Table S1. The relative gene expression levels were calculated using the $2^{-\Delta\Delta CT}$ formula [40].

3. Results

3.1. Identification and Characterization of Ethylene Signaling Pathway Genes in Strawberry

To extensively identify FaETRs and FaEIN3/EINs in the ethylene signaling transduction pathway from the strawberry, both HMM and the local BLAST program were used to scan the strawberry genome database (v1.0.a1). The two gene families had 29 putative members in total, including 15 *FaETRs* and 14 *FaEIN3/EINs*. Their sequence feature is analyzed in Tables 1 and 2. The length of FaETR and FaEIN3/EIN proteins ranged from 553 to 774 and 232 to 797 aa with molecular weight varying from 62,261.50 to 87,072.14 and 26,951.89 to 88,258.07 Da and theoretical pI varying from 5.83 to 7.25 and 5.09 to 6.20, respectively. Instability indexes of most FaETRs and all FaEIN3/EINs were above 40 except for ETR4, ETR5, ETR6 and ETR7, indicating that only ETR4-7 were stable proteins. Subcellular localization analysis predicted that all FaETRs and FaEIN3/EINs were respectively localized to the endoplasmic reticulum (ER) and the nucleus.

Table 1. Information of FaETR genes identified in strawberry.

Name	Protein Size	MW (Da)	Theoretical pI	Instability Index	Aliphatic Index	GRAVY	Subcellular Localization
FaETR1	741	83,123.00	6.97	49.44	108.49	0.120	ER
FaETR2	741	83,037.96	6.82	49.72	109.01	0.135	ER
FaETR3	741	83,162.03	7.15	49.57	108.10	0.106	ER
FaETR4	633	70,886.33	6.11	36.77	105.34	0.109	ER
FaETR5	633	70,856.21	5.83	36.41	105.34	0.104	ER
FaETR6	633	70,805.21	6.04	36.21	105.04	0.117	ER
FaETR7	553	62,261.50	5.92	37.50	108.05	0.126	ER
FaETR8	741	83,152.04	7.14	49.01	108.35	0.116	ER

Table 1. *Cont.*

Name	Protein Size	MW (Da)	Theoretical pI	Instability Index	Aliphatic Index	GRAVY	Subcellular Localization
FaETR9	765	84,496.32	6.66	40.47	103.20	0.132	ER
FaETR10	765	84,601.37	6.49	40.22	102.30	0.123	ER
FaETR11	765	84,544.39	6.49	40.37	102.30	0.124	ER
FaETR12	774	86,848.73	6.31	40.41	97.07	0.017	ER
FaETR13	769	86,328.16	6.80	44.43	97.20	0.013	ER
FaETR14	774	87,072.14	7.25	44.55	96.45	0.002	ER
FaETR15	774	86,920.84	6.41	42.38	97.07	0.016	ER

Table 2. Information of FaEIN3/EIN genes identified in strawberry.

Name	Protein Size	MW (Da)	Theoretical pI	Instability Index	Aliphatic Index	GRAVY	Subcellular Location
FaEIN1	616	69,685.51	5.41	47.26	61.09	−0.690	nucleus
FaEIN2	618	70,050.05	5.47	47.89	62.14	−0.677	nucleus
FaEIN3	618	69,906.89	5.52	46.86	61.50	−0.675	nucleus
FaEIN4	617	69,732.66	5.53	47.91	62.40	−0.668	nucleus
FaEIN5	602	67,753.36	5.09	49.95	65.76	−0.633	nucleus
FaEIN6	602	67,664.21	5.17	50.03	64.80	−0.657	nucleus
FaEIN7	593	66,308.72	5.62	56.76	71.21	−0.732	nucleus
FaEIN8	593	66,332.72	5.50	53.83	70.56	−0.732	nucleus
FaEIN9	797	88,258.07	6.13	51.47	77.06	−0.533	nucleus
FaEIN10	590	66,162.56	5.68	55.08	70.42	−0.749	nucleus
FaEIN11	449	50,979.55	5.10	43.40	77.71	−0.607	nucleus
FaEIN12	232	26,951.89	5.94	55.40	85.30	−0.674	nucleus
FaEIN13	408	46,862.97	5.11	46.53	73.55	−0.682	nucleus
FaEIN14	275	31,453.79	6.20	43.51	61.35	−0.589	nucleus

3.2. Chromosomal Localization and Phylogenetic Tree

To determine the distribution of *FaETR* and *FaEIN3/EIN* genes on chromosomes in strawberry, a chromosome map was constructed according to the genome annotation. The 13 *FaETR* genes were evenly located on 13 chromosomes, with the remaining two genes (*FaETR12* and *FaETR15*) localized on chromosome 6-3. Chromosomes Fvb1-2, Fvb1-3 and Fvb1-4 had two members of *FaEIN3/EINs*, while the other eight chromosomes only had one member (Figure 1). To unravel the evolutionary history of the FaETR or FaEIN3/EIN gene family from strawberry and to help in their classification, FaETRs and FaEIN3/EINs from different species were used to build an unrooted phylogenetic tree. The phylogenetic trees showed that FaETR families were divided into two clades. Eight FaETRs (FaETR1-8) in strawberry, together with AtERS1 (AT2G40940) and AtETR1 (AT1G66340) in Arabidopsis and SlETR2 (Solyc07g056580), SlETR1 (Solyc12g011330) and SlETR3 (Solyc09g075440) in tomato, belonged to Class 1. Seven FaETRs (FaETR9-15) in strawberry, together with AtETR2 (AT3G23150), AtERS2 (AT1G04310) and AtEIN4 (AT3G04580) in Arabidopsis and SlETR6 (Solyc09g089610), SlETR4 (Solyc06g053710), SlETR7 (Solyc05g055070) and SlETR5 (Solyc11g006180) in tomato, belonged to Class 2. FaEIN3/EIN families were divided into three classes. Class 1 had four FaEIN3/EINs that were clustered with AtEIL3 (AT1G73730), Class 2 had three FaEIN3/EINs that were clustered with AtEIL5 (AT5G65100), AtEIL4 (AT5G10120) and AtEIL2 (AT5G21120), and Class 3 had seven FaEIN3/EINs that were clustered with AtEIN3 (AT3G20770) and ATEIL1 (AT2G27050) (Figure 2).

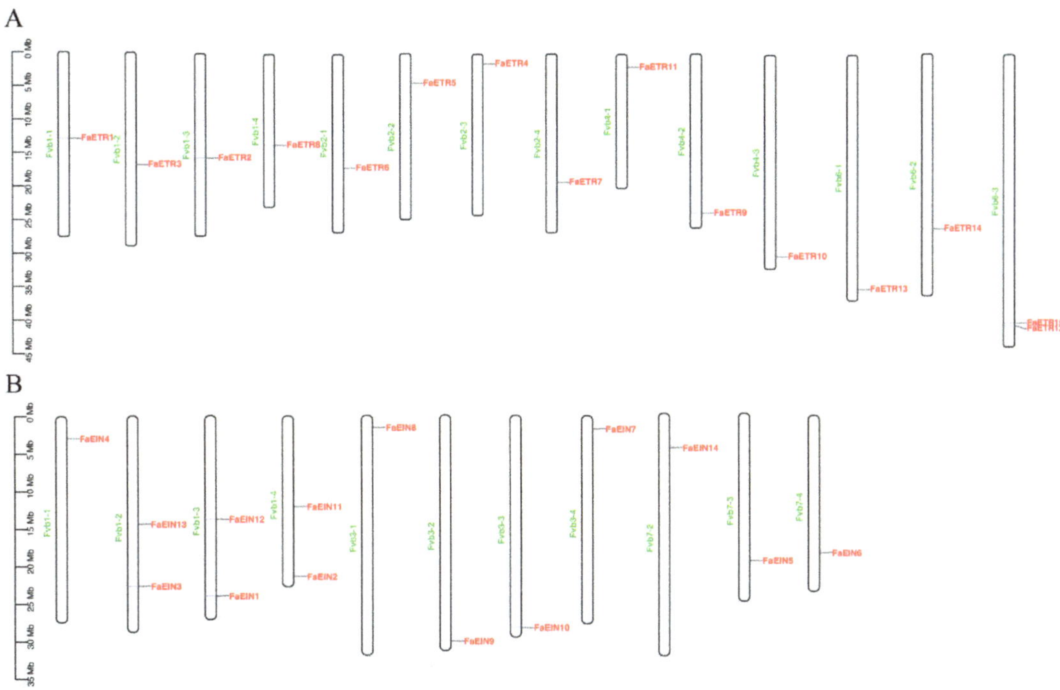

Figure 1. Chromosome distribution of *FaETR genes* (**A**) and *FaEIN3/EIN* genes (**B**) in strawberry.

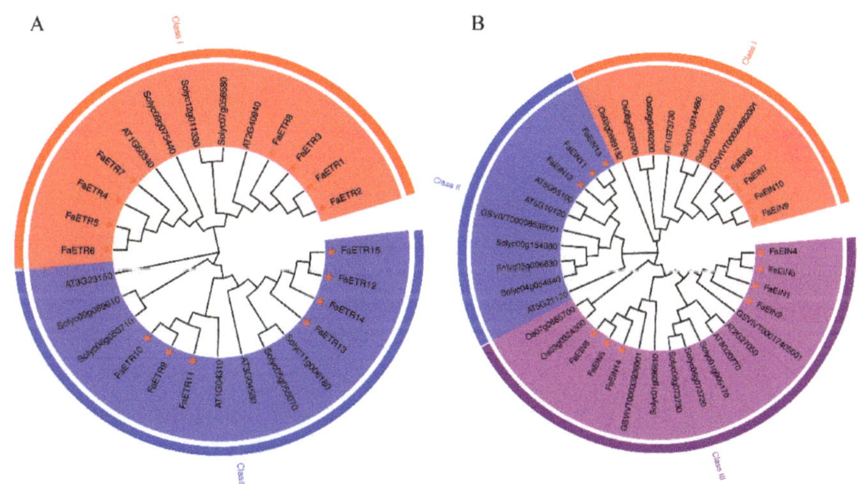

Figure 2. Phylogenetic analysis of FaETR family (**A**) and FaEIN3/EIN family (**B**) among different plant species. Red stars indicated the FaETRs or FaEIN3/EINs in strawberry. Fa, *Fragaria × ananassa;* At, *Arabidopsis thaliana;* Solyc, *Solanum lycopersicum;* Os, *Oryza sativa;* GSVIVT, *Vitis vinifera.*

3.3. Gene Structure and Conserved Motif

To better understand the gene structural characteristics of the ethylene signaling pathway genes, their intron/exon arrangement was analyzed. All *FaETRs* had at least one intron and up to six exons. *FaETR1-8* clustered into Class 1 contained the most exons, while *FaETR9-15* clustered into Class 2 only had two exons. Seven *FaEIN3/EINs* had no intron, and other members presented the break sequences because of the distribution of different size introns. The conserved motifs of FaETR and FaEIN3/EIN proteins were analyzed using the MEME tool. The closely related proteins in the same clade exhibited similar motif compositions. Remarkably, the similarity in exon–intron structure and motif distribution supported the results from the phylogenetic analysis of genes, indicating that their function was both conserved and diversified (Figure 3).

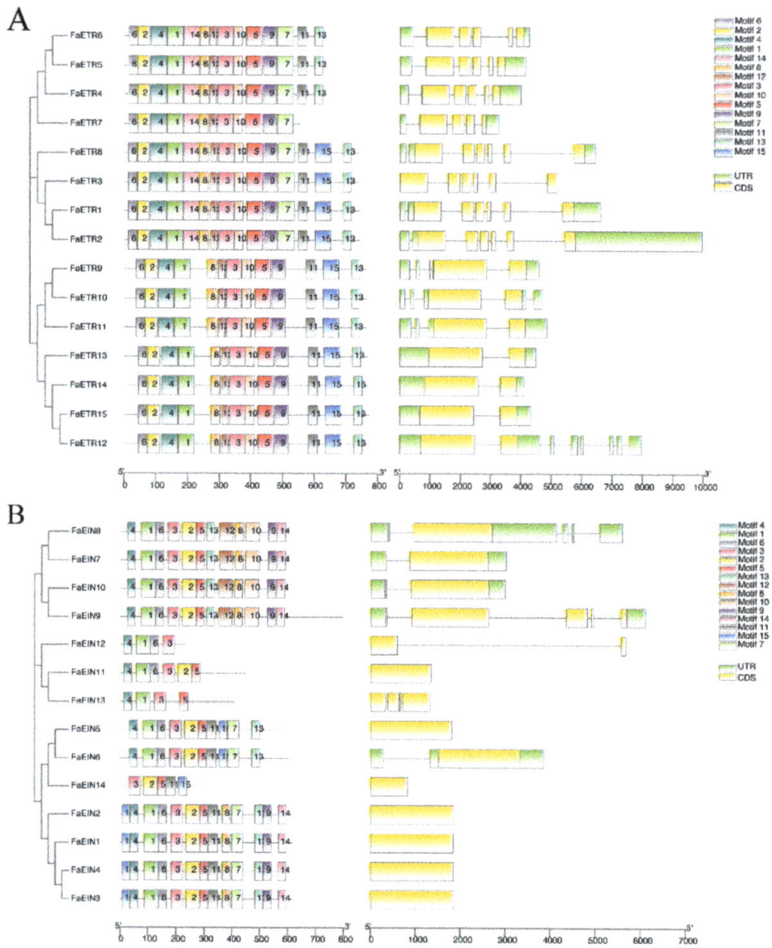

Figure 3. Gene structure and conserved motif of FaETRs (**A**) and FaEIN3/EINs (**B**). Exons and introns are indicated by filled yellow boxes and single lines, respectively. Conserved motifs are indicated by different color boxes numbered 1–15.

3.4. Protein–Protein Interaction Network Analysis

A protein–protein interaction network was constructed to predict the molecular interaction of FaETRs and FaEIN3/EINs in the ethylene signaling pathway, based on their orthologous genes in diploid strawberry (*Fragaria × vesca*). It was predicted that FaETRs interacted with CTR1 and RTE1 and FaEIN3/EINs interacted with EIN2, ERF1B, EBF1 and MAPKs. All proteins are important components in the ethylene signaling pathway (Figure 4).

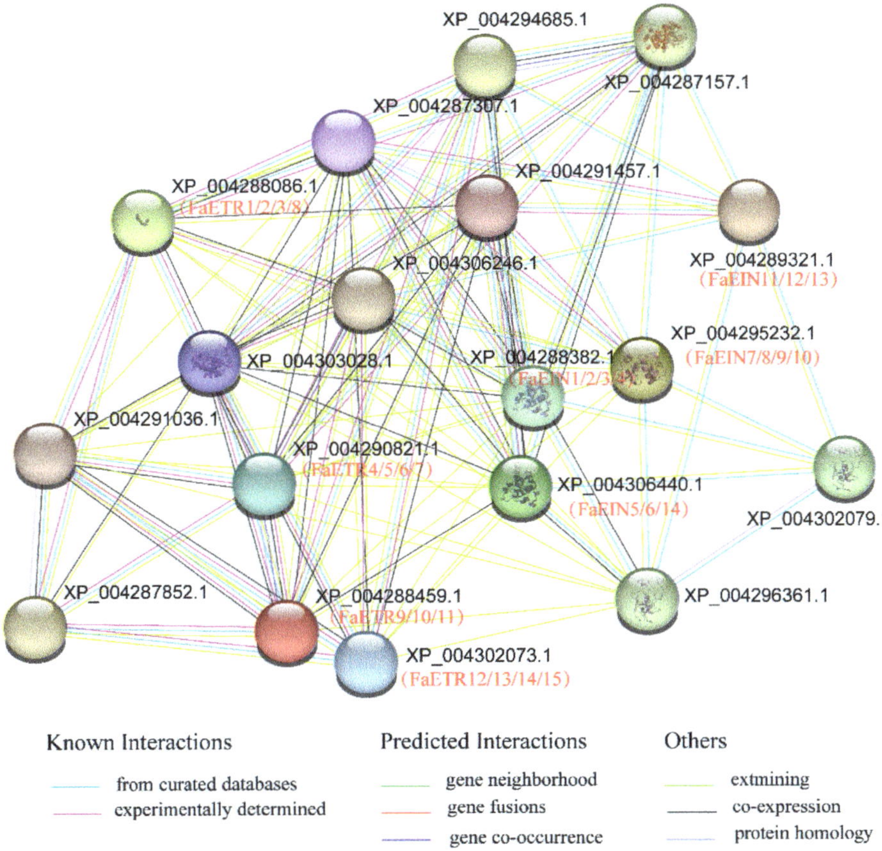

Figure 4. A protein–protein interaction network for FaETRs and FaEIN3/EINs in ethylene signaling pathway. XP_004303028.1, Serine/threonine-protein kinase CTR1-like (CTR1); XP_004287307.1, EIN3-binding F-box protein 1-like (EBF1); XP_004291457.1, EIN3-binding F-box protein 1-like (EBF1); XP_004291036.1, Protein REVERSION-TO-ETHYLENE SENSITIVITY1-like (RTE1); XP_004306246.1, Ethylene-insensitive protein 2-like (EIN2); XP_004287852.1, Protein RTE1-HOMOLOG-like (RTE1); XP_004294685.1, Mitogen-activated protein kinase 3-like (MAPK3); XP_004287157.1, Mitogen-activated protein kinase homolog MMK1-like (MAPK1); XP_004296361.1, Ethylene-responsive transcription factor 1B-like (ERF1B); XP_004302079.1, Ethylene-responsive transcription factor 1B-like (ERF1B). The red color text was indicated as the corresponding putative orthologous genes in the octoploid strawberry.

3.5. Expression Profiles during Fruit Ripening

To identify the *FaETR* and *FaEIN3/EIN* genes involved in fruit ripening, their expression patterns during the fruit development of three strawberry cultivars were examined. As shown in Figure 5A, most *FaETRs* had higher expression levels in the cultivar 'Benihoppe'. Interestingly, *FaETRs* were mainly expressed at the early stage of fruit development in three cultivars. *FaEIN2*, *FaEIN4*, *FaEIN7* and *FaEIN10* had the highest expression at the IR stage of 'Benihoppe'. *FaEIN3*, *FaEIN6*, *FaEIN8* and *FaEIN14* exhibited higher expression over fruit development in 'Snow Princess' than the other two cultivars. Moreover, *FaEIN5* and *FaEIN12* had the highest expression at the IR stage of 'Snow Princess' (Figure 5B). The transcript levels of *FaETR* and *FaEIN3/EIN* genes in the strawberry fruit exposed to ABA and ABA biosynthesis blocker nordihydroguaiaretic acid (NDGA) treatments were profiled. The results showed that only a couple of *FaETRs* (*FaETR2* and *FaETR13*) and *FaEINs*

(*FaEIN2* and *FaEIN7*) could be induced by ABA and inhibited by NDGA (Figure 5C,D). In addition, qRT-PCR showed that the expression patterns of *FaETR2*, *FaETR13*, *FaEIN2* and *FaEIN7* were consistent with the transcriptome data (Figure 6).

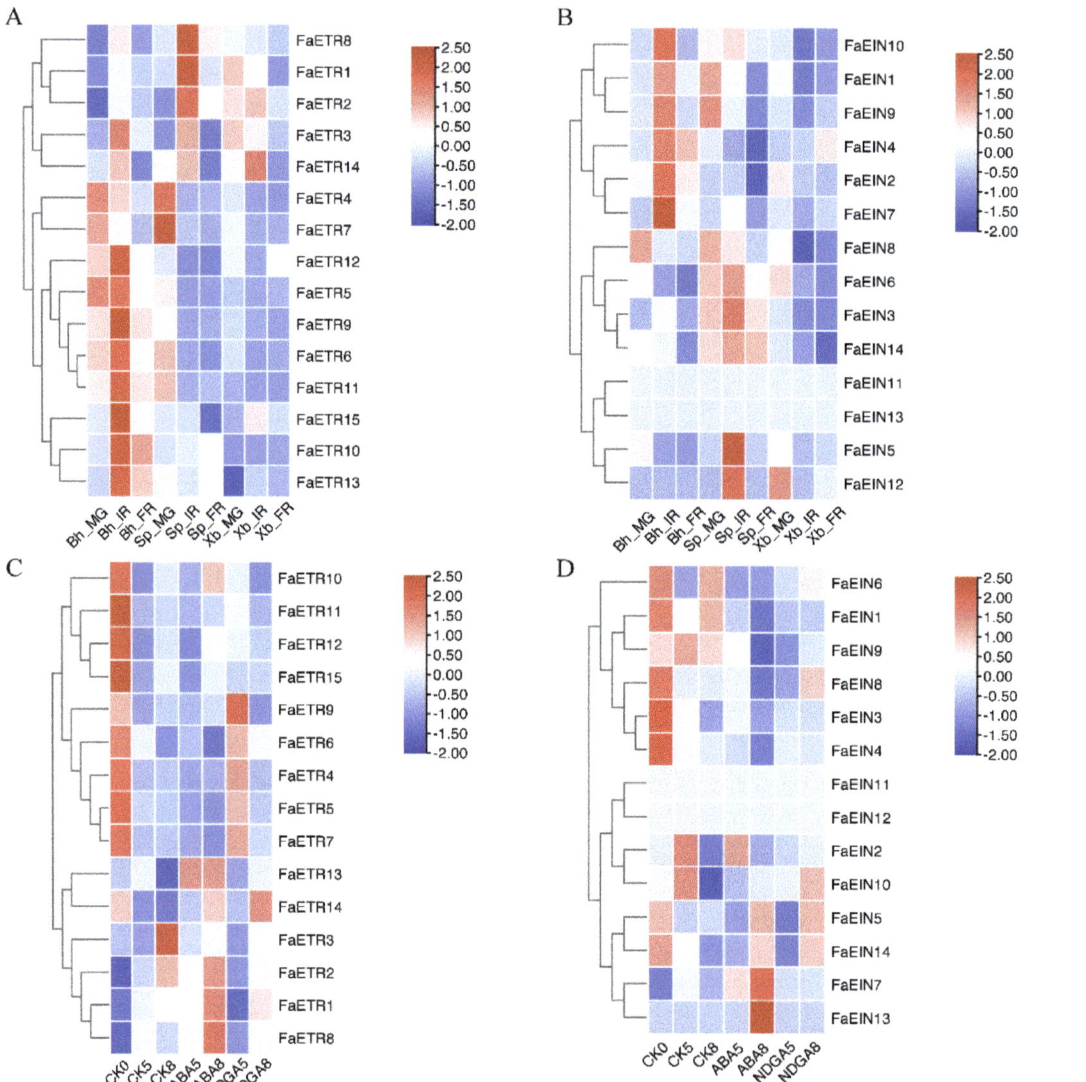

Figure 5. Expression levels of *FaETR* and *FaEIN3/EIN* genes in strawberries. (**A**) Expression levels of *FaETRs* in three strawberry cultivars at different fruit developmental stages. (**B**) Expression levels of *FaEINs* in three strawberry cultivars at different fruit developmental stages. (**C**) Expression levels of *FaETRs* in strawberries after ABA and NDGA treatment. (**D**) Expression levels of *FaEINs* in strawberries after ABA and NDGA treatment. CK0, strawberries treated with distilled water on day 0; CK5, ABA5 and NDGA5, strawberries treated with distilled water, ABA and NDGA on day 5, respectively; CK8, strawberries treated with distilled water, ABA and NDGA on day 8, respectively.

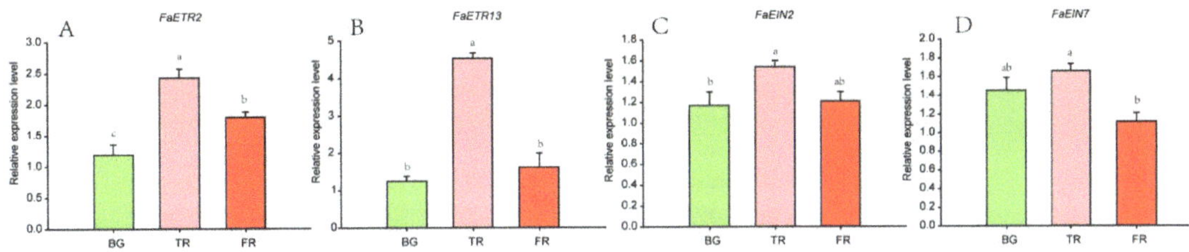

Figure 6. Expression levels of selected *FaETRs* and *FaEIN3/EINs* in 'Xiaobai' strawberry. (**A**) Expression levels of *FaETR2* at different fruit developmental stages. (**B**) Expression levels of *FaETR13* at different fruit developmental stages. (**C**) Expression levels of *FaEIN2* at different fruit developmental stages. (**D**) Expression levels of *FaEIN7* at different fruit developmental stages.BG, big green; TR, turn red; FR, full red. The data are presented as mean ± standard error (SE) analyzed using IBM SPSS Statistics 23.0 and one-way ANOVA. Duncan's multiple-range test was employed to determine the differences in gene expression levels among different fruit developmental stages at the significance level of $p \leq 0.05$.

4. Discussion

A growing body of evidence suggests that ethylene affects color evolution, nutrient accumulation and other ripening-related processes of non-climacteric fruits. For instance, ethylene treatment accelerates grape fruit ripening by promoting coloration and ethylene production, increasing phytochemical contents and decreasing chlorophyll and titratable acid contents [41,42]. The exogenous ethylene-triggered coloration in the fruit peel increases with the ripening of Navelate orange, since ethylene stimulates carotenoid accumulation and reduces the chloroplastic carotenoid contents [43]. In litchi, ethylene plays a more important role in chlorophyll degradation in comparison with abscisic acid [44]. During blueberry ripening and storage, the role of ethylene in pigment and texture changes is genotype-dependent [45,46]. Nevertheless, the progress concerning the ethylene-induced molecular change in non-climacteric ripening is still slow, although extensive literature research has been reported on ethylene regulating non-climacteric ripening from physiological aspects. In addition, multiple research studies have spotlighted ethylene biosynthesis, while few studies have systematically focused on ethylene signal transduction.

Strawberry is a model non-climacteric fruit crop with a lack of ethylene burst during the ripening process. It has been documented that ethylene could affect some aspects of strawberry ripening. Therefore, the identification of ethylene signaling pathway genes in strawberry is necessary for fruit ripening study and breeding. In this study, a total of 29 ethylene signal transduction genes including 15 *FaETRs* and 14 *FaEIN3/EINs* were characterized from the octoploid strawberry genome, which was more than the number existing in the genomes of many fruit species: four *ETRs* and five *EIN3/EINs* in woodland strawberry [13,47], four *ETRs* and four *EIN3/EINs* in mulberry [48], four *ETRs* and four *EIN3/EINs* in peach [49], seven *ETRs* and nine *EIN3/EINs* in tomato [13,50], four *ETRs* and four *EIN3/EINs* in grape [51] and other fruit trees. This finding suggested that the *FaETR* and *FaEIN3/EIN* gene family has undergone an expansion in the octoploid strawberry. However, diverse expression patterns were observed between members of the *FaETR* and *FaEIN3/EIN* gene family, suggesting that there are different regulatory systems between the members of these gene families affecting their expression levels [52,53]. The ETRs from strawberry, Arabidopsis and tomato in the phylogenetic tree were divided into two subfamilies, which was consistent with previous reports [25,48]. The ETR classification in subfamily 1 or subfamily 2 relies on the non-functional HATPase_c domain present in the subfamily 2 members [25]. FaETRs and FaEIN3/EINs in the same cluster shared similar conserved motifs and exon–intron structures, while there existed differences in gene structure and motifs among different clusters. Multiple research projects have suggested that structural diversity is the primary resource for the evolution of multigene families [36,54,55]. Also, based on phylogeny analysis, it seems that the expansion and diversity among *ETR* and *EIN3/EIN* gene family members occur more after the divergence of monocot and dicot [56].

To investigate the potential function of *FaETR* and *FaEIN3/EIN* genes during strawberry fruit development and ripening, RNAseq datasets that were produced and validated in previous studies [37,38] were used to analyze the expression changes of these genes. The results showed that most *FaETRs* had a higher expression at the onset of fruit ripening in three strawberry cultivars, in line with the results of several studies on other non-climacteric fruits like grape [25,57] and citrus [25]. It has been reported that *ETR* genes seem to be more responsive in younger strawberries than in older ones during fruit ripening, since they showed a great expression increase between stages large green and white [35]. EIN3/EINs belong to a small family of transcription factors that activate the downstream components (ethylene-responsive factors and other downstream genes) of the ethylene signaling pathway [58–60]. Specific *FaEIN3/EINs* showed high expression levels at the initial red stage of the strawberry fruit in a certain cultivar, suggesting they might play a role during fruit ripening. Also, Ma et al. [61,62] proposed that LcEIL2 and LcEIL3 regulated ethylene-activated litchi fruitlet abscission by binding to *LcCEL2/8* and *LcPG1/2* in a cell wall remodeling process and *LcACS1/4/7* and *LcACO2/3* in the ethylene biosynthetic pathway. Mu et al. (2016) [57] demonstrated that the expression levels of most of the genes involved in ethylene biosynthesis and signaling had the same changing trend in non-climacteric fruits (grape and strawberry): they reached a maximum in the fruit development and then decreased, indicating that the mechanism of ethylene production and perception occurs in these fruits prior to ripening.

It has been well-documented that the ABA signaling pathway plays a central role in non-climacteric fruit ripening, including strawberry, grape and sweet cherry [63,64]. Multiple lines of evidence have shown that ethylene and ABA worked in synergy during non-climacteric fruit ripening; ethylene signaling could be involved in adjusting non-climacteric fruit ripening by merging into the ABA signaling pathway [20,21,44]. Jiang et al. (2003) [65] reported that exogenous ABA accelerated fruit coloring and softening through the up-regulation of ethylene production and PAL activity. Here, the expression levels of two *FaETRs* (*FaETR2* and *FaETR13*) and two *FaEINs* (*FaEIN2* and *FaEIN7*) were induced by ABA and inhibited by NDGA, indicating they might integrate ethylene and ABA signaling to regulate strawberry fruit ripening.

5. Conclusions

In summary, a total of 15 FaETRs and 14 FaEIN3/EINs in the ethylene signaling pathway from the cultivated strawberry were screened. Subsequently, a systematic analysis, including the chromosome location, evolutionary relationship, gene structure, conserved motif and protein–protein interaction network, was performed. Phylogenetic analysis showed that FaETRs and FaEIN3/EIL genes were respectively divided into two and three classes. The closely related proteins in the same class exhibited similar gene structure and motif composition. Furthermore, the expression profiles of *FaETRs* and *FaEIN3/EIL* genes during fruit ripening and in response to abscisic acid indicated that some of them in the ethylene signaling pathway probably participated in the strawberry ripening process. This work provides a landscape of strawberry ethylene signaling pathway gene families and a basis for further studies on strawberry breeding.

Supplementary Materials: The following supporting information can be downloaded at https://www.mdpi.com/article/10.3390/agronomy13071930/s1, Table S1: Primers used for qRT-PCR.

Author Contributions: Conceptualization, H.T. and Y.Z. (Yunting Zhang); software, X.G. and C.G.; validation, M.D. and Y.C.; methodology, Y.Z. (Yunting Zhang) and C.G.; formal analysis, Y.L. (Yuanxiu Lin) and Y.W.; visualization, Y.Z. (Yunting Zhang) and C.G.; writing—original draft preparation, Y.Z. (Yunting Zhang), M.D. and X.G.; writing—review and editing, Y.Z. (Yong Zhang), Q.C., Y.L. (Ya Luo) and X.W.; funding acquisition, H.T. and Y.Z. (Yunting Zhang). All authors have read and agreed to the published version of the manuscript.

Funding: This research was funded by the Double Support Project of Discipline Construction of Sichuan Agricultural University (03573134) and the Natural Science Foundation of Sichuan Province (2023NSFSC1243).

Institutional Review Board Statement: Not applicable.

Informed Consent Statement: Not applicable.

Data Availability Statement: Not applicable.

Conflicts of Interest: The authors declare no conflict of interest.

References

1. Bleecker, A.B.; Kende, H. Ethylene: A gaseous signal molecule in plants. *Annu. Rev. Cell Dev. Biol.* **2000**, *16*, 1–18. [CrossRef] [PubMed]
2. Merchante, C.; Alonso, J.M.; Stepanova, A.N. Ethylene signaling: Simple ligand, complex regulation. *Curr. Opin. Plant Biol.* **2013**, *16*, 554–560. [CrossRef] [PubMed]
3. Husain, T.; Fatima, A.; Suhel, M.; Singh, S.; Sharma, A.; Prasad, S.M.; Singh, V.P. A brief appraisal of ethylene signaling under abiotic stress in plants. *Plant Signal. Behav.* **2020**, *15*, 1782051. [CrossRef] [PubMed]
4. Broekgaarden, C.; Caarls, L.; Vos, I.A.; Pieterse, C.M.; Van Wees, S.C. Ethylene: Traffic controller on hormonal crossroads to defense. *Plant Physiol.* **2015**, *169*, 2371–2379. [CrossRef]
5. Yang, S.F.; Hoffman, N.E. Ethylene biosynthesis and its regulation in higher plants. *Annu. Rev. Plant Physiol.* **1984**, *35*, 155–189. [CrossRef]
6. Guo, H.; Ecker, J.R. The ethylene signaling pathway: New insights. *Curr. Opin. Plant Biol.* **2004**, *7*, 40–49. [CrossRef]
7. Binder, B.M. Ethylene signaling in plants. *J. Biol. Chem.* **2020**, *295*, 7710–7725. [CrossRef]
8. Schaller, G.E. Ethylene and the regulation of plant development. *BMC Biol.* **2012**, *10*, 9. [CrossRef]
9. Liu, Y.; Tang, M.; Liu, M.; Su, D.; Chen, J.; Gao, Y.; Bouzayen, M.; Li, Z. The molecular regulation of ethylene in fruit ripening. *Small Methods* **2020**, *4*, 1900485. [CrossRef]
10. Chen, T.; Qin, G.; Tian, S. Regulatory network of fruit ripening: Current understanding and future challenges. *New Phytol.* **2020**, *228*, 1219–1226. [CrossRef]
11. Zhu, X.; Zhu, Q.; Zhu, H. Towards a better understanding of fruit ripening: Crosstalk of hormones in the regulation of fruit ripening. *Front. Plant Sci.* **2023**, *14*, 1173877. [CrossRef]
12. Yin, X.-R.; Chen, K.-S.; Allan, A.C.; Wu, R.-M.; Zhang, B.; Lallu, N.; Ferguson, I.B. Ethylene-induced modulation of genes associated with the ethylene signalling pathway in ripening kiwifruit. *J. Exp. Bot.* **2008**, *59*, 2097–2108. [CrossRef]
13. Jourda, C.; Cardi, C.; Mbéguié-A-Mbéguié, D.; Bocs, S.; Garsmeur, O.; D'Hont, A.; Yahiaoui, N. Expansion of banana (*Musa acuminata*) gene families involved in ethylene biosynthesis and signalling after lineage-specific whole-genome duplications. *New Phytol.* **2014**, *202*, 986–1000. [CrossRef]
14. Tieman, D.M.; Ciardi, J.A.; Taylor, M.G.; Klee, H.J. Members of the tomato *LeEIL* (*EIN3-like*) gene family are functionally redundant and regulate ethylene responses throughout plant development. *Plant J.* **2001**, *26*, 47–58. [CrossRef]
15. Barickman, T.C.; Kopsell, D.A.; Sams, C.E. Abscisic acid increases carotenoid and chlorophyll concentrations in leaves and fruit of two tomato genotypes. *J. Am. Soc. Hortic. Sci.* **2014**, *139*, 261–266. [CrossRef]
16. Sun, L.; Yuan, B.; Zhang, M.; Wang, L.; Cui, M.; Wang, Q.; Leng, P. Fruit-specific rnai-mediated suppression of *SlNCED1* increases both lycopene and β-carotene contents in tomato fruit. *J. Exp. Bot.* **2012**, *63*, 3097–3108. [CrossRef]
17. Mou, W.; Li, D.; Bu, J.; Jiang, Y.; Khan, Z.U.; Luo, Z.; Mao, L.; Ying, T. Comprehensive analysis of aba effects on ethylene biosynthesis and signaling during tomato fruit ripening. *PLoS ONE* **2016**, *11*, e0154072. [CrossRef]
18. Wu, Q.; Bai, J.; Tao, X.; Mou, W.; Luo, Z.; Mao, L.; Ban, Z.; Ying, T.; Li, L. Synergistic effect of abscisic acid and ethylene on color development in tomato (*Solanum lycopersicum* L.) fruit. *Sci. Hortic.* **2018**, *235*, 169–180. [CrossRef]
19. Mou, W.; Li, D.; Luo, Z.; Li, L.; Mao, L.; Ying, T. SlAREB1 transcriptional activation of nor is involved in abscisic acid-modulated ethylene biosynthesis during tomato fruit ripening. *Plant Sci.* **2018**, *276*, 239–249. [CrossRef] [PubMed]
20. Kou, X.; Feng, Y.; Yuan, S.; Zhao, X.; Wu, C.; Wang, C.; Xue, Z. Different regulatory mechanisms of plant hormones in the ripening of climacteric and non-climacteric fruits: A review. *Plant Mol. Biol.* **2021**, *107*, 477–497. [CrossRef]
21. Bai, Q.; Huang, Y.; Shen, Y. The physiological and molecular mechanism of abscisic acid in regulation of fleshy fruit ripening. *Front. Plant Sci.* **2021**, *11*, 619953. [CrossRef]
22. Chervin, C.; El-Kereamy, A.; Roustan, J.-P.; Latché, A.; Lamon, J.; Bouzayen, M. Ethylene seems required for the berry development and ripening in grape, a non-climacteric fruit. *Plant Sci.* **2004**, *167*, 1301–1305. [CrossRef]
23. Gong, Y.; Fan, X.; Mattheis, J.P. Responses of 'Bing'and 'Rainier' sweet cherries to ethylene and 1-methylcyclopropene. *J. Am. Soc. Hortic. Sci.* **2002**, *127*, 831–835. [CrossRef]
24. Cherian, S.; Figueroa, C.R.; Nair, H. 'Movers and shakers' in the regulation of fruit ripening: A cross-dissection of climacteric versus non-climacteric fruit. *J. Exp. Bot.* **2014**, *65*, 4705–4722. [CrossRef] [PubMed]
25. Chen, Y.; Grimplet, J.; David, K.; Castellarin, S.D.; Terol, J.; Wong, D.C.; Luo, Z.; Schaffer, R.; Celton, J.-M.; Talon, M. Ethylene receptors and related proteins in climacteric and non-climacteric fruits. *Plant Sci.* **2018**, *276*, 63–72. [CrossRef] [PubMed]

26. Li, L.; Lichter, A.; Chalupowicz, D.; Gamrasni, D.; Goldberg, T.; Nerya, O.; Ben-Arie, R.; Porat, R. Effects of the ethylene-action inhibitor 1-methylcyclopropene on postharvest quality of non-climacteric fruit crops. *Postharvest Biol. Technol.* **2016**, *111*, 322–329. [CrossRef]
27. Gao, J.; Zhang, Y.; Li, Z.; Liu, M. Role of ethylene response factors (ERFs) in fruit ripening. *Food Qual. Saf.* **2020**, *4*, 15–20. [CrossRef]
28. Xie, X.-L.; Shen, S.-L.; Yin, X.-R.; Xu, Q.; Sun, C.-D.; Grierson, D.; Ferguson, I.; Chen, K.-S. Isolation, classification and transcription profiles of the AP2/ERF transcription factor superfamily in citrus. *Mol. Biol. Rep.* **2014**, *41*, 4261–4271. [CrossRef] [PubMed]
29. Kuang, J.-F.; Chen, J.-Y.; Luo, M.; Wu, K.-Q.; Sun, W.; Jiang, Y.-M.; Lu, W.-J. Histone deacetylase HD2 interacts with ERF1 and is involved in longan fruit senescence. *J. Exp. Bot.* **2012**, *63*, 441–454. [CrossRef]
30. Jia, H.-F.; Chai, Y.-M.; Li, C.-L.; Lu, D.; Luo, J.-J.; Qin, L.; Shen, Y.-Y. Abscisic acid plays an important role in the regulation of strawberry fruit ripening. *Plant Physiol.* **2011**, *157*, 188–199. [CrossRef]
31. Villarreal, N.M.; Bustamante, C.A.; Civello, P.M.; Martínez, G.A. Effect of ethylene and 1-MCP treatments on strawberry fruit ripening. *J. Sci. Food Agric.* **2010**, *90*, 683–689. [CrossRef]
32. Villarreal, N.M.; Marina, M.; Nardi, C.F.; Civello, P.M.; Martínez, G.A. Novel insights of ethylene role in strawberry cell wall metabolism. *Plant Sci.* **2016**, *252*, 1–11. [CrossRef] [PubMed]
33. Balogh, A.; Koncz, T.; Tisza, V.; Kiss, E.; Heszky, L. The effect of 1-MCP on the expression of several ripening-related genes in strawberries. *HortScience* **2005**, *40*, 2088–2090. [CrossRef]
34. Sun, J.-H.; Luo, J.-J.; Tian, L.; Li, C.-L.; Xing, Y.; Shen, Y.-Y. New evidence for the role of ethylene in strawberry fruit ripening. *J. Plant Growth Regul.* **2013**, *32*, 461–470. [CrossRef]
35. Trainotti, L.; Pavanello, A.; Casadoro, G. Different ethylene receptors show an increased expression during the ripening of strawberries: Does such an increment imply a role for ethylene in the ripening of these non-climacteric fruits? *J. Exp. Bot.* **2005**, *56*, 2037–2046. [CrossRef]
36. Zhang, Y.; Guo, C.; Deng, M.; Li, S.; Chen, Y.; Gu, X.; Tang, G.; Lin, Y.; Wang, Y.; He, W.; et al. Genome-wide analysis of the erf family and identification of potential genes involved in fruit ripening in octoploid strawberry. *Int. J. Mol. Sci.* **2022**, *23*, 10550. [CrossRef]
37. Li, D.; Li, L.; Luo, Z.; Mou, W.; Mao, L.; Ying, T. Comparative transcriptome analysis reveals the influence of abscisic acid on the metabolism of pigments, ascorbic acid and folic acid during strawberry fruit ripening. *PLoS ONE* **2015**, *10*, e0130037. [CrossRef]
38. Zhao, F.; Li, G.; Hu, P.; Zhao, X.; Li, L.; Wei, W.; Feng, J.; Zhou, H. Identification of basic/helix-loop-helix transcription factors reveals candidate genes involved in anthocyanin biosynthesis from the strawberry white-flesh mutant. *Sci. Rep.* **2018**, *8*, 2721. [CrossRef]
39. Chen, Q.; Yu, H.; Wang, X.; Xie, X.; Yue, X.; Tang, H. An alternative cetyltrimethylammonium bromide-based protocol for rna isolation from blackberry (*Rubus* L.). *Genet. Mol. Res.* **2012**, *11*, 1773–1782. [CrossRef]
40. Livak, K.J.; Schmittgen, T.D. Analysis of relative gene expression data using real-time quantitative PCR and the $2^{-\Delta\Delta CT}$ method. *Methods* **2001**, *25*, 402–408. [CrossRef]
41. Li, Z.; Chen, C.; Zou, D.; Li, J.; Huang, Y.; Zheng, X.; Tan, B.; Cheng, J.; Wang, W.; Zhang, L.; et al. Ethylene accelerates grape ripening via increasing *VvERF75*-induced ethylene synthesis and chlorophyll degradation. *Fruit Res.* **2023**, *3*, 3. [CrossRef]
42. Chaudhary, P.R.; Jayaprakasha, G.; Patil, B.S. Ethylene degreening modulates health promoting phytochemicals in Rio red grapefruit. *Food Chem.* **2015**, *188*, 77–83. [CrossRef]
43. Rodrigo, M.J.; Zacarias, L. Effect of postharvest ethylene treatment on carotenoid accumulation and the expression of carotenoid biosynthetic genes in the flavedo of orange (*Citrus sinensis* L. Osbeck) fruit. *Postharvest Biol. Technol.* **2007**, *43*, 14–22. [CrossRef]
44. Wang, H.; Huang, H.; Huang, X. Differential effects of abscisic acid and ethylene on the fruit maturation of *Litchi chinensis* Sonn. *Plant Growth Regul.* **2007**, *52*, 189–198. [CrossRef]
45. Farneti, B.; Khomenko, I.; Ajelli, M.; Emanuelli, F.; Biasioli, F.; Giongo, L. Ethylene production affects blueberry fruit texture and storability. *Front. Plant Sci.* **2022**, *13*, 813863. [CrossRef] [PubMed]
46. Costa, D.V.; Almeida, D.P.; Pintado, M. Effect of postharvest application of ethylene on the profile of phenolic acids and anthocyanins in three blueberry cultivars (*Vaccinium corymbosum*). *J. Sci. Food Agric.* **2018**, *98*, 5052–5061. [CrossRef]
47. Cao, Y.; Han, Y.; Meng, D.; Li, D.; Jin, Q.; Lin, Y.; Cai, Y. Genome-wide analysis suggests high level of microsynteny and purifying selection affect the evolution of *EIN3/EIL* family in Rosaceae. *PeerJ* **2017**, *5*, e3400. [CrossRef]
48. Liu, C.; Zhao, A.; Zhu, P.; Li, J.; Han, L.; Wang, X.; Fan, W.; Lü, R.; Wang, C.; Li, Z.; et al. Characterization and expression of genes involved in the ethylene biosynthesis and signal transduction during ripening of mulberry fruit. *PLoS ONE* **2015**, *10*, e0122081. [CrossRef]
49. Wang, X.; Ding, Y.; Wang, Y.; Pan, L.; Niu, L.; Lu, Z.; Cui, G.; Zeng, W.; Wang, Z. Genes involved in ethylene signal transduction in peach (*Prunus persica*) and their expression profiles during fruit maturation. *Sci. Hortic.* **2017**, *224*, 306–316. [CrossRef]
50. Liu, M.; Pirrello, J.; Chervin, C.; Roustan, J.-P.; Bouzayen, M. Ethylene control of fruit ripening: Revisiting the complex network of transcriptional regulation. *Plant Physiol.* **2015**, *169*, 2380–2390. [CrossRef]
51. Chervin, C.; Deluc, L. Ethylene signalling receptors and transcription factors over the grape berry development: Gene expression profiling. *Vitis* **2010**, *49*, 129–136.
52. Yaghobi, M.; Heidari, P. Genome-wide analysis of aquaporin gene family in triticum turgidum and its expression profile in response to salt stress. *Genes* **2023**, *14*, 202. [CrossRef] [PubMed]

53. Hashemipetroudi, S.H.; Arab, M.; Heidari, P.; Kuhlmann, M. Genome-wide analysis of the laccase (LAC) gene family in *Aeluropus littoralis*: A focus on identification, evolution and expression patterns in response to abiotic stresses and ABA treatment. *Front. Plant Sci.* **2023**, *14*, 1112354. [CrossRef]
54. Zhang, Y.; Ye, Y.; Jiang, L.; Lin, Y.; Gu, X.; Chen, Q.; Sun, B.; Zhang, Y.; Luo, Y.; Wang, Y.; et al. Genome-wide characterization of snf1-related protein kinases (snrks) and expression analysis of snrk1. 1 in strawberry. *Genes* **2020**, *11*, 427. [CrossRef] [PubMed]
55. Liu, H.; Xiong, J.-S.; Jiang, Y.-T.; Wang, L.; Cheng, Z.-M. Evolution of the R2R3-MYB gene family in six Rosaceae species and expression in woodland strawberry. *J. Integr. Agric.* **2019**, *18*, 2753–2770. [CrossRef]
56. Ahmadizadeh, M.; Rezaee, S.; Heidari, P. Genome-wide characterization and expression analysis of fatty acid desaturase gene family in *Camelina sativa*. *Gene Rep.* **2020**, *21*, 100894. [CrossRef]
57. Mu, Q.; Wang, B.; Leng, X.; Sun, X.; Shangguan, L.; Jia, H.; Fang, J. Comparison and verification of the genes involved in ethylene biosynthesis and signaling in apple, grape, peach, pear and strawberry. *Acta Physiol. Plant.* **2016**, *38*, 44.
58. Chen, C.; Zhang, M.; Zhang, M.; Yang, M.; Dai, S.; Meng, Q.; Lv, W.; Zhuang, K. ETHYLENE-INSENSITIVE 3-LIKE 2 regulates β-carotene and ascorbic acid accumulation in tomatoes during ripening. *Plant Physiol.* **2023**, *192*, kiad151. [CrossRef]
59. Peng, J.; Li, Z.; Wen, X.; Li, W.; Shi, H.; Yang, L.; Zhu, H.; Guo, H. Salt-induced stabilization of EIN3/EIL1 confers salinity tolerance by deterring ROS accumulation in Arabidopsis. *PLoS Genet.* **2014**, *10*, e1004664. [CrossRef]
60. Ma, X.; Li, C.; Yuan, Y.; Zhao, M.; Li, J. Xyloglucan endotransglucosylase/hydrolase genes *LcXTH4/7/19* are involved in fruitlet abscission and are activated by LcEIL2/3 in litchi. *Physiol. Plant.* **2021**, *173*, 1136–1146. [CrossRef]
61. Ma, X.; Ying, P.; He, Z.; Wu, H.; Li, J.; Zhao, M. The LcKNAT1-LcEIL2/3 regulatory module is involved in fruitlet abscission in litchi. *Front. Plant Sci.* **2022**, *12*, 3258. [CrossRef] [PubMed]
62. Ma, X.; Yuan, Y.; Wu, Q.; Wang, J.; Li, J.; Zhao, M. LcEIL2/3 are involved in fruitlet abscission via activating genes related to ethylene biosynthesis and cell wall remodeling in litchi. *Plant J.* **2020**, *103*, 1338–1350. [CrossRef] [PubMed]
63. Li, B.-J.; Grierson, D.; Shi, Y.; Chen, K.-S. Roles of abscisic acid in regulating ripening and quality of strawberry, a model non-climacteric fruit. *Hortic. Res.* **2022**, *9*, uhac089. [CrossRef] [PubMed]
64. Wang, W.; Fan, D.; Hao, Q.; Jia, W. Signal transduction in non-climacteric fruit ripening. *Hortic. Res.* **2022**, *9*, uhac190. [CrossRef] [PubMed]
65. Jiang, Y.; Joyce, D.C. ABA effects on ethylene production, PAL activity, anthocyanin and phenolic contents of strawberry fruit. *Plant Growth Regul.* **2003**, *39*, 171–174. [CrossRef]

agronomy

MDPI

Article

S-nitrosylation of *SlAPX* Is Involved in Alleviating Oxidative Damage in Transgenic Tobacco under Nitrate Stress

Chuntao Lv [1], Yuanlin Liang [1], Manqi Wang [1], Kunzhi Li [1], Xudong Sun [2,]* and Huini Xu [1,]*

[1] Faculty of Life Science and Technology, Kunming University of Science and Technology, Jingming South Street, Kunming 650224, China; lchuntao16@163.com (C.L.)

[2] Key Laboratory for Plant Diversity and Biogeography of East Asia, Kunming Institute of Botany, Chinese Academy of Sciences, Kunming 650201, China

* Correspondence: sunxudong@mail.kib.ac.cn (X.S.); xuhn@kust.edu.cn (H.X.)

Abstract: Nitric oxide (NO) modulates plant response by post-translationally modifying proteins, mainly through S-nitrosylation. Ascorbate peroxidase (APX) in the ascorbate-glutathione (AsA-GSH) cycle participates in the removal of hydrogen peroxide (H_2O_2). However, the relationship between S-nitrosylation and the role of tomato APX (*SlAPX*) under nitrate stress is still unclear. In this study, the enzyme activity, mRNA expression, and S-nitrosylation level of *SlAPX* were significantly increased in tomato roots after nitrate treatment. *SlAPX* protein could be S-nitrosylated by S-nitrosoglutathione in vitro, and APX activity was significantly increased after S-nitrosylation. The *SlAPX* overexpressed tobacco plants grew better than the wild type (WT) plants under nitrate stress. Meanwhile, the transgenic plants showed lower reactive oxygen species and malondialdehyde content, higher APX, monodehydroascorbate reductase, glutathione reductase activities, ascorbic acid/dehydroascorbic acid, and reduced glutathione/oxidized glutathione ratio, proline, and soluble sugar contents than those in the WT plants under nitrate treatment. Moreover, overexpressed transgenic seeds showed higher tolerance to methyl viologen induced oxidative stress compared with the WT. The NO accumulation and S-nitrosylation APX level were higher in transgenic plants than in WT plants after nitrate stress treatment. Our results provide novel insights into the mechanism of *SlAPX* modulation excess nitrate stress tolerance involving the S-nitrosylation modification.

Keywords: ascorbate peroxidase; NO; AsA-GSH cycle; tomato

Citation: Lv, C.; Liang, Y.; Wang, M.; Li, K.; Sun, X.; Xu, H. S-nitrosylation of *SlAPX* Is Involved in Alleviating Oxidative Damage in Transgenic Tobacco under Nitrate Stress. *Agronomy* **2023**, *13*, 1322. https://doi.org/10.3390/agronomy13051322

Academic Editors: Aisheng Xiong, Lijun Ou and Guanglong Wang

Received: 16 March 2023
Revised: 17 April 2023
Accepted: 18 April 2023
Published: 9 May 2023

1. Introduction

Salt stress is one of the major abiotic stresses affecting plant growth and global productivity [1,2]. Salt stress seriously affects the normal growth and development of plants, with low seed germination rate, leaf wilt, and yellowing, and it decreases photosynthesis metabolic capacity, eventually resulting in a decrease in yield [1,3–7]. When subjected to salt stress, plant cells produce excess reactive oxygen species (ROS), such as hydrogen peroxide (H_2O_2), hydroxyl radical, singlet oxygen, and superoxide [8–11]. Accumulation of ROS can interfere with cellular redox and oxidation, leading to DNA damage, membrane protein polymerization, lipid peroxidation, and inactivation of cell enzymes, ultimately destroying membrane structure and leading to cell dysfunction [12–14].

As sessile organisms, plants have evolved non-enzymatic and enzymatic scavenging ROS mechanisms to adapt to saline environments [15–17]. The enzymes that remove ROS in the plants are superoxide dismutase (SOD), catalase (CAT), peroxidase (POD), ascorbate peroxidase (APX), single monodehydroascorbate reductase (MDHAR), dehydroascorbate reductase (DHAR), glutathione reductase (GR), and certain nonenzymatic antioxidants [18–21]. Among them, the highly efficient ascorbate-glutathione (AsA-GSH) cycle formed by APX, MDHAR, DHAR, and GR is a crucial antioxidant system involved in the removal of intracellular ROS under plant development and stress conditions [22–24]. The APX uses

AsA as an electron donor to reduce the H_2O_2 to water. Monodehydroascorbate (MDHA) and dehydroascorbate (DHA) were reduced by MDHAR and DHAR to AsA, respectively, with GSH, and, finally, glutathione disulfide (GSSG) was restored to GSH. Past studies have shown that APX played an important role in alleviating plant abiotic stresses by regulating the H_2O_2 contents [25].

Nitric oxide (NO) is a ubiquitous bioactive gas molecule. In plants, NO is involved in the regulation of various physiological processes, including seed germination, root development, stomatal closure, flowering, hormonal signaling, gene expression and regulation, stress response, and programmed cell death. In abiotic stresses, NO helps plant cells to maintain their homeostasis and alleviate various stresses. NO plays its role mainly through S-nitrosylation—that is, the NO group binds to the residues of the protein cysteine [26]. Various abiotic stress conditions significantly regulate the S-nitrosylation of proteins [27]. S-nitrosylation of Cys20 and Cys147 of potato DHAR1 reduced DHAR activity [28]. Previous research showed that APX enzyme activity can be mediated by S-nitrosylation [29]. The S-nitrosylation site of pea APX was Cys32, and the S-nitrosylation increased APX activity [30,31]. Cys32 of Arabidopsis cytosolic APX1 was a S-nitrosylation site, and S-nitrosylation enhanced its enzymatic activity [32]. However, the cytosolic APX decreased its activity after S-nitrosylation in tobacco [33]. Proteomic analysis also identified APX as a target protein of S-nitrosylation in Arabidopsis [34]. There is no report about S-nitrosylation modification on tomato APX.

Over-utilization of chemical fertilizer has caused secondary salinization in Chinese greenhouses. The excessively accumulated anion in the soil of the greenhouse is nitrate (NO_3^-) [35,36]. NO_3^- excess is common in greenhouse soils, imposing environmental risks and degrading vegetable quality. Shi et al. suggested that excess NO_3^- stress to plants might share the similar defense pathways with NaCl stress [37]. To gain more insight into the mechanism of the tomato *SlAPX* under nitrate stress, we investigated the response of *SlAPX* overexpressed tobacco plants under excess nitrate stress. Meanwhile, we also investigated the possibility of the regulation of APX activity by S-nitrosylation.

2. Materials and Methods

2.1. Plant Materials and Treatments

Tomato (*Solanum lycopersicum* L.) seeds were surface sterilized with 55 °C sterile water for 10 min, and they were germinated in vermiculite. Plants were then hydroponically grown for 4 weeks in plastic tanks containing 4 L of aerated nutrient solution, including $Ca(NO_3)_2{\cdot}4H_2O$ 590 mg·L^{-1}, KNO_3 404 mg·L^{-1}, $KH_2PO_4$136 mg·L^{-1}, $MgSO_4{\cdot}7H_2O$ 246 mg·L^{-1}, EDTA·Na_2-Fe 40 mg·L^{-1}, H_3BO_3 2.86 mg·L^{-1}, $MnSO_4{\cdot}4H_2O$ 2.13 mg·L^{-1}, $ZnSO_4{\cdot}7H_2O$ 0.22 mg·L^{-1}, $CuSO_4{\cdot}5H_2O$ 0.08 mg·L^{-1}, and $(NH_4)_6Mo_7O_{24}{\cdot}4H_2O$ 0.02 mg·L^{-1}. The nutrient solution pH was adjusted to 6.0–6.5 by the addition of 98% (*w/v*) H_2SO_4. The experiment was conducted in the greenhouse of Kunming University of Technology under natural conditions. The temperature was 23–28 °C during the day and 13–18 °C at night, with a photoperiod of 12–14 h. The six–week–old tomato seedlings were treated with normal nutrient solution (CK), normal nutrient solution adding 100 mM nitrate (NO_3^-), normal nutrient solution adding 100 μM NO donor sodium nitroprusside (SNP), and normal nutrient solution adding 100 mM nitrate + 100 μM SNP (NO_3^-+SNP) for 24 h.

2.2. Gene Expression Analysis

The RNA was extracted using the TRIzol Reagent (Takara, Dalian, China) and detected by 1% agarose gel. For qRT-PCR, reverse transcription of RNA was carried out according to the instruction of the SYBR® PrimeScript™ RT-PCR Kit II (Takara, Dalian, China). qRT-PCR was performed using the iCycler iQ Real-time PCR detection system (Bio-Rad, Hercules, CA, USA). qRT-PCR was performed in three technical repetitions with complementary DNAs (cDNAs) synthesized from three biological replicates. The relative expression of specific genes was quantified using the $2^{-\Delta\Delta Ct}$ method. Tomato *Actin* was used as inner control for qRT-PCR analysis. These primer sequences are listed in Supplemental Table S1.

2.3. Plasmid Construction and Overexpressed Tobacco Characterization

The full length of the *SlAPX1* (accession no. NM_001247853) cDNA sequence amplified from tomato was 753 bp. The cDNA was amplified with the gene specific primers with restriction enzyme sites, *SlAPX*-F-*Bam*HI (5′-CGGGGGTACCGGATCCATGGGTAAGTGCTATCCTACTGT-3′), and *SlAPX*-R-*Bam*HI (5′-TCAGAATTCGGATCCTTAAG CTTCAGCAAATCCC-3′) using the Pfu DNA polymerase (Vazyme, Najing, China). The PCR fragment was ligated into the binary plant vector pRI101-GFP (Takara, Dalian, China) with the ClonExpress II one-step cloning kit (Vazyme, Najing, China). The pRI101-GFP plasmid was transformed into the *Agrobacterium tumefaciens* LBA4404 [38]. Transgenic tobacco was obtained with the leaf plate method [39].

To confirm the *SlAPX* gene integration into the tobacco genome, genomic DNA was extracted from young leaves using the cetyltrimethylammonium bromide (CTAB) method [40]. The binary plant vector pRI101-*SlAPX*-GFP and wild-type (WT) genomic DNA were used as positive and negative controls, respectively. The transgenic plants were then identified by qRT-PCR and Western blot analysis.

2.4. Analysis of Transgenic Plants under Nitrate Stress

To test the nitrate tolerance of tobacco seedlings, transgenic and WT tobacco plants were grown in vermiculite and peat-filled pots (1:1). Six-week-old tobacco seedlings were watered with 50 mL of water as a control, or 50 mL of 150 mM nitrate solution (provided by same mol of KNO_3 and $Ca(NO_3)_2$) for 14 d for the nitrate treatment group. The experiment was also conducted in the greenhouse of Kunming University of Technology under natural conditions. The leaves were taken from 3 plants (n = 3), immediately frozen in liquid nitrogen, and stored at −80 °C until use.

2.5. Analysis of Transgenic Plants under Methyl Viologen (MV) Stress

Seeds from transgenic and WT tobacco plants were soaked in sterile water at 55 °C for 30 min, and then sterilised with 4% NaClO for 20 min before being washed three times with sterile water. After that, the seeds were sown on Murashige Skoog (MS) agar plates (9 cm) with 0 and 20 μM MV, and the survival rate was analyzed after 10 days.

To test the MV tolerance of tobacco seedlings, transgenic and WT tobacco plants were grown in vermiculite and peat-filled pots (1:1). Tobacco plants were treated with water (control) and 50 μM MV solution for 5 days. The phenotype was then photographed.

2.6. Endogenous ROS and NO Fluorescence Analysis

The tomato root tips were cut into 0.5~1 cm pieces, rinsed with pure water, and then soaked in EP tubes with either 2 μM H_2DCFDA solution for ROS accumulation analysis [41] or 5 μM DAF-FM DA for NO accumulation analysis [42]. The EP tubes were then exposed in darkness for 30 min and the root tips were stained. The root tips were washed 3 times for 15 min each with 20 mM HEPES-KOH (pH 7.8) buffer solution. Finally, the root tips of the washed tomato seedlings were placed under the epifluorescence microscope (model DMI6000B; Leica, Solms, Germany) to observe the accumulation of ROS (excitation 488 nm; emission 525 nm) or NO (excitation 485 nm; emission 515 nm).

2.7. Antioxidant Enzyme Activities Analysis

0.2 g leaves were grinded in an ice bath in precooled mortar. 1 mL of enzyme extract (50 mmol·L^{-1}, PH 7.8, phosphate buffer, 1 mmol·L^{-1} EDTA, 1 mmol·L^{-1} AsA and 1% PVP) was added and then centrifuged at 12,000 rpm for 20 min at 4 °C. The resulting supernatant was finally collected for enzymatic activity analysis. The decrease in absorbance at 290 nm when the ascorbate was oxidized was used to determine APX activity, as described by Nakano and Asada [43]. The MDHAR activity was evaluated by monitoring the change in absorbance at 340 nm, as described by Miyake and Asada [44]. The rate of NADPH oxidation was used to calculate GR activity [45].

2.8. AsA, DHA, GSH, GSSG, Proline and Soluble Sugar Contents Analysis

The reduced AsA, oxidized DHA, GSH, and GSSH were analyzed according to Jiang and Zhang [46]. Free proline contents were analyzed using the ninhydrin assay as previously described [47]. The material was extracted for 15 min at 4 °C with 3% (*w/v*) sulfosalicylic acid. The free proline was determined using the supernatant after centrifugation. The soluble sugar content was determined using the method of Yemm and Willis [48].

2.9. Lipid Peroxidation Level and H_2O_2 Contents Analysis

The generation of malondialdehyde (MDA) was estimated using the thiobarbituric acid reaction method to measure lipid peroxidation [49]. H_2O_2 contents were analyzed using the method of Gay and Gebicki [50].

2.10. Measurement of Protein Expression

Protein expression level was analyzed with the Western blot, following the method of Bai et al. [51]. SDS-PAGE was used to separate the proteins, which were then transferred to PVDF membranes. APX peptides were used to immunize white mice to obtain an antibody for APX detection.

2.11. The Biotin Switch Approach to Detect S-nitrosylation Protein

Detection of S-nitrosylation proteins was performed by the methods of Jaffrey and Snyder, with minor modifications [52]. 1 g plant was soaked in HEN buffer to extract soluble protein (containing1 mM EDTA, 0.1 mM neocuproine, 25 mM HEPES-NaOH, pH 7.7). The homogenate was centrifuged at 10,000× *g* for 20 min and the supernatant was measured and adjusted the protein concentration to 1 mg·mL^{-1}. All protein samples were incubated at 50 °C for 20 min with regular vortexing after being treated with 25 mM methylmethanethiosulfonate (MMTS) and 2.5% sodium dodecyl sulfate (SDS). The residual MMTS was eliminated by the acetone precipitation method. At 37 °C for 1.5 h, sodium ascorbate and biotin-HPDP were added, and the excess biotin and ascorbate were removed by acetone precipitation. SDS-PAGE was used to electrophorese the biotin–labeled S–nitrosylation protein, and the protein was then submitted to a western blot examination with the APX antibody.

2.12. Purification of Recombinant SlAPX Protein

The *SlAPX* fragment was ligated into the prokaryotic expression vector pDE1 vector (Beijing KoSo Biotechnology Co., Ltd., Beijing, China) with the ClonExpress II one-step cloning kit (Vazyme, Nanjing, China). The pDE1-*SlAPX1* was transformed into the Rosetta (DE3) plysS *E. coli* Strain (Ybscience, Beijing, China). The *SlAPX* gene recombinant expressing strain was inoculated into 25 mL LB medium at a ratio of 1:100 and incubated at 37 °C, 220 rpm·min^{-1} for about 4 to 6 h (OD 600 ≈ 0.5), with IPTG added at a final concentration of 1 mM and induced at 37 °C and 28 °C, respectively. A total of 2 mL bacterial fluid after induction at 0, 2, 4, and 6 h was centrifuged at 12,000 rpm·min^{-1} for 1 min, and bacteria were collected. The bacteria were resuspended in an appropriate amount of PBS, broken for 15 min by sonication, and then centrifuged at 12,000 rpm·min^{-1} for 1 min, and the supernatant was then collected and the precipitate was resuspended in an appropriate volume of PBS. A total of 10 µL of supernatant and 10 µL of precipitation resuspension were collected for SDS-PAGE analysis. Protein was purified using the MagneHis™ Protein Purification System (Promega, Madison, WI, USA) kit. Target proteins were eluted with 20 mM phosphate buffer containing 100 mM NaCl and a different concentration gradient of imidazole, and 10 µL samples were taken for 12% SDS-PAGE analysis.

2.13. Determination of S-nitrosylation and Activity of SlAPX Protein by GSNO Treatment In Vitro

In order to analyze the S-nitrosylation level of *SlAPX* protein in vitro, the purified *SlAPX* protein was treated with or without GSNO for 30 min, and analyzed by biotin conversion method with or without AsA. The *SlAPX* activity was analyzed after GSNO treatment.

2.14. Statistical Analysis

Each sample was statistically analyzed and the data was presented as mean ± standard errors (SE) of three independent experiments. One-way analysis of variance (ANOVA) was performed on the data. (p-values < 0.05 are summarized with one asterisk, and p-values < 0.01 are summarized with two asterisks).

3. Results

3.1. Effect of Nitrate Stress and NO Treatment on the Activity, Expression, and S-nitrosylation Level of SlAPX in Tomato

To determine the activity, expression, and S-nitrosylation level of *SlAPX* in tomato, 6-week-old tomato seedlings were treated with nitrate and NO donor SNP for 24 h. The activity of APX in roots increased significantly by 282.5% after 24 h nitrate treatment, compared with CK (Figure 1A). The APX enzyme activity in the roots increased by 14.85% after exogenous application of SNP compared to nitrate stress treatment alone, respectively. The *SlAPX* gene expression in tomato seedling roots was significantly ($p < 0.05$) increased by nitrate as well as by nitrate and SNP co-treatment compared with CK (Figure 1B). Compared with CK, *SlAPX* gene expression in tomato seedling roots increased 2.96-fold after nitrate treatment and 3.47-fold after nitrate and SNP treatment. Compared with CK, S-nitrosylation level of *SlAPX* protein was significantly increased in roots after nitrate treatment (Figure 1C). The result showed that nitrate stress enhanced S-nitrosylation of *SlAPX* in tomato roots.

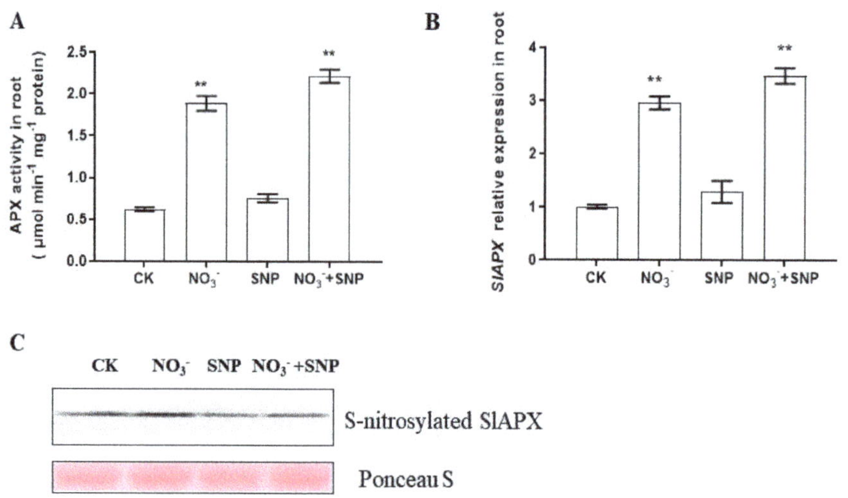

Figure 1. Effect of nitrate stress and SNP treatment on the enzyme activities, expression profile, and S-nitrosylation level of *SlAPX*. 6-week-old tomato seedlings were treated with control (CK), 100 mM nitrate, 100 µM SNP, or 100 mM nitrate + 100 µM SNP for 24 h. Subsequently, *SlAPX* enzyme activities (**A**) relative mRNA expression (**B**) and S-nitrosylation level (**C**) were analyzed. p-values < 0.01 are summarized with two asterisks.

3.2. S-nitrosylation of SlAPX In Vitro and the Effect of S-nitrosylation on APX Activity

Strains with a recombinant expression of the *SlAPX* gene were grown in LB medium, and expression was induced using IPTG. Cultures at 0, 2, 4, and 6 h after 28 °C and 37 °C induction were analyzed by SDS-PAGE. The results showed that the induced *SlAPX* recombinant protein size was approximately 28 kD, as expected (Figure S1A). SDS-PAGE analysis indicated the expression of homogeneous target proteins in the supernatant and precipitate, but mainly concentrated in the supernatant (Figure S1B). The supernatant was purified using a Ni^{2+} NTA affinity column, and 100 and 150 mM of imidazole eluates were eluted, with a clear single band of about 28 kD, indicating a high purity of *SlAPX* protein (Figure S1C).

The purified *SlAPX* protein was treated with or without 1 mM S-nitrosylation agent GSNO for 30 min, and then the S-nitrosylation of *SlAPX* protein in the presence or absence of AsA was analyzed. The results showed that S-nitrosylation of *SlAPX* protein occurred only when GSNO and AsA were added at the same time (Figure 2A).

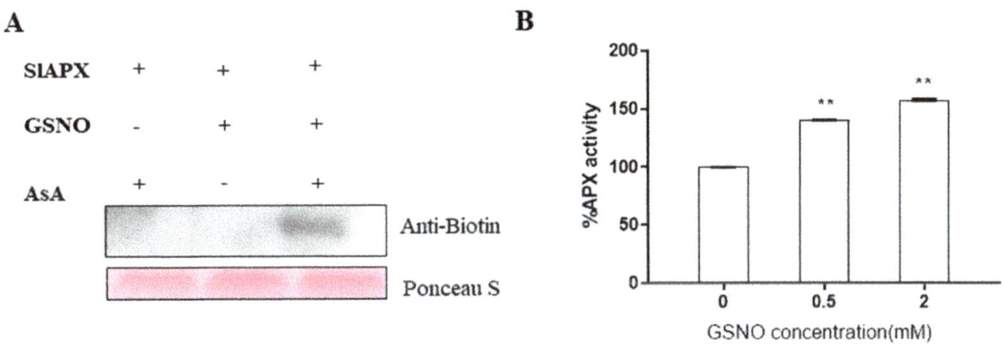

Figure 2. Effect of GSNO treatment on the S-nitrosylation level and enzyme activity of *SlAPX* protein. (**A**): The *SlAPX* protein was treated with or without 1 mM GSNO for 30 min, and then the S-nitrosylation level of *SlAPX* protein in vitro was measured with or without AsA. (**B**): Effect of 0, 0.5, and 2 mM GSNO treatment on the activity of *SlAPX* protein. p values < 0.01 are summarized with two asterisks.

To investigate the activity of *SlAPX* after S-nitrosylation, the recombinant *SlAPX* protein were purified and treated with GSNO, and *SlAPX* activity was measured. *SlAPX* activity increased significantly with increasing GSNO concentration (Figure 2B). The *SlAPX* activity increased by 40.14% and 57.45% after incubation with 0.5 and 2 mM of GSNO at room temperature for 30 min, respectively. These results suggested that tomato *SlAPX* activity might be increased after S-nitrosylation.

3.3. The Characterization of Overexpressed SlAPX Transgenic Tobacco

We identified 12 positive overexpressed *SlAPX* transgenic lines by kanamycin screening. Genomic PCR showed the target band in the transgenic plants with specific primers for NPTII (Figure 3A). qRT-PCR, Western blot, and enzyme activities were then conducted. The qRT-PCR results showed that the *SlAPX* gene overexpressed transgenic lines of OE2, OE3, and OE4increased 3.64, 3.75, and 3.47 times, respectively, compared with WT (Figure 3B). The results of Western blot showed that APX protein expression was significantly increased in OE2, OE3, and OE4 in the transgenic plants compared with WT (Figure 3C). The APX activity was significantly increased in OE2, OE3, and OE4 compared with WT (Figure 3D). The above results indicated that *SlAPX* successfully overexpressed in tobacco.

3.4. Overexpression of SlAPX Reduced Oxidative Damage in Tobacco under Nitrate Stress

The phenotype of WT leaves was smaller and yellower than that of overexpressed transgenic plants after nitrate treatment (Figure 4A). After nitrate treatment, ROS fluorescence accumulation was dramatically lower in the *SlAPX* overexpressed transgenic lines than that in the WT (Figure 4B). Under normal growth, the H_2O_2 content of the transgenic plants was lower than that of the WT; after nitrate treatment, the OE2, OE3, and OE4 transgenic plants have a significantly lower H_2O_2 content than WT (Figure 4C). Similar to the H_2O_2 content, MDA contents increased in both WT and transgenic tobacco after nitrate treatment, and the contents in transgenic tobacco was significantly lower than in WT plants (Figure 4D).

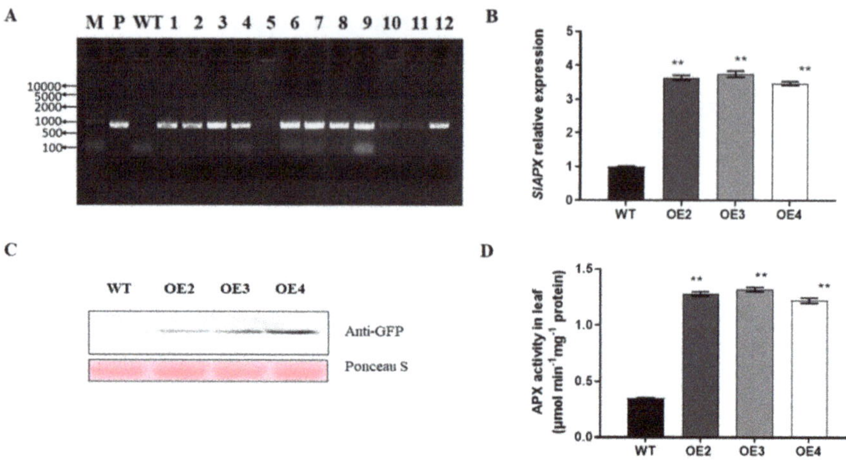

Figure 3. The genomic PCR (**A**), relative expression (**B**), protein expression (**C**), and APX activity (**D**) analysis in *SlAPX* transgenic tobacco plants. OE2, OE3, and OE4 were three different *SlAPX* overexpressed transgenic lines. p values < 0.01 are summarized with two asterisks.

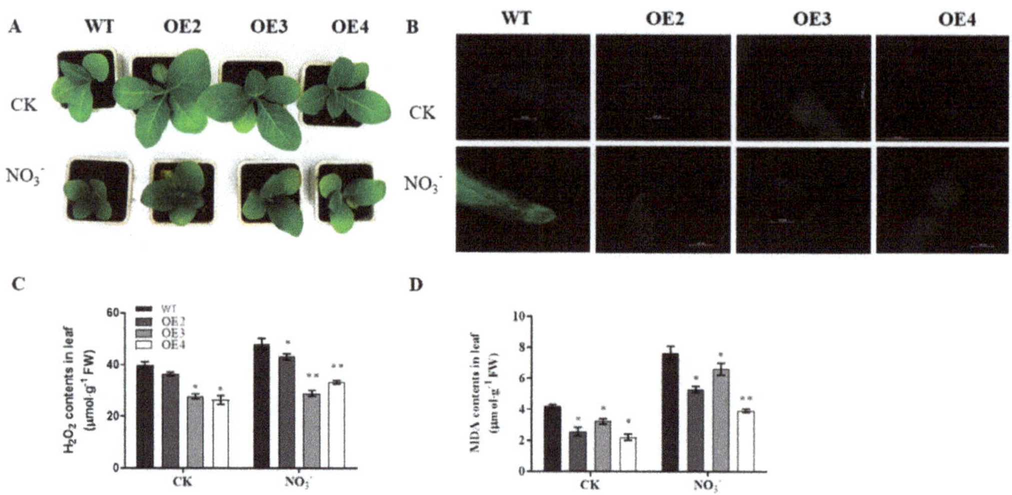

Figure 4. Effects of nitrate stress on the phenotype (**A**), ROS fluoresence (**B**), H_2O_2 contents (**C**), and MDA contents (**D**) in *SlAPX* overexpressing and WT plants. Transgenic and WT tobacco plants were watered with 50 mL of water as a control, or 50 mL of 150 mM nitrate solution once every 2 days for 2 weeks. Scale bar = 100 μM. p values < 0.05 are summarized with one asterisk, and p values < 0.01 are summarized with two asterisks.

3.5. *SlAPX Overexpressed Tobacco Plants Had Increased Oxidative Stress Tolerance*

To further investigate the tolerance of *SlAPX* overexpressing tobacco to oxidative stress, seeds of transgenic lines were transferred to MS medium containing MV induced oxidative stress, and MS medium without MV was used as a control (Figure 5A). There was no significant difference in seed germination between the WT and transgenic plants on the MS medium without MV. On MS medium containing 20 μM MV, the growth of WT and overexpressed tobacco was significantly inhibited and the leaves showed photobleaching, but the survival rate of transgenic tobacco was significantly higher than that of WT (Figure 5B).

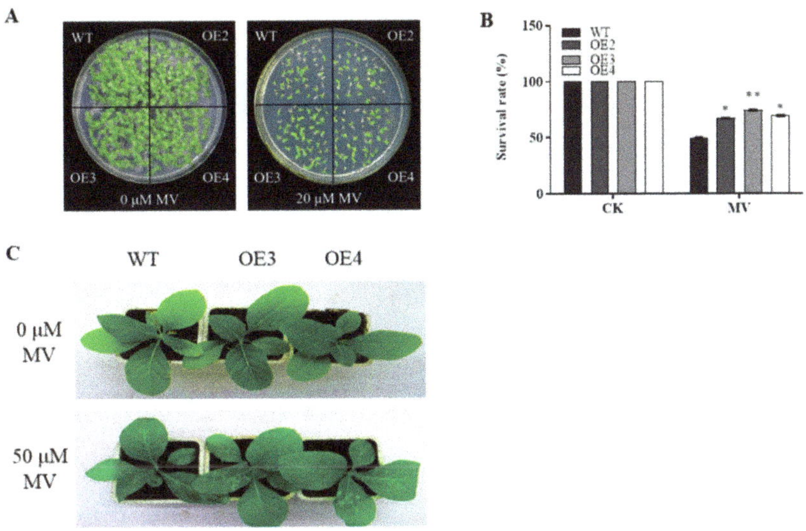

Figure 5. Effects of MV stress on the seed germination and seedling growth in *SlAPX* overexpressing and WT plants. (**A**): The phenotype of *SlAPX* overexpressing and WT seeds sown on MS agar plates with 0 or 20 μM MV. (**B**): The survival rate was assayed after 10 days of MV treatment. (**C**): Tobacco plants were sprayed with water or 50 μM MV solution for 5 days. The phenotype was then observed. p values < 0.05 are summarized with one asterisk, and p values < 0.01 are summarized with two asterisks.

To investigate the oxidative stress tolerance of transgenic plants seedlings, leaves were sprayed with solutions containing 0 μM (control) or 50 μM MV for 5 days. There was no significant difference in growth between WT and transgenic plants in the control. After treatment with 50 μM MV solution, the leaves of WT plants showed severe damage, while the leaves of transgenic plants showed less damage symptoms (Figure 5C).

3.6. SlAPX Overexpressed Plants Had Higher Antioxidant Enzyme Activities, and the AsA/DHA, GSH/GSSG Ratio under Nitrate Stress

Under nitrate treatment, the mRNA expression level and activity of APX were significantly higher in transgenic plants than in WT plants (Figure 6A,B). Activities of antioxidant enzymes in the AsA-GSH cycle were then analyzed. As shown in Figure 6C,D, the activities of MDHAR, and GR were significantly higher in transgenic plants than in WT plants under normal conditions. After nitrate treatment, MDHAR and GR activities in overexpressed plants were significantly higher than those in WT. The AsA/DHA and GSH/GSSG ratio of *SlAPX* overexpressed plants was significantly higher than that of WT plants under control and nitrate stress conditions (Figure 6E,F).

3.7. SlAPX Overexpressed Plants Had Enhanced Accumulation of Osmotic Substance under Nitrate Stress

As shown in Figure 7A, proline contents were significantly higher in transgenic lines than in WT plants under normal conditions and nitrate stress. There is no significant difference in soluble sugar contents between transgenic and WT plants under control. After nitrate treatment, the soluble sugar contents in transgenic plants were significantly higher than WT plants (Figure 7B). We then analyzed the mRNA expression of several osmotic stress marker genes by qRT-PCR, including *Ntosmotin*, *NtP5CS*, *NtDREB*, and *NtLEA5*. As shown in Figure 7C–F, there was no significant difference between the WT and transgenic lines under normal conditions. The mRNA transcript levels of *NtP5CS*, *NtDREB2*, *Ntosmotin*, and *NtLEA5* genes were dramatically increased after nitrate stress treatment.

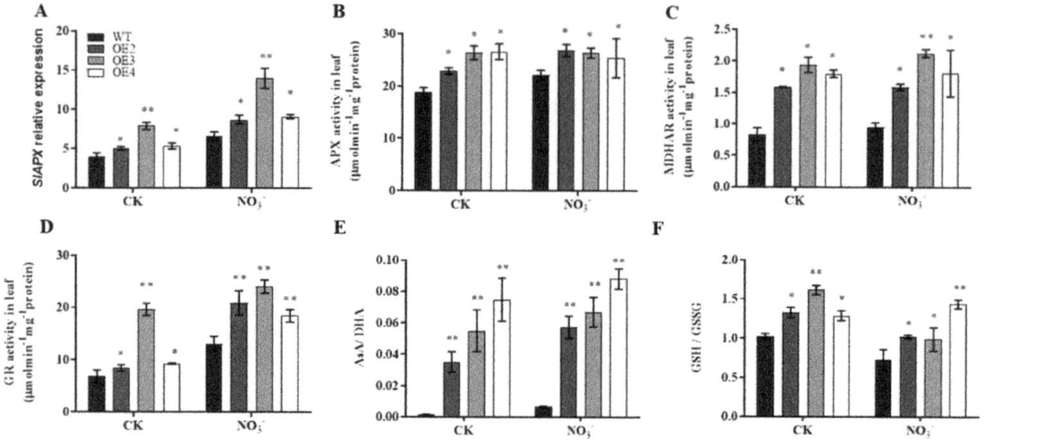

Figure 6. Effects of nitrate treatment on the key antioxidant enzymes activities and AsA/DHA, GSH/GSSG ratio in *SlAPX* transgenic and WT plants. Transgenic and WT tobacco plants were watered with 50 mL of water as a control, or 50 mL of 150 mM nitrate solution once every 2 days for 2 weeks. (**A**): Relative expression of *SlAPX*. (**B–D**): Activities of APX, MDHAR, GR. (**E**,**F**):AsA/DHA and GSH/GSSG ratio. p values < 0.05 are summarized with one asterisk, and p values < 0.01 are summarized with two asterisks.

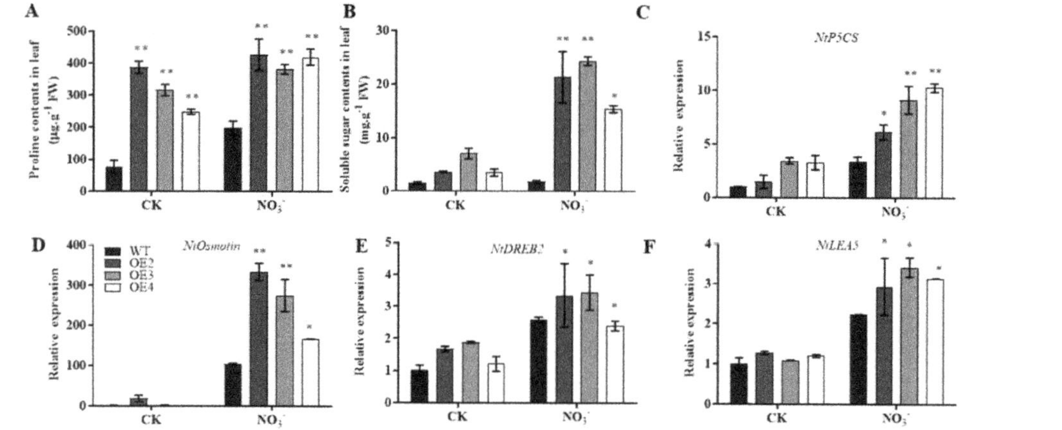

Figure 7. Effects of nitrate stress on proline (**A**) and soluble sugar (**B**) contents and expression of *NtP5CS* (**C**), *NtOsmotin* (**D**), *NtDREB* (**E**), *NtLEA5* (**F**) in *SlAPX* overexpressing and WT plants. Transgenic and WT tobacco plants were watered with 50 mL of water as a control, or 50 mL of 150 mM nitrate solution once every 2 days for 2 weeks. p values < 0.05 are summarized with one asterisk, and p values < 0.01 are summarized with two asterisks.

3.8. SlAPX Overexpressed Transgenic Plants Had Higher S-nitrosylation Level of APX under Nitrate Stress

To investigate whether *SlAPX* was S-nitrosylated under nitrate stress, the NO accumulation was first assayed. There was no dramatical difference in NO accumulation between WT and *SlAPX* overexpressed plants in the control (Figure 8A). When tobacco seedlings were exposed to nitrate stress, NO content increased significantly in WT and transgenic plants, especially in the latter. As shown in Figure 8B, the S-nitrosylation APX levels of overexpressed transgenic plants were significantly higher than those of WT plants under normal growth and nitrate treatment.

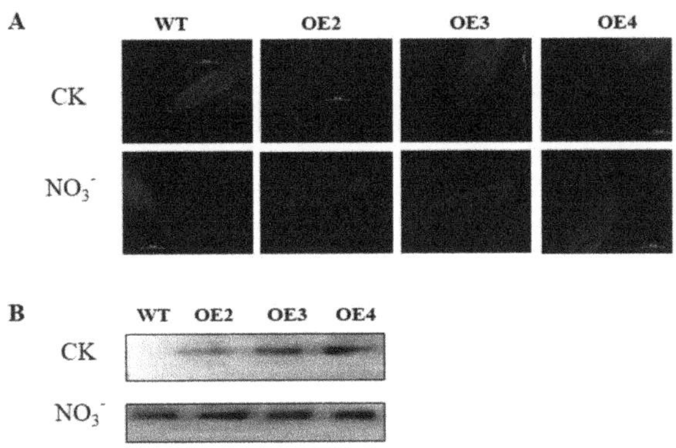

Figure 8. Effect of nitrate stress on the NO fluorescence (**A**) and S-nitrosylation APX level (**B**) in *SlAPX* overexpressed and WT tobacco. Transgenic and WT tobacco plants were watered with 50 mL of water as a control, or 50 mL of 150 mM nitrate solution once every 2 days for 2 weeks. The NO accumulation and S-nitrosylation level of APX were then analysed.

4. Discussion

APX, one of the important enzymes in the AsA-GSH cycle, is involved in regulating H_2O_2 levels under plant growth and stress conditions [53–56]. Salinity caused an increase in the activity of APX in sugarcane [57]. Compared with salt stress, the enzyme activity of APX was significantly increased in wheat with SNP and salt stress [58]. In this study, the activity and transcription levels of *SlAPX* were significantly increased after nitrate treatment (Figure 1A,B).

Overexpression of APX has been reported to enhance plant tolerance under environmental stress. The transgenic plants overexpressing the Populus *PpAPX* gene increased APX activity under salt stress, and they also significantly increased salt tolerance during the vegetative period [59]. Overexpressing the *LmAPX* gene of *Lycium chinense* Mill. in tobacco showed high APX activity under salt stress, which improved the salt tolerance of tobacco [60]. The transgenic *Arabidopsis* plants overexpressing *Oncidium OgCytAPX1* showed highly efficient ROS scavenging activity and salt tolerance [61]. In this study, under nitrate stress, the growth of *SlAPX* overexpressed transgenic tobacco was better than WT, indicating that *SlAPX* overexpression enhanced nitrate stress tolerance of tobacco (Figure 4A).

Salt stress leads to excessive production of ROS in plants, leading to oxidative stress, thus affecting the growth and development of the plant. MDA content, as a product of lipid peroxidation, was used to evaluate the extent of oxidative damage [62]. APX plays a key role in stress regulation responses by enhancing ROS clearance and maintaining ROS homeostasis, regulating H_2O_2 content, and reducing MDA content. *Populus euphratica PeAPX2* can promote the accumulation of cAPX under salt stress, scavenging ROS and reducing the content of H_2O_2 and MDA, thus enhancing the salt tolerance of poplar [63]. In our study, the ROS, H_2O_2, and MDA contents in *SlAPX* overexpressed transgenic tobacco plants were lower than WT plants, suggesting that the oxidative damage caused by excessive nitrate was lower in overexpressed plants than WT (Figure 4). Visual assessment of transgenic and control lines exposed to MV confirmed that overexpression of Populus *PpAPX* minimized leaf damage, indicating that *APX* transgenic plants improved oxidative damage in response to abiotic stress [59]. The transgenic tobacco plants overexpressing *APX* showed increased tolerance to oxidative stress caused by application of MV [64]. In our study, the survival rate of *SlAPX* overexpressed plants was higher than WT plants under MV treatment (Figure 5), indicating that *SlAPX* when overexpressed showed higher tolerance to oxidative stress than WT.

Our study showed that the *SlAPX* transgenic plants have higher antioxidant enzyme activity of APX, MDHAR, and GR to enhance nitrate stress tolerance (Figure 6). The alteration in AsA and GSH production may contribute to increased resistance to abiotic and biotic stresses [65]. Under drought stress, the AsA and DHA contents of *CytAPX* overexpressed tobacco increased, the AsA/DHA ratio decreased, and the GSH/GSSG ratio increased [66]. In a sensitive genotype of pea under NaCl stress, the total AsA content in the soluble fraction decreased, with a slightly lower AsA/DHA ratio, and a 60% decrease in the glutathione pool compared with the control group [19]. Our study found that AsA/DHA and GSH/GSSG ratios were significantly higher in *SlAPX* overexpressing plants under control and nitrate stress conditions (Figure 6F,G), indicating that the transgenic plants had more reducing agents to clear ROS.

Our study found that under normal conditions and after nitrate stress treatment, the contents of proline in transgenic tobacco were significantly higher than those in WT plants (Figure 7). Osmotin proteins have been shown to be induced in plants in response to abiotic and biotic stresses, involved in protecting plants from these stresses [67]. Tobacco *osmotin* transgenic plants showed significantly increased accumulation of free proline compared to WT plants, resulting in better stress tolerance [68]. The transgenic tomato plants carrying the tobacco *osmotin* gene exhibited enhanced salt stress tolerance compared with WT plants, indicating that overexpression of the *osmotin* gene enhanced salt tolerance in transgenic tomato plants [69]. A bifunctional enzyme, delta(1)-pyrrolin-5-carboxylate synthetase (P5CS), controls the glutamate pathway in proline biosynthesis in plants and positively regulates plant response to salt stress [70,71]. Overexpression of *P5CS* in soybean transgenic lines increased the content of proline and showed high salt tolerance [72]. Previous studies have confirmed that dehydration responsive element binding protein (DREB) is a transcription factor that responds to salt stress by enhancing transcription expression and activating salt tolerance related genes in plants [73]. Under salt stress, transgenic tobacco plants with the soybean *GmDREB6* gene increased the transcription levels of *GmDREB6* and *NtP5CS* genes, thus improving salt tolerance [74]. Late embryogenetic abundant (LEA) proteins play an important role in plant responses to abiotic stresses as osmotic regulatory materials and protective materials for cell membrane structure [75]. In our study, the expression levels of *Ntosmotin*, *NtP5CS*, *NtDREB2*, and *NtLEA5* genes in *SlAPX* transgenic tobacco were significantly higher than WT plants after nitrate stress treatment, suggesting that *SlAPX* may regulate osmotic potential to enhance nitrate stress tolerance (Figure 7).

NO, as a signaling molecule, regulates plant growth and development mainly through S-nitrosylation [26]. Proteomic studies have shown that APX acts as an S-nitrosylation target protein [34]. Studies have shown that S-nitrosylation increased APX activity [30]. After treatment of *Antiaris toxicaria* with NO gas, the S-nitrosylation of APX increased its enzyme activity and contributed to seed drying [76]. Proteomic analysis of *Arabidopsis* roots showed that cytoplasmic APX (APX1) could undergo S-nitrosylation, and the activity of recombinant APX1 was increased after S-nitrosylation [77]. In the present study, the APX S-nitrosylation level was increased in tomato after nitrate treatment (Figure 1C). *SlAPX* was S-nitrosylated by GSNO treatment and its activity was increased (Figure 2). In addition, *SlAPX* overexpressed transgenic tobacco plants had higher S-nitrosylation level of APX and activity when compared to WT (Figure 8B). Our results suggest that APX are regulated by S-nitrosylation, thus highlighting the close involvement of interactions between NO metabolism and antioxidant enzymes related to ROS metabolism in stress tolerance. The key cysteine cite of S-nitrosylation will be studied further in the future.

5. Conclusions

SlAPX protein was S-nitrosylated in tomato under nitrate stress and in vitro. The growth of *SlAPX* overexpressed plants was significantly better than that of WT under nitrate stress, with lower ROS accumulation, higher AsA/DHA ratio and antioxidant enzyme activities, and higher proline and soluble sugar contents. These results suggest that *SlAPX* is involved in mitigating oxidative damage under nitrate stress related to the S-nitrosylation of APX.

Supplementary Materials: The following supporting information can be downloaded at: https://www.mdpi.com/article/10.3390/agronomy13051322/s1, Figure S1: Induction and purification of tomato *SlAPX* recombinant protein prokaryotically; Table S1: Specific primers for qRT-PCR amplification.

Author Contributions: H.X. and X.S. designed the project. C.L. wrote the manuscript. C.L., Y.L. and M.W. conducted the experiments and analyzed the data. K.L. helped in the writing of the article. All authors have read and agreed to the published version of the manuscript.

Funding: This research was funded by the National Natural Science Foundation of China (grant no. 32260753) and the Yunnan Ten Thousand Talents Plan: Young & Elite Talents Project.

Data Availability Statement: The data presented in this study are available on request from the corresponding author.

Conflicts of Interest: The authors declare no conflict of interest.

References

1. Shahid, M.A.; Sarkhosh, A.; Khan, N.; Balal, R.M.; Ali, S.; Rossi, L.; Gómez, C.; Mattson, N.; Nasim, W.; Garcia-Sanchez, F. Insights into the Physiological and Biochemical Impacts of Salt Stress on Plant Growth and Development. *Agronomy* **2020**, *10*, 938. [CrossRef]
2. Vaidyanathan, H.; Sivakumar, P.; Chakrabarty, R.; Thomas, G. Scavenging of reactive oxygen species in NaCl-stressed rice (*Oryza sativa* L.)—Differential response in salt-tolerant and sensitive varieties. *Plant Sci.* **2003**, *165*, 1411–1418. [CrossRef]
3. Quan, R.; Lin, H.; Mendoza, I.; Zhang, Y.; Cao, W.; Yang, Y.; Shang, M.; Chen, S.; Pardo, J.M.; Guo, Y. SCABP8/CBL10, a putative calcium sensor, interacts with the protein kinase SOS2 to protect *Arabidopsis* shoots from salt stress. *Plant Cell* **2007**, *19*, 1415–1431. [CrossRef]
4. Park, H.J.; Kim, W.Y.; Yun, D.J. A role for GIGANTEA: Keeping the balance between flowering and salinity stress tolerance. *Plant Signal. Behav.* **2013**, *8*, e24820. [CrossRef]
5. Alharby, H.F.; Al-Zahrani, H.S.; Hakeem, K.R.; Iqbal, M. Identification of physiological and biochemical markers for salt (NaCl) stress in the seedlings of mungbean [*Vigna radiata* (L.) Wilczek] genotypes. *Saudi J. Biol. Sci.* **2019**, *26*, 1053–1060. [CrossRef]
6. van Zelm, E.; Zhang, Y.; Testerink, C. Salt Tolerance Mechanisms of Plants. *Annu. Rev. Plant Biol.* **2020**, *71*, 403–433. [CrossRef]
7. Khan, N.; Bano, A. Effects of exogenously applied salicylic acid and putrescine alone and in combination with rhizobacteria on the phytoremediation of heavy metals and chickpea growth in sandy soil. *Int. J. Phytoremediat.* **2018**, *20*, 405–414. [CrossRef]
8. Foyer, C.H.; Lelandais, M.; Kunert, K.J. Photooxidative stress in plants. *Physiol. Plant.* **1994**, *92*, 696–717. [CrossRef]
9. Mittler, R. Oxidative stress, antioxidants and stress tolerance. *Trends Plant Sci.* **2002**, *7*, 405–410. [CrossRef]
10. Perez-Lopez, U.; Robredo, A.; Lacuesta, M.; Sgherri, C.; Munoz-Rueda, A.; Navari-Izzo, F.; Mena-Petite, A. The oxidative stress caused by salinity in two barley cultivars is mitigated by elevated CO_2. *Physiol. Plant.* **2009**, *135*, 29–42. [CrossRef]
11. Tanou, G.; Job, C.; Rajjou, L.; Arc, E.; Belghazi, M.; Diamantidis, G.; Molassiotis, A.; Job, D. Proteomics reveals the overlapping roles of hydrogen peroxide and nitric oxide in the acclimation of citrus plants to salinity. *Plant J.* **2009**, *60*, 795–804. [CrossRef]
12. Zheng, C.; Jiang, D.; Liu, F.; Dai, T.; Liu, W.; Jing, Q.; Cao, W. Exogenous nitric oxide improves seed germination in wheat against mitochondrial oxidative damage induced by high salinity. *Environ. Exp. Bot.* **2009**, *67*, 222–227. [CrossRef]
13. Zhu, J.-K. Regulation of ion homeostasis under salt stress. *Curr. Opin. Plant Biol.* **2003**, *6*, 441–445. [CrossRef]
14. Fan, H.-F.; Du, C.-X.; Guo, S.-R. Nitric oxide enhances salt tolerance in cucumber seedlings by regulating free polyamine content. *Environ. Exp. Bot.* **2013**, *86*, 52–59. [CrossRef]
15. Asada, K. The water-water cycle as alternative photon and electron sinks. *Philos. Trans. R. Soc. Lond. Ser. B Biol. Sci.* **2000**, *355*, 1419–1431. [CrossRef]
16. Willekens, H.; Chamnongpol, S.; Davey, M.; Schraudner, M.; Langebartels, C.; Van Montagu, M.; Inze, D.; Van Camp, W. Catalase is a sink for H_2O_2 and is indispensable for stress defence in C3 plants. *EMBO J.* **1997**, *16*, 4806–4816. [CrossRef]
17. Bowler, C.; Montagu, M.V.; Inze, D. Superoxide Dismutase and Stress Tolerance. *Annu. Rev. Plant Biol. Physiol. Plant Mol. Biol.* **1992**, *43*, 83–116. [CrossRef]
18. Schafer, F.Q.; Wang, H.P.; Kelley, E.E.; Cueno, K.L.; Martin, S.M.; Buettner, G.R. Comparing beta-carotene, vitamin E and nitric oxide as membrane antioxidants. *Biol. Chem.* **2002**, *383*, 671–681. [CrossRef]
19. Hernandez, J.A.; Jiménez, A.; Mullineaux, P.; Sevilia, F. Tolerance of pea (*Pisum sativum* L.) to long-term salt stress is associated with induction of antioxidant defences. *Plant Cell Environ.* **2001**, *23*, 853–862. [CrossRef]
20. Kukreja, S.; Nandwal, A.S.; Kumar, N.; Sharma, S.K.; Sharma, S.K.; Unvi, V.; Sharma, P.K. Plant water status, H_2O_2 scavenging enzymes, ethylene evolution and membrane integrity of Cicer arietinum roots as affected by salinity. *Biol. Plant.* **2005**, *49*, 305–308. [CrossRef]
21. Hernández, J.A.; Campillo, A.; Jiménez, A.; Alarcón, J.J.; Sevilla, F. Response of antioxidant systems and leaf water relations to NaCl stress in pea plants. *New Phytol.* **1999**, *141*, 241–251. [CrossRef]
22. Shigeoka, S.; Ishikawa, T.; Tamoi, M.; Miyagawa, Y.; Takeda, T.; Yabuta, Y.; Yoshimura, K. Regulation and function of ascorbate peroxidase isoenzymes. *J. Exp. Bot.* **2002**, *53*, 1305–1319. [CrossRef]

23. Noctor, G.; Foyer, C.H. Ascorbate and Glutathione: Keeping Active Oxygen under Control. *Annu. Rev. Plant Physiol. Plant Mol. Biol.* **1998**, *49*, 249–279. [CrossRef]
24. Asada, K. Ascorbate peroxidase—A hydrogen peroxide-scavenging enzyme in plants. *Physiol. Plant.* **1992**, *85*, 235–241. [CrossRef]
25. Hasanuzzaman, M.; Nahar, K.; Gill, S.S.; Alharby, H.F.; Razafindrabe, B.H.; Fujita, M. Hydrogen Peroxide Pretreatment Mitigates Cadmium-Induced Oxidative Stress in *Brassica napus* L.: An Intrinsic Study on Antioxidant Defense and Glyoxalase Systems. *Front. Plant Sci.* **2017**, *8*, 115. [CrossRef]
26. Astier, J.; Lindermayr, C. Nitric oxide-dependent posttranslational modification in plants: An update. *Int. J. Mol. Sci.* **2012**, *13*, 15193–15208. [CrossRef]
27. Jain, P.; Bhatla, S.C. Molecular mechanisms accompanying nitric oxide signalling through tyrosine nitration and S-nitrosylation of proteins in plants. *Funct. Plant Biol.* **2018**, *45*, 70–82. [CrossRef]
28. Kato, H.; Takemoto, D.; Kawakita, K. Proteomic analysis of S-nitrosylated proteins in potato plant. *Physiol. Plant.* **2013**, *148*, 371–386. [CrossRef]
29. Gross, F.; Durner, J.; Gaupels, F. Nitric oxide, antioxidants and prooxidants in plant defence responses. *Front. Plant Sci.* **2013**, *4*, 419. [CrossRef]
30. Begara-Morales, J.C.; Sánchez-Calvo, B.; Chaki, M.; Valderrama, R.; Mata-Pérez, C.; López-Jaramillo, J.; Padilla, M.N.; Carreras, A.; Corpas, F.J.; Barroso, J.B. Dual regulation of cytosolic ascorbate peroxidase (APX) by tyrosine nitration and S-nitrosylation. *J. Exp. Bot.* **2014**, *65*, 527–538. [CrossRef]
31. Begara-Morales, J.C.; Sánchez-Calvo, B.; Chaki, M.; Mata-Pérez, C.; Valderrama, R.; Padilla, M.N.; López-Jaramillo, J.; Luque, F.; Corpas, F.J.; Barroso, J.B. Differential molecular response of monodehydroascorbate reductase and glutathione reductase by nitration and *S*-nitrosylation. *J. Exp. Bot.* **2015**, *66*, 5983–5996. [CrossRef] [PubMed]
32. Yang, H.; Mu, J.; Chen, L.; Feng, J.; Hu, J.; Li, L.; Zhou, J.-M.; Zuo, J. *S*-nitrosylation positively regulates ascorbate peroxidase activity during plant stress responses. *Plant Physiol.* **2015**, *167*, 1604–1615. [CrossRef] [PubMed]
33. de Pinto, M.C.; Locato, V.; Sgobba, A.; Romero-Puertas, M.D.C.; Gadaleta, C.; Delledonne, M.; De Gara, L. *S*-nitrosylation of ascorbate peroxidase is part of programmed cell death signaling in tobacco Bright Yellow-2 cells. *Plant Physiol.* **2013**, *163*, 1766–1775. [CrossRef] [PubMed]
34. Fares, A.; Rossignol, M.; Peltier, J.-B. Proteomics investigation of endogenous S-nitrosylation in Arabidopsis. *Biochem. Biophys. Res. Commun.* **2011**, *416*, 331–336. [CrossRef]
35. Ju, X.; Kou, C.; Christie, P.; Dou, Z.; Zhang, F. Changes in the soil environment from excessive application of fertilizers and manures to two contrasting intensive cropping systems on the North China Plain. *Environ. Pollut* **2007**, *145*, 497–506. [CrossRef]
36. Yang, X.; Wang, X.; Wei, M.; Hikosaka, S.; Goto, E. Response of Ammonia Assimilation in Cucumber Seedlings to Nitrate Stress. *J. Plant Biol.* **2010**, *53*, 173–179. [CrossRef]
37. Shi, Q.-H.; Zhu, Z.-J.; Khalid, A.-A.; Liu, H.-Y.; Yu, J.-Q. [Effects of iso-osmotic salt stress on the activities of antioxidative enzymes, H+-ATPase and H+-PPase in tomato plants]. *Zhi Wu Sheng Li Yu Fen Zi Sheng Wu Xue Xue Bao* **2004**, *30*, 311–316.
38. Hoekema, A.; Hirsch, P.R.; Hooykaas, P.J.J.; Schilperoort, R.A. A binary plant vector strategy based on separation of vir- and T-region of the Agrobacterium tumefaciens Ti-plasmid. *Nature* **1983**, *303*, 179–180. [CrossRef]
39. Horsch, R.B.; Fry, J.E.; Hoffmann, N.L.; Wallroth, M.; Eichholtz, D.; Rogers, S.G.; Fraley, R.T. A simple and general method for transferring genes into plants. *Science* **1985**, *227*, 1229–1231. [CrossRef]
40. Porebski, S.; Bailey, L.G.; Baum, B.R. Modification of a CTAB DNA extraction protocol for plants containing high polysaccharide and polyphenol components. *Plant Mol. Biol. Rep.* **1997**, *15*, 8–15. [CrossRef]
41. Mazel, A.; Leshem, Y.; Tiwari, B.S.; Levine, A. Induction of salt and osmotic stress tolerance by overexpression of an intracellular vesicle trafficking protein AtRab7 (AtRabG3e). *Plant Physiol.* **2004**, *134*, 118–128. [CrossRef] [PubMed]
42. Zhao, M.-G.; Chen, L.; Zhang, L.-L.; Zhang, W.-H. Nitric reductase-dependent nitric oxide production is involved in cold acclimation and freezing tolerance in Arabidopsis. *Plant Physiol.* **2009**, *151*, 755–767. [CrossRef]
43. Nakano, Y.; Asada, K. Hydrogen Peroxide is Scavenged by Ascorbate-specific Peroxidase in Spinach Chloroplasts. *Plant Cell Physiol.* **1981**, *22*, 867–880.
44. Miyake, C.; Asada, K. Thylakoid-Bound Ascorbate Peroxidase in Spinach Chloroplasts and Photoreduction of Its Primary Oxidation Product Monodehydroascorbate Radicals in Thylakoids. *Plant Cell Physiol.* **1992**, *33*, 541–553.
45. Foyer, C.H.; Halliwell, B. The presence of glutathione and glutathione reductase in chloroplasts: A proposed role in ascorbic acid metabolism. *Planta* **1976**, *133*, 21–25. [CrossRef]
46. Jiang, M.; Zhang, J. Effect of abscisic acid on active oxygen species, antioxidative defence system and oxidative damage in leaves of maize seedlings. *Plant Cell Physiol.* **2001**, *42*, 1265–1273. [CrossRef]
47. Bates, L.S.; Waldren, R.P.; Teare, I.D. Rapid determination of free proline for water-stress studies. *Plant Soil* **1973**, *39*, 205–207. [CrossRef]
48. Yemm, E.W.; Willis, A.J. The estimation of carbohydrates in plant extracts by anthrone. *Biochem. J.* **1954**, *57*, 508–514. [CrossRef]
49. Rao, K.V.M.; Sresty, T.V. Antioxidative parameters in the seedlings of pigeonpea (*Cajanus cajan* (L.) Millspaugh) in response to Zn and Ni stresses. *Plant Sci.* **2000**, *157*, 113–128.
50. Gay, C.A.; Gebicki, J.M. Measurement of protein and lipid hydroperoxides in biological systems by the ferric-xylenol orange method. *Anal. Biochem.* **2003**, *315*, 29–35. [CrossRef]

51. Bai, X.; Long, J.; He, X.; Yan, J.; Chen, X.; Tan, Y.; Li, K.; Chen, L.; Xu, H. Overexpression of spinach non-symbiotic hemoglobin in *Arabidopsis* resulted in decreased NO content and lowered nitrate and other abiotic stresses tolerance. *Sci. Rep.* **2016**, *6*, 26400. [CrossRef] [PubMed]
52. Jaffrey, S.R.; Snyder, S.H. The biotin switch method for the detection of *S*-nitrosylated proteins. *Sci. STKE* **2001**, *2001*, pl1. [CrossRef] [PubMed]
53. Jiménez, A.; Hernández, J.A.; Pastori, G.; del Río, L.A.; Sevilla, F. Role of the ascorbate-glutathione cycle of mitochondria and peroxisomes in the senescence of pea leaves. *Plant Physiol.* **1998**, *118*, 1327–1335. [CrossRef] [PubMed]
54. Gomez, J.M.; Jiménez, A.; Olmos, E.; Sevilla, F. Location and effects of long-term NaCl stress on superoxide dismutase and ascorbate peroxidase isoenzymes of pea (*Pisum sativum* cv. Puget) chloroplasts. *J. Exp. Bot.* **2004**, *55*, 119–130. [CrossRef] [PubMed]
55. Palma, J.M.; Jiménez, A.; Sandalio, L.M.; Corpas, F.J.; Lundqvist, M.; Gómez, M.; Sevilla, F.; del Río, L.A. Antioxidative enzymes from chloroplasts, mitochondria, and peroxisomes during leaf senescence of nodulated pea plants. *J. Exp. Bot.* **2006**, *57*, 1747–1758. [CrossRef] [PubMed]
56. Leterrier, M.; Del Río, L.A.; Corpas, F.J. Cytosolic NADP-isocitrate dehydrogenase of pea plants: Genomic clone characterization and functional analysis under abiotic stress conditions. *Free Radic. Res.* **2007**, *41*, 191–199. [CrossRef]
57. Willadino, L.; de Oliveira Filho, R.A.; da Silva Junior, E.A.; Neto, A.G.; Camara, T.R. Estresse salino em duas variedades de cana-de-açúcar: Enzimas do sistema antioxidativo e fluorescência da clorofila. *Rev. Ciência Agronômica* **2011**, *42*, 417–422. [CrossRef]
58. Hasanuzzaman, M.; Hossain, M.A.; Fujita, M. Nitric oxide modulates antioxidant defense and the methylglyoxal detoxification system and reduces salinity-induced damage of wheat seedlings. *Plant Biotechnol. Rep.* **2011**, *5*, 353–365. [CrossRef]
59. Li, Y.-J.; Hai, R.-L.; Du, X.-H.; Jiang, X.-N.; Lu, H. Over-expression of a Populus peroxisomal ascorbate peroxidase (PpAPX) gene in tobacco plants enhances stress tolerance. *Plant Breed.* **2009**, *128*, 404–410. [CrossRef]
60. Wu, G.; Wang, G.; Ji, J.; Gao, H.; Guan, W.; Wu, J.; Guan, C.; Wang, Y. Cloning of a cytosolic ascorbate peroxidase gene from *Lycium chinense* Mill. and enhanced salt tolerance by overexpressing in tobacco. *Gene* **2014**, *543*, 85–92. [CrossRef]
61. Chin, D.-C.; Kumar, R.S.; Suen, C.-S.; Chien, C.-Y.; Hwang, M.-J.; Hsu, C.-H.; Xuhan, X.; Lai, Z.X.; Yeh, K.-W. Plant Cytosolic Ascorbate Peroxidase with Dual Catalytic Activity Modulates Abiotic Stress Tolerances. *iScience* **2019**, *16*, 31–49. [CrossRef]
62. Wang, Y.; Li, J.; Wang, J.; Li, Z. Exogenous H_2O_2 improves the chilling tolerance of manilagrass and mascarenegrass by activating the antioxidative system. *Plant Growth Regul.* **2010**, *61*, 195–204. [CrossRef]
63. He, F.; Niu, M.-X.; Feng, C.-H.; Li, H.-G.; Su, Y.; Su, W.-L.; Pang, H.; Yang, Y.; Yu, X.; Wang, H.-L.; et al. PeSTZ1 confers salt stress tolerance by scavenging the accumulation of ROS through regulating the expression of PeZAT12 and PeAPX2 in Populus. *Tree Physiol.* **2020**, *40*, 1292–1311. [CrossRef] [PubMed]
64. Yabuta, Y.; Motoki, T.; Yoshimura, K.; Takeda, T.; Ishikawa, T.; Shigeoka, S. Thylakoid membrane-bound ascorbate peroxidase is a limiting factor of antioxidative systems under photo-oxidative stress. *Plant J.* **2002**, *32*, 915–925. [CrossRef]
65. Xu, S.; Li, J.; Zhang, X.; Wei, H.; Cui, L. Effects of heat acclimation pretreatment on changes of membrane lipid peroxidation, antioxidant metabolites, and ultrastructure of chloroplasts in two cool-season turfgrass species under heat stress. *Environ. Exp. Bot.* **2006**, *56*, 274–285. [CrossRef]
66. Faize, M.; Burgos, L.; Faize, L.; Piqueras, A.; Nicolas, E.; Barba-Espin, G.; Clemente-Moreno, M.J.; Alcobendas, R.; Artlip, T.; Hernandez, J.A. Involvement of cytosolic ascorbate peroxidase and Cu/Zn-superoxide dismutase for improved tolerance against drought stress. *J. Exp. Bot.* **2011**, *62*, 2599–2613. [CrossRef] [PubMed]
67. Patade, V.Y.; Khatri, D.; Kumari, M.; Grover, A.; Gupta, S.M.; Ahmed, Z. Cold tolerance in Osmotin transgenic tomato (*Solanum lycopersicum* L.) is associated with modulation in transcript abundance of stress responsive genes. *Springerplus* **2013**, *2*, 117. [CrossRef] [PubMed]
68. Goyary, D. Transgenic Crops, and their Scope for Abiotic Stress Environment of High Altitude: Biochemical and Physiological Perspectives. *DRDO Sci. Spectrum* **2009**, 195–201.
69. Goel, D.; Singh, A.K.; Yadav, V.; Babbar, S.B.; Bansal, K.C. Overexpression of osmotin gene confers tolerance to salt and drought stresses in transgenic tomato (*Solanum lycopersicum* L.). *Protoplasma* **2010**, *245*, 133–141. [CrossRef]
70. Szekely, G.; Abraham, E.; Cseplo, A.; Rigo, G.; Zsigmond, L.; Csiszar, J.; Ayaydin, F.; Strizhov, N.; Jasik, J.; Schmelzer, E.; et al. Duplicated P5CS genes of Arabidopsis play distinct roles in stress regulation and developmental control of proline biosynthesis. *Plant J.* **2008**, *53*, 11–28. [CrossRef]
71. Funck, D.; Baumgarten, L.; Stift, M.; von Wirén, N.; Schönemann, L. Differential Contribution of P5CS Isoforms to Stress Tolerance in Arabidopsis. *Front. Plant Sci.* **2020**, *11*, 565134. [CrossRef]
72. Nguyen, Q.H.; Vu, L.T.K.; Nguyen, L.T.N.; Pham, N.T.T.; Nguyen, Y.T.H.; Le, S.V.; Chu, M.H. Overexpression of the GmDREB6 gene enhances proline accumulation and salt tolerance in genetically modified soybean plants. *Sci. Rep.* **2019**, *9*, 19663. [CrossRef]
73. Agarwal, P.K.; Agarwal, P.; Reddy, M.K.; Sopory, S.K. Role of DREB transcription factors in abiotic and biotic stress tolerance in plants. *Plant Cell Rep.* **2006**, *25*, 1263–1274. [CrossRef]
74. Tu, T.Q.; Vaciaxa, P.; Lo, T.T.M.; Nguyen, N.H.; Pham, N.T.T.; Nguyen, Q.H.; Do, P.T.; Nguyen, L.T.N.; Nguyen, Y.T.H.; Chu, M.H. GmDREB6, a soybean transcription factor, notably affects the transcription of the NtP5CS and NtCLC genes in transgenic tobacco under salt stress conditions. *Saudi J. Biol. Sci.* **2021**, *28*, 7175–7181. [CrossRef]
75. Banerjee, A.; Roychoudhury, A. Group II late embryogenesis abundant (LEA) proteins: Structural and functional aspects in plant abiotic stress. *Plant Growth Regul.* **2015**, *79*, 1–17. [CrossRef]

76. Bai, X.; Yang, L.; Tian, M.; Chen, J.; Shi, J.; Yang, Y.; Hu, X. Nitric oxide enhances desiccation tolerance of recalcitrant *Antiaris toxicaria* seeds via protein S-nitrosylation and carbonylation. *PLoS ONE* **2011**, *6*, e20714. [CrossRef]
77. Correa-Aragunde, N.; Foresi, N.; Delledonne, M.; Lamattina, L. Auxin induces redox regulation of ascorbate peroxidase 1 activity by S-nitrosylation/denitrosylation balance resulting in changes of root growth pattern in Arabidopsis. *J. Exp. Bot.* **2013**, *64*, 3339–3349. [CrossRef]

agronomy

MDPI

Article

The γ-Aminobutyric Acid (GABA) Synthesis Gene Regulates the Resistance to Water Core-Induced Hypoxia Stress for Pear Fruits

Xiao Liu, Hao Ma, Jing Liu, Donghe Liu and Chunlei Wang *

College of Horticulture and Landscape Architecture, Yangzhou University, Yangzhou 225009, China
* Correspondence: wangcl@yzu.edu.cn

Abstract: Watercore is a physiological disorder which often occurs in *Rosaceae* fruits, and it causes hypoxia stress, promoting fruit decay. γ-aminobutyric acid (GABA) was reported as being involved in different abiotic stresses, and glutamate decarboxylase (GAD) is the key enzyme of GABA synthesis in plants. Our previous transcriptome analysis found that *PpGAD2* was significantly induced in watercore fruit; however, the mechanism through which *PpGAD2* regulates watercore-induced hypoxia stress resistance in pears is unclear. The present study found that the fruit pulp ethanol, malondialdehyde (MDA) and H_2O_2 content was significantly inhibited by exogenous GABA. The transcript abundance of *PpGAD2* was significantly higher than that of other *PpGADs* in watercore fruit or healthy fruit. Tissue expression showed that the content of *PpGAD2* in mature fruit was higher than in young fruit. Moreover, subcellular localization showed that *PpGAD2* was located in the cytoplasm. Transient overexpression assays suggested that *PpGAD2* had a role in GABA synthesis. Several CML (calmodulin–like) genes were also significantly increased in watercore fruit. Moreover, *PpWRKY53* was significantly induced in watercore fruit, and the GUS activity assay showed that *PpWRKY53* can significantly increase the activity of the *PpGAD2* promoter. Taken together, these results demonstrate that *PpGAD2* played an important role in GABA synthesis to increase plants' resistance to hypoxia stress, and its activity may be affected by *PpWRKY53* and several watercore-induced CML genes.

Keywords: pear; watercore; γ-aminobutyric; transcriptome; glutamate decarboxylase

Citation: Liu, X.; Ma, H.; Liu, J.; Liu, D.; Wang, C. The γ-Aminobutyric Acid (GABA) Synthesis Gene Regulates the Resistance to Water Core-Induced Hypoxia Stress for Pear Fruits. *Agronomy* **2023**, *13*, 1062. https://doi.org/10.3390/agronomy13041062

Academic Editor: Spyridon A. Petropoulos

Received: 17 February 2023
Revised: 2 April 2023
Accepted: 3 April 2023
Published: 6 April 2023

1. Introduction

Watercore is a very serious physiological disorder that frequently occurs in *Rosaceae* fruits, such as pear, apple and peach [1]. The symptoms of watercore are translucent fruit flesh and watery tissue. Our previous study suggested that watery tissue increased anaerobic respiration and caused the fruit to suffer from hypoxia stress [2]. Under hypoxia stress, plants experience a derangement of the cellular energy metabolism, causing a decrease in the pH in the cytoplasm, the inhibition of the ion transport pathway or nutrient acquisition, and the accumulation of toxic products from anaerobic respiration and reactive oxygen species (ROS) [3]. Pear watercore is most common in Japanese and Korean pear varieties with high internal quality. In recent years, Japanese and Korean sand pear varieties have been widely cultivated in Asia, and with the upgrading of cultivation technology, the proportion of high-quality fruit has increased significantly, leading to the increasing occurrence of sand pear watercore in production. However, watercore during pear production has not received enough attention, and it has caused great economic losses. Studying and understanding the hypoxia stress resistance mechanism in watercore fruit is urgent.

γ-aminobutyric acid (GABA) is a four-carbon non–protein amino acid. GABA has received a lot of research in the human and medical fields. For instance, GABA is an important neuroactive inhibitor in the nervous system, with effects such as reducing blood pressure, treating insomnia, alleviating anxiety, enhancing immunity and increasing memory [4]. In plants, many studies have shown that hypoxia can restore membrane potential and prevent an ROS-induced ion homeostasis imbalance; improve the defense ability of reactive oxygen species; remove chloroplast hydroxyl radicals and stabilize and protect chloroplast thylakoids; consume protons in cells to regulate the pH of the cytoplasm and alleviate the acidosis caused by hypoxia stress; and improve carbon recovery in the TCA cycle and promote the production of ATP [5]. Salvatierra et al. found that exogenous GABA application transiently increased the hypoxia tolerance of a sensitive genotype *Prunus* rootstock root [6]. GABA also regulates phenolic compound accumulation and enhances the antioxidant system in germinated hull-less barley under NaCl stress [7]. In plants, the GABA shunt pathway plays a key role in many regulatory mechanisms under stress conditions, such as endogenous signaling response molecules or metabolite accumulation [8]. The GABA shunt pathway is widely distributed in eukaryotes and prokaryotes and is an important metabolic pathway for GABA. In the cytoplasm, glutamate is first catalyzed by GAD for the irreversible production of GABA, GABA is subsequently converted into succinate acid by GABA transaminase (GABA-T). Succinate acid enters into the mitochondria and is catalyzed by means of succinate hemialdehyde dehydrogenase (SSADH) oxidation. Subsequently, succinate enters into the TCA cycle. Polyamine degradation is also one of the pathways used to generate GABA [9]. Compared with the GABA shunt pathway, polyamine degradation produced less GABA content. To date, GAD has been reported in many crops that respond to various stresses. For instance, the cotton *GhGAD6* responds to cadmium stress and increases the GABA content to relieve cadmium stress-induced oxidative damage [10]. Moreover, the activation of glutamate decarboxylase (GAD) requires the involvement of Ca^{2+}. Among many signal transduction pathways, Ca^{2+} is a multifunctional second messenger that regulates and activates many downstream responses of plants to various stresses. When plants are subjected to abiotic stress, Ca^{2+} in the cytoplasm will accumulate rapidly, resulting in the formation of a concentration difference of Ca^{2+} inside and outside the cell, which then generates a Ca^{2+} signal [11]. Different external stress stimuli induce plants to produce corresponding specific Ca^{2+} signals. Therefore, different calcium signal receptor proteins are required to recognize, decode and transmit the signals to the downstream, so that the downstream effector factors respond to the sensed specific Ca^{2+} signals, and then cause physiological and biochemical changes in plants to adapt to external adversity [11]. At present, Ca^{2+} receptor proteins in plants are mainly composed of three families: calmodulin (CaM)/calmodulin-like protein (CML), Ca^{2+}-dependent protein kinase (CDPK), and calmodulin B-like protein (CBL). CaM is ubiquitous in eukaryotes, while CML, CDPK and CBL are unique to plants and some protozoa. CaM itself has no catalytic activity; however, after binding with Ca^{2+}, it can interact with a downstream CaM-binding protein (CaMBP) to activate the function of these interacting proteins, and GAD is the CaMBP [12]. Our previous study reported that in watercore fruit, several GAD genes were significantly induced to up-regulated [2]. These *PpGADs* may play a key role in regulating fruit to resist hypoxia stress.

Many stress-associated transcription factors, such as WRKY, NAC, ERF, etc., were found to vary in expression level under hypoxia stress. For instance, transcription factors including MYB, MYB-related, bZIP, bHLH and WRKYs employ calcium signaling and sugar metabolism pathways to induce resistance to hypoxia stress in cucumber seedlings [13]. Tang et al. observed that the interplay between ERF members and the two WRKYs increased the adaptation to hypoxia stress in *Arabidopsis* induced by submergence [14]. In persimmon fruit, *DkNAC7* regulates de-astringency by activating *DkERF9* and *DkPDC2*, encoding pyruvate decarboxylase under hypoxia conditions [15]. In addition, many transcription factors belonging to IAA, WRKY, HB, and ZIPs demonstrated a higher expression in low-oxygen-concentration apples [16]. In the current study, we found that the *PpGAD2* promoter contained the W-box cis-element and that a WRKY gene was significantly upregulated in watercore fruit. Whether this WRKY gene can regulate the expression of *PpGAD2* needs to be further studied. 'Akibae' pear is a good-fruit-quality sand pear cultivar; however, it is highly susceptible to watercore. Here, we tried to dissect the molecular mechanism of a key GAD gene, *PpGAD2*, in regulating the resistance to watercore-induced hypoxia stress by participating in GABA synthesis, which might help us to rich the preventive measures to avoid watercore in pears.

2. Materials and Methods

2.1. Plant Materials and Treatment

The three cultivars ('Akibae', 'Aikansui' and 'Housui') were grafted onto *Pyrus calleryana*. The watercore fruits of 'Akibae' were collected about 125 days after flowering (DAF). 'Akibae' fruits treated with GABA at a concentration of 5 mM 100 days after flowering (DAF) were sampled two weeks later. Young leaves (third to fourth leaf from the top of the plant), mature leaves (second to third leaf from the bottom of the shoot), young fruits (60 DAF) and mature fruits (105 DAF) were collected for tissue gene expression. In each fruit sample, five healthy and uniform in appearance fruits were randomly collected from the outer crown of the tree with three replicates. All samples were immediately sampled and frozen in liquid nitrogen and stored at −80 °C for further analysis.

2.2. Measurement of Ethanol, H_2O_2, MDA, GABA Content and GAD Activity

The fruits ethanol content was measured using a test kit from Suzhou Comin Biotechnology. Ethanol was oxidized and dehydrogenated into acetaldehyde under the catalysis of ethanol dehydrogenase. At the same time, NAD was reduced to produce NADH, which caused WST-8 to turn orange under the action of 1-mPMS. We weighed out about 0.5 g of tissue, added 1 mL of distilled water for homogenization, centrifuged at 8000× *g* and 25 °C for 10 min, and removed the supernatant for testing. Then, we added the reaction solution in sequence. The content of ethanol was measured by measuring the change in the absorbance value at 450 nm.

The fruits H_2O_2 content was measured using a test kit from Suzhou Comin Biotechnology. H_2O_2 and titanium sulfate formed a yellow titanium peroxide complex with characteristic absorption at 415 nm. We weighed out about 0.5 g of tissue, added 1 mL of distilled water for homogenization, centrifuged at 8000× *g* and 25 °C for 10 min, and removed the supernatant for testing. Then, we added the reaction solution according to the order of the instructions. We allowed it to stand at room temperature for 5 min, poured it into a cuvette, and measured the absorbance value at 415 nm.

Fruits GAD activity was determined using a test kit from Shanghai Enzyme-linked Biotechnology. The purified plant GAD antibody was used to coat the microporous plate to prepare the solid-phase antibody. Firstly, we diluted the standard solution in the kit and created a standard curve. Then, according to the instructions, we performed the processes of temperature incubation, solution preparation, washing, and enzyme color development, the absorbance was measured with an enzyme marker at the wavelength of 450 nm, and the activity concentration of plant GAD in the sample was calculated through the standard curve.

The fruits' GABA content was determined using a test kit from Shanghai Enzyme-linked Biotechnology. GABA reacted with hypochlorite and phenol in the alkaline solution to produce a blue-green substance. We weighed out about 0.1 g of the tissue sample and add it into the mortar, added 1 mL of the extract, homogenized it on ice, centrifuged it at 12,000 rpm and 4 °C or room temperature for 10 min, and removed the supernatant for testing. Then, we added the reaction solution in sequence, evenly mixed it, placed it in a boiling water bath (95–100 °C) for 10 min, and then placed it in an ice bath which was increased to room temperature to show the blue-green color. Then, adding 200 μL to a 96-well plate, the content of GABA in the sample was obtained by detecting the value of the colored substance at a 645 nm wavelength.

Fruits MDA was measured according to Li [17]. The content of malondialdehyde was determined by means of the thiobarbituric acid method. MDA can react with sulfur under acidic and high temperature conditions. The reaction of barbituric acid (TBA) produces reddish-brown trimethylene, which has its maximum light absorption at 532 nm and its minimum light absorption at 600 nm.

2.3. RNA Extraction and Sequencing

The detailed RNA-seq analysis process for watercore fruit and healthy fruit can be found in our previous work [18]. The high quality clean data was mapped to the reference genome of *Pyrus bretschneideri*. All RNA-seq data were uploaded to the National Center for Biotechnology Information (NCBI) Gene Expression Omnibus (GEO) database (GEO accession number: GSE164987). Differential gene expression levels were computed using the fragments per kilobase of exons per million mapped reads (RPKM) method. We used the DESeq2 package to analyze the differential gene expression. The differentially expressed genes (DEGs) were selected using the following criteria: $|\log_2{}^{\text{foldchange}}| \geq 1$ and corrected $p < 0.05$. Gene Ontology (GO) enrichment analysis and Kyoto Encyclopedia of Genes and Genomes (KEGG) enrichment analysis were performed using the cluster Profiler package. GO categories and KEGG pathways with false discovery rate (FDR) q values ≤ 0.05 were considered significantly enriched.

2.4. Quantitative Real-Time PCR (qRT-PCR) Analysis

Total RNA of pear fruit skin was extracted using the CTAB method [19]. First-strand cDNA was synthesized using the qPCR HiScript3 RT superMix kit from Vazyme Biotechnology, Nanjing, China. qRT-PCR was determined with a Bio-Rad CFX96 instrument (Bio-Rad, Waltham, MA, USA) by using ChamQ SYBR qPCR Master Mix (Vazyme Biotechnolog, Nanjing, China). Gene-specific primers were designed using the Primer5 (v5.0) software (Table S1), and the specificity and quality of each primer pair were checked through melting curve analysis and sequencing. We used the Livak [20] method to calculate relative gene expression levels.

2.5. Subcellular Localization of PpGAD2

The full CDS of *PpGAD2* was ligated into the pCAMBIA1301-GFP vector. If sequence verification demonstrated it was accurate, it was then used for tobacco leaf transient transformation by means of *Agrobacterium*-mediated transformation according to our previously described methods [21]. Leaves transformed with a noncoding sequence vector were used as the control (CK).

2.6. Transient Transformation of Pear Fruits

To establish *PpGAD2*'s role in GABA synthesis, the *PpGAD2*-overexpressing vectors were mobilized into the *A. tumefaciens* strain *EHA105* and then used to transiently transform three kinds of pear fruits ('Akibae', 'Aikansui' and 'Housui'). Fruits transformed with a noncoding sequence vector were used as the control (CK). Transformed fruits were collected 5 days later for further analysis.

2.7. GUS Activity Assay

The *PpGAD2* promoter was ligated into the pCAMBIA1301-GUS vector. The CDS of *PpWRKY53* was cloned into the binary pCAMBIA1301 vector. The fusion constructs and the positive control were separately transferred into the *A. tumefaciens* strain EHA105 by means of heat shock. *A. tumufaciens*-mediated transformation was used for transient GUS expression in tobacco leaves. The leaves' GUS activity was tested using the test kit (Shanghai Enzyme-linked Biotechnology).

2.8. Statistical Analysis

Error bars indicate the standard errors (SEs) of the means. The data in the experiment were compared by means of one-way analysis of variance (SPSS 17.0). Differences were determined using the Tukey test ($p < 0.05$) and indicated using different letters. Bar graphs were drawn using the scientific software of GraphPad Prism 7.0 (San Diego, CA, USA).

3. Results

3.1. Effect of Exogenous GABA Treatment on Pear Fruit

To investigate the effect of exogenous GABA treatment on 'Akibae' pear fruit, 5 mM GABA was administered to the fruits. After the treatment with exogenous GABA, the ethanol, MDA and H_2O_2 content was significantly decreased compared with watercore fruit (Figure 1). Specifically, the ethanol, MDA and H_2O_2 content decreased by 40.38%, 68.99% and 41.96%, respectively.

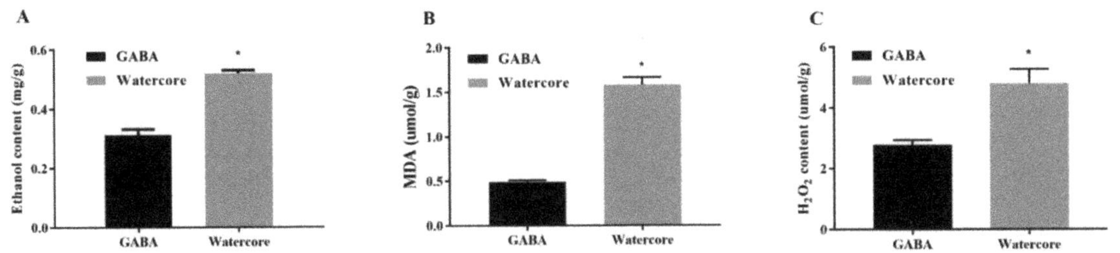

Figure 1. Analysis the content of ethanol (**A**), MDA (**B**) and H_2O_2 (**C**) in the watercore and GABA treatment of 'Akibae' fruits. The asterisk on the bars represent standard errors from three independent replicates. * $p < 0.05$.

3.2. Expression Pattern of PpGADs in Watercore Fruit

In our previous research, the expression of three *PpGADs* was significantly increased, which correlated with the GABA content in watercore fruit [2]. To further evaluate the transcript abundance of *PpGADs*, we used *PpGAD1* in healthy fruit as a control (CK) and compared the *PpGADs* expression level again. The result show that the transcript abundance of *PpGAD2* was significantly higher than that of *PpGAD1* and *PpGAD3* both in healthy fruit and in watercore fruit, and so *PpGAD2* was selected for this study (Figure 2).

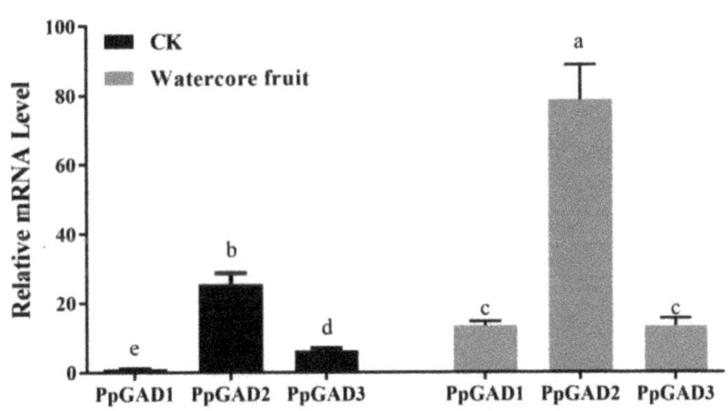

Figure 2. Analysis of *PpGADs'* expression in the 'Akibae' watercore fruits. CK refers to healthy fruit. The lower case letter on the bars represent standard errors from three independent replicates.

3.3. Tissue-Specific Expression of the PpGAD2

Different tissues from the 'Akibae' pear, including young leaves, mature leaves, mature fruits and young fruits, were used to determine the expression levels of *PpGAD2*. The results show that the *PpGAD2* gene is highly expressed in leaves, and it is predominantly expressed in young leave. Moreover, *PpGAD2* expression in mature fruits was higher than that in young fruits (Figure 3).

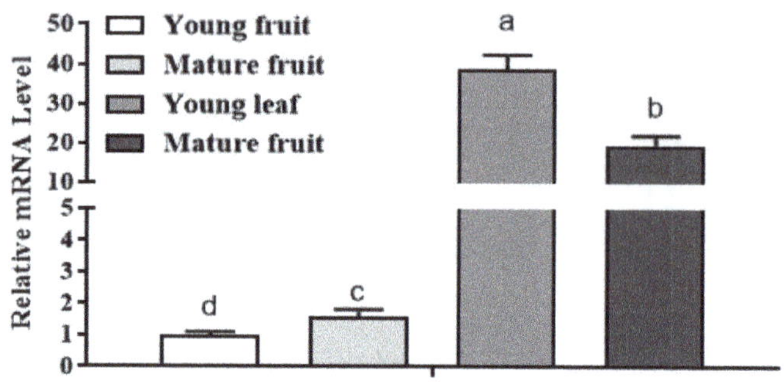

Figure 3. Analysis of *PpGAD2* expression in the different tissues of 'Akibae'. Young leaves (third to fourth leaf from the top of leaf), mature leaves (second to third leaf from the bottom of shoot), young fruit (60 DAF) and mature fruit (105 DAF). The lower case letter on the bars represent standard errors from three independent replicates.

3.4. *PpGAD2* Was Localized in Cytoplasm

To further validate the potential location of *PpGAD2*, the subcellular *PpGAD2* was investigated in tobacco leaves. After observation with confocal microscopy, GFP signals showed that *PpGAD2* was located in the cytoplasmic area (Figure 4).

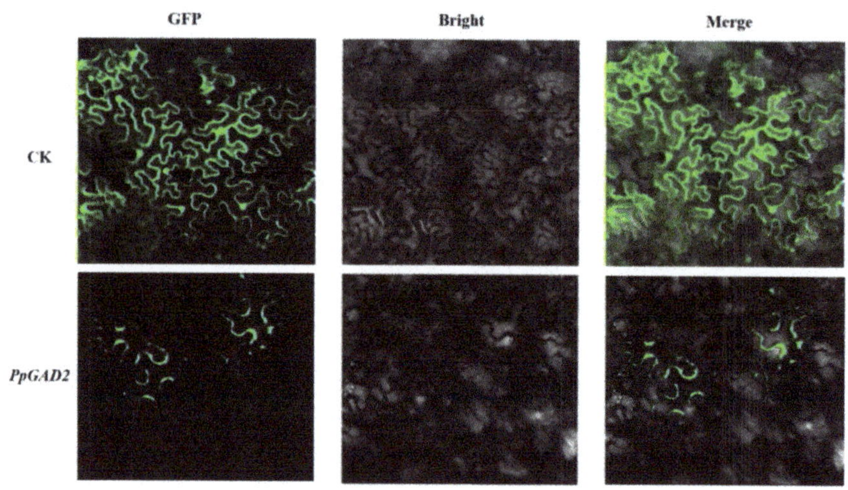

Figure 4. Subcellular localization of *PpGAD2* in tobacco leaves. The images from left to right are green fluorescent signals from GFP, bright fields, and merged, respectively. Leaves transformed with a noncoding sequence vector were used as the control (CK).

3.5. *Transient Overexpression of PpGAD2 in Pear Fruits*

To further validate the roles of *PpGAD2* in GABA synthesis, *PpGAD2* overexpression constructs were agroinfiltrated into pear fruits. Besides to 'Akibae', 'Hosui' and 'Aikansui' were also used to further verify gene function. The transcript level of *PpGAD2* was significantly increased in the three pear cultivars. After the transient transformation of *PpGAD2* in the three cultivars of pear fruits, they also had a significantly increased GABA content and GAD activity (Figure 5).

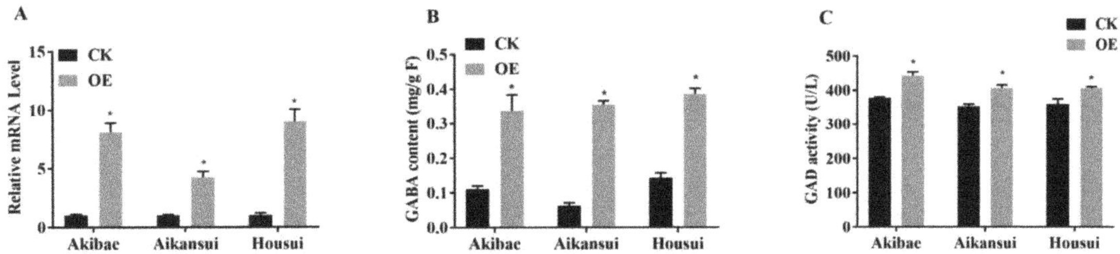

Figure 5. Transient overexpression of *PpGAD2* in three pear cultivars ('Aikansui', 'Akibae' and 'Hosui'). (**A–C**) refer to *PpGAD2* gene expression, GABA content and GAD activity, respectively. The asterisk on the bars represent standard errors from three independent replicates when compared to CK of each variety itself. * $p < 0.05$. CK and OE refer to vector only and *PpGAD2* over-expression, respectively.

3.6. Transcriptome Changes in Calmodulin-Related Genes in Watercore Fruit

As shown in Table 1, an analysis of DEGs found that four calmodulin-like genes were significantly upregulated in watercore fruit (LOC108865441, LOC103960085, LOC103939317 and LOC103953910) (Table 1).

Table 1. Expression levels of calmodulin-like genes in watercore fruit using RNA-Seq.

Gene Description	FC	ID
calmodulin-like protein 3	5.31	LOC108865441
calmodulin-like protein 11	5.03	LOC103960085
calmodulin-like protein 7-1	3.31	LOC103939317
calmodulin-like protein 7-2	1.40	LOC103953910

3.7. PpWRKY53 Regulate the PpGAD2 Promoter Activity

According to the RNA-Seq analysis of watercore fruit, a significantly up expressed WRKY gene (Fold change: 5.58) was identified as *PpWRKY53* (ID: LOC103943771). The regulation of the *PpGAD2* promoter using *PpWRKY53* was determined in tobacco after co-transformation. The results show that the *PpGAD2* exhibited increased GUS activity compared to CK, which indicated that *PpWRKY53* may activate GUS transcription, as driven by the *PpGAD2* promoter (Figure 6).

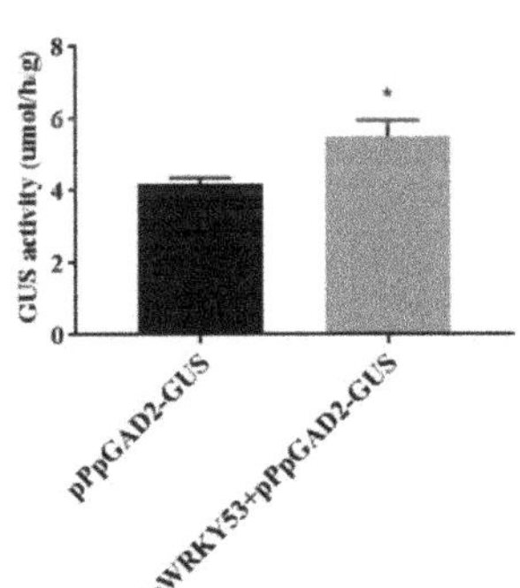

Figure 6. The change of *PpGAD2* promoter-gus activity combine with *PpWRKY53*. The asterisk on the bars represent standard errors from three independent replicates. * $p < 0.05$.

4. Discussion

Plant stress, often caused by different environmental conditions exceeding the tolerance limit, leads to lower plant reproduction rates due to increased ROS production [22].

hypoxia stress mainly occurs in natural environments when the root system is waterlogged or the fruit experiences physiological hypoxia. Due to the differences in the tissues of fleshy fruits, O_2 diffusion resistance will be caused, which creates an obvious oxygen gradient in the pulp tissues [23]. Long-term water soaking may aggravate the formation of an oxygen gradient and cause the insufficient exchange of CO_2 and O_2, which can cause fruit hypoxia. In production, GABA application is an effective approach that relieves multiple types of abiotic stress by inhibiting ROS generation [24]. In the current study, the content of ethanol, MDA and H_2O_2 was significantly inhibited by GABA (Figure 1). Thus, our study suggested that GABA can alleviate watercore-induced hypoxia stress.

GAD genes showed different expression patterns in various plant tissues and organs. In citrus, Liu et al. found that the transcript of *CsGAD1* was mainly expressed in flowers, while *CsGAD2* was mainly expressed in fruit juice sacs [25]. In soybean, the expression of *GmGAD4* and *GmGAD5* was detected in cotyledons, whereas *GmGAD1* and *GmGAD3* were mostly expressed in hypocotyls and roots [26]. In the current study, we found that *PpGAD2* was mainly expressed in the leaves compared with young and mature fruits. However, the expression of *PpGAD2* in mature fruits was significantly higher than in young fruits, which may suggest that *PpGAD2* plays a more important role in mature fruits than in young fruits. Sand pears are mainly cultivated in the region south of the Yangtze River in China and mature in July to September. The fruit growth and development season in this region is vulnerable to extreme conditions, such as high temperature, high humidity or drought. Therefore, the high expression of genes in the mature stage may also improve the resistance of fruit. In plants, GAD catalyzes the irreversible conversion of glutamate to GABA, and GAD exists in the cytosolic system. A previous study found the *SlGAD2* and *SGAD3* expression levels are positively correlated with the GABA content during tomato fruit development [27]. In *SlGAD* mutant plants, the fruit GABA content was significantly decreased, suggesting that GABA biosynthesis in tomato fruits involves the decarboxylation of glutamate by GAD enzymes [28]. Rajani et al. reported the overexpression of *AtGAD1* in maize, with the intent of increasing the synthesis of GABA [29]. In this study, the GABA content and GAD activity were significantly increased in the three sand pear cultivars when *PpGAD2* was overexpressed. Together with the subcellular localization analysis of *PpGAD2* (Figure 4), we suggest that *PpGAD2* plays a key role in regulating GABA synthesis. Moreover, CaMs/CaMs-like is an important Ca^{2+} sensor that functions in plant resistance, interacting with GADs. *Arabidopsis* GAD1, 2, and 4 contain CaM-binding domains, and all have the potential to interact with CaMs [30]. *AtCML8* interacts with *AtGAD4* and functions by regulating GABA accumulation in *Arabidopsis* defense induced by (Z)-3-hexenol [31]. In the current study, we found that several CML genes were significantly induced in watercore fruit, suggesting that they may interact with *PpGAD2* to respond to fruit hypoxia stress. The validation of the interaction mechanism of *PpGAD2* and CMLs requires further study in the future.

In recent years, the transcription regulation mechanisms of GADs have been reported. One previous study found that *OsMYB55* binds to the promoter regions of *OsGAD3*, and *OsMYB55* overexpression resulted in an accumulation of GABA and resistance to high temperatures in rice [32]. *FaMYB5* could repress the transcription levels of *FaGAD* and influences strawberry citric acid accumulation [33]. WRKY TFs contain a specific signature with a highly conserved WRKY domain and are involved in many biological functions, such as regulating plant growth or development, producing secondary metabolites and increasing tolerance to different abiotic stresses [34]. To date, there is no direct evidence to prove that WRKY genes directly regulate the expression of GAD. Mirabella et al. found that the GAD4 gene was up-regulated in wrky40 and wrky6 mutant *Arabidopsis*, induced by E-2-hexenal [35]. In the current study, we found that WRKY genes were significantly upregulated in watercore fruit, which was consistent with the expression of *PpGAD2*. The GUS activity assay also showed that they can regulate the expression of *PpGAD2*. Together with the results above, it is suggested that *PpWRKY53*'s response to hypoxia stress signals is regulated by *PpGAD2* expression, increasing GABA synthesis, which improves fruit resistance to ROS.

5. Conclusions

Watercore is a physiological disorder that leads to pear fruits suffering from hypoxia stress, seriously affecting the internal quality and storage capacity of pears. GABA is a non-protein amino acid that can improve plant resistance. Therefore, it is necessary to study the regulation of GABA on sand pear resistance to watercore-induced hypoxia stress. This study proved that GABA can improve the resistance of fruits to hypoxia stress. Gene function verification showed that *PpGAD2* is the key gene that catalyzes GABA synthesis. Its expression is directly regulated by the transcription factor *PpWRKY53*. Several *PpCML* genes were also significantly increased in watercore fruit which may interact with *PpGAD2* and thus improve GAD activity in pear; however, these need to be further studied in the future. Our study represents a valuable gene resource in providing a theoretical basis for watercore-preventative technologies and resistant breeding.

Supplementary Materials: The following supporting information can be downloaded at: https://www.mdpi.com/article/10.3390/agronomy13041062/s1, Table S1: List of primers.

Author Contributions: X.L. and C.W. designed the research. J.L., H.M. and D.L. performed the experiment and analyzed the data. X.L. wrote the manuscript. All authors have read and agreed to the published version of the manuscript.

Funding: This work was supported by the Postgraduate Research & Practice Innovation Program of Jiangsu Province (Yangzhou University) (No. SJCX21_1615) and Yangzhou City's Green and Golden Phoenix Program.

Institutional Review Board Statement: Not applicable.

Informed Consent Statement: Not applicable.

Data Availability Statement: Data are available by contacting the authors.

Conflicts of Interest: The authors declare no conflict of interest.

References

1. Rahimi, V.; Esmaielpour, B.; Heydari, H.R. *Abiotic Stress Biology in Horticultural Plants*; Springer: Berlin/Heidelberg, Germany, 2016.
2. Liu, X.; Liu, D.H.; Chen, T.; Zhang, J.; Wang, C.L. Watercore Pear Fruit Respiration Changed and Accumulated gamma-Aminobutyric Acid (GABA) in Response to Inner Hypoxia Stress. *Genes* **2022**, *13*, 977. [CrossRef] [PubMed]
3. Cukrov, D. Progress toward Understanding the Molecular Basis of Fruit Response to Hypoxia. *Plants* **2018**, *7*, 78. [CrossRef] [PubMed]
4. Hepsomali, P.; Groeger, J.A.; Nishihira, J.; Scholey, A. Effects of Oral Gamma-Aminobutyric Acid (GABA) Administration on Stress and Sleep in Humans: A Systematic Review. *Front. Neurosci.* **2020**, *14*, 923. [CrossRef] [PubMed]
5. Takayama, M.; Ezura, H. How and why does tomato accumulate a large amount of GABA in the fruit? *Front. Plant Sci.* **2015**, *6*, 612. [CrossRef]
6. Salvatierra, A.; Pimentel, P.; Almada, R.; Hinrichsen, P. Exogenous GABA application transiently improves the tolerance to root hypoxia on a sensitive genotype of Prunus rootstock. *Environ. Exp. Bot.* **2016**, *125*, 52–66. [CrossRef]
7. Ma, Y.; Wang, P.; Wang, M.; Sun, M.; Gu, Z.; Yang, R. GABA mediates phenolic compounds accumulation and the antioxidant system enhancement in germinated hulless barley under NaCl stress. *Food Chem.* **2019**, *270*, 593–601. [CrossRef] [PubMed]
8. Ansari, M.I.; Jalil, S.U.; Ansari, S.A.; Hasanuzzaman, M. GABA shunt: A key-player in mitigation of ROS during stress. *Plant Growth Regul.* **2021**, *94*, 131–149. [CrossRef]
9. Bouché, N.; Fromm, H. GABA in plants: Just a metabolite? *Trends Plant Sci.* **2004**, *9*, 110–115. [CrossRef]
10. Huang, H.; He, Y.; Cui, A.; Sun, L.; Han, M.; Wang, J.; Rui, C.; Lei, Y.; Liu, X.; Xu, N.; et al. Genome-wide identification of GAD family genes suggests *GhGAD6* functionally respond to Cd^{2+} stress in cotton. *Front. Genet.* **2022**, *13*, 965058. [CrossRef]
11. Kudla, J.; Becker, D.; Grill, E.; Hedrich, R.; Schumacher, K. Advances and current challenges in calcium signaling. *New Phytol.* **2018**, *218*, 414–431. [CrossRef]
12. Costa, A.; Luoni, L.; Marrano, C.A.; Hashimoto, K.; Koester, P.; Giacometti, S.; De Michelis, M.I.; Kudla, J.; Bonza, M.C. Ca^{2+}-dependent phosphoregulation of the plasma membrane Ca^{2+}-ATPase ACA8 modulates stimulus-induced calcium signatures. *J. Exp. Bot.* **2017**, *68*, 3215–3230. [CrossRef] [PubMed]

13. He, L.; Yan, J.; Ding, X.; Jin, H.; Zhang, H.; Cui, J.; Zhou, Q.; Yu, J. Integrated analysis of transcriptome and microRNAs associated with exogenous calcium-mediated enhancement of hypoxia tolerance in cucumber seedlings (*Cucumis sativus* L.). *Front. Plant Sci.* **2023**, *13*, 994268. [CrossRef] [PubMed]
14. Tang, H.; Bi, H.; Liu, B.; Lou, S.; Song, Y.; Tong, S.; Chen, N.; Jiang, Y.; Liu, J.; Liu, H. WRKY33 interacts with WRKY12 protein to up-regulate RAP2.2 during submergence induced hypoxia response in *Arabidopsis thaliana*. *New Phytol.* **2021**, *229*, 106–125. [CrossRef]
15. Jin, R.; Zhu, Q.G.; Shen, X.Y.; Wang, M.M.; Jamil, W.; Grierson, D.; Yin, X.R.; Chen, K.S. *DkNAC7*, a novel high-CO_2/hypoxia-induced NAC transcription factor, regulates persimmon fruit de-astringency. *PLoS ONE* **2018**, *13*, e0194326. [CrossRef]
16. Cukrov, D.; Zermiani, M.; Brizzolara, S.; Cestaro, A.; Licausi, F.; Luchinat, C.; Santucci, C.; Tenori, L.; Van Veen, H.; Zuccolo, A.; et al. Extreme hypoxia Conditions Induce Selective Molecular Responses and Metabolic Reset in Detached Apple Fruit. *Front. Plant Sci.* **2016**, *7*, 146. [CrossRef]
17. Li, H.-H. *Principles and Techniques of Plant Physiological Biochemical Experimental*; Higher Education Press: Beijing, China, 2000.
18. Liu, X.; Fan, H.M.; Liu, D.H.; Liu, J.; Shen, Y.; Zhang, J.; Wei, J.; Wang, C.L. Transcriptome and Metabolome Analyses Provide Insights into the Watercore Disorder on "Akibae" Pear Fruit. *Int. J. Mol. Sci.* **2021**, *22*, 4911. [CrossRef]
19. Zhang, J.; Zhang, Y.F.; Zhang, P.F.; Bian, Y.H.; Liu, Z.Y.; Zhang, C.; Liu, X.; Wang, C.L. An integrated metabolic and transcriptomic analysis reveals the mechanism through which fruit bagging alleviates exocarp semi-russeting in pear fruit. *Tree Physiol.* **2021**, *41*, 1306–1318. [CrossRef]
20. Livak, K.J.; Schmittgen, T.D. Analysis of relative gene expression data using real-time quantitative PCR and the 2(T)(-Delta Delta C) method. *Methods* **2001**, *25*, 402–408. [CrossRef] [PubMed]
21. Liu, X.; Shen, Y.; Liu, D.H.; Liu, J.; Zhang, J.; Wei, J.; Wang, C.L. A sorbitol transporter gene plays specific role in the occurrence of watercore by modulating the level of intercellular sorbitol in pear. *Plant Sci.* **2022**, *317*, 111179. [CrossRef] [PubMed]
22. Lee, S.; Park, C.-M. Regulation of reactive oxygen species generation under drought conditions in arabidopsis. *Plant Signal. Behav.* **2012**, *7*, 599–601. [CrossRef]
23. Mori, K.; Beauvoit, B.P.; Biais, B.; Chabane, M.; Allwood, J.W.; Deborde, C.; Maucourt, M.; Goodacre, R.; Cabasson, C.; Moing, A.; et al. Central Metabolism Is Tuned to the Availability of Oxygen in Developing Melon Fruit. *Front. Plant Sci.* **2019**, *10*, 594. [CrossRef] [PubMed]
24. Kaspal, M.; Kanapaddalagamage, M.H.; Ramesh, S.A. Emerging Roles of gamma Aminobutyric Acid (GABA) Gated Channels in Plant Stress Tolerance. *Plants* **2021**, *10*, 2178. [CrossRef]
25. Liu, X.; Hu, X.M.; Jin, L.F.; Shi, C.Y.; Liu, Y.Z.; Peng, S.A. Identification and transcript analysis of two glutamate decarboxylase genes, *CsGAD1* and *CsGAD2*, reveal the strong relationship between *CsGAD1* and citrate utilization in citrus fruit. *Mol. Biol. Rep.* **2014**, *41*, 6253–6262. [CrossRef]
26. Hyun, T.K.; Eom, S.H.; Jeun, Y.C.; Han, S.H.; Kim, J.S. Identification of glutamate decarboxylases as a gamma-aminobutyric acid (GABA) biosynthetic enzyme in soybean. *Ind. Crops Prod.* **2013**, *49*, 864–870. [CrossRef]
27. Akihiro, T.; Koike, S.; Tani, R.; Tominaga, T.; Watanabe, S.; Iijima, Y.; Aoki, K.; Shibata, D.; Ashihara, H.; Matsukura, C.; et al. Biochemical mechanism on GABA accumulation during fruit development in tomato. *Plant Cell Physiol.* **2008**, *49*, 1378–1389. [CrossRef] [PubMed]
28. Takayama, M.; Koike, S.; Kusano, M.; Matsukura, C.; Saito, K.; Ariizumi, T.; Ezura, H. Tomato Glutamate Decarboxylase Genes *SlGAD2* and *SlGAD3* Play Key Roles in Regulating gamma-Aminobutyric Acid Levels in Tomato (*Solanum lycopersicum*). *Plant Cell Physiol.* **2015**, *56*, 1533–1545. [CrossRef] [PubMed]
29. Rajani, M.S.; Bedair, M.F.; Li, H.; Duff, S.M.G. Phenotypic effects from the expression of a deregulated *AtGAD1* transgene and GABA pathway suppression mutants in maize. *PLoS ONE* **2021**, *16*, e0259365.
30. Shelp, B.J.; Bozzo, G.G.; Zarei, A.; Simpson, J.P.; Trobacher, C.P.; Allan, W.L. Strategies and tools for studying the metabolism and function of gamma-aminobutyrate in plants. II. Integrated analysis. *Botany* **2012**, *90*, 781–793. [CrossRef]
31. Jiao, C.; Guo, Z.; Gong, J.; Zuo, Y.; Li, S.; Vanegas, D.; McLamore, E.S.; Shen, Y. CML8 and GAD4 function in (Z)-3-hexenol-mediated defense by regulating gamma-aminobutyric acid accumulation in *Arabidopsis*. *Plant Physiol. Biochem.* **2022**, *186*, 135–144. [CrossRef]
32. El-kereamy, A.; Bi, Y.M.; Ranathunge, K.; Beatty, P.H.; Good, A.G.; Rothstein, S.J. The Rice R2R3-MYB Transcription Factor *OsMYB55* Is Involved in the Tolerance to High Temperature and Modulates Amino Acid Metabolism. *PLoS ONE* **2012**, *7*, e52030. [CrossRef]
33. Liu, Y.; Zhu, L.; Yang, M.; Xie, X.; Sun, P.; Fang, C.; Zhao, J. R2R3-MYB transcription factor *FaMYB5* is involved in citric acid metabolism in strawberry fruits. *J. Plant Physiol.* **2022**, *277*, 153789. [CrossRef] [PubMed]
34. Ramamoorthy, R.; Jiang, S.Y.; Kumar, N.; Venkatesh, P.N.; Ramachandran, S. A comprehensive transcriptional profiling of the WRKY gene family in rice under various abiotic and phytohormone treatments. *Plant Cell Physiol.* **2008**, *49*, 865–879. [CrossRef] [PubMed]
35. Mirabella, R.; Rauwerda, H.; Allmann, S.; Scala, A.; Spyropoulou, E.A.; de Vries, M.; Boersma, M.R.; Breit, T.M.; Haring, M.A.; Schuurink, R.C. WRKY40 and WRKY6 act downstream of the green leaf volatile E-2-hexenal in *Arabidopsis*. *Plant J.* **2015**, *83*, 1082–1096. [CrossRef] [PubMed]

agronomy

MDPI

Article

Combined Morphological and Palynological Classification for *Hibiscus syriacus* L. (Malvaceae): Construction of the Diagnostic Classification Framework and Implications of Pollen Morphological Variation on Fruiting

Fen Xiao [1,2], Xiaohong Wang [1,2,*], Yun Jiang [3], Chulin Chen [1], Jiajia Chen [1], Jingwen Zhang [1] and Yafeng Wen [1,2]

1 College of Landscape Architecture, Central South University of Forestry & Technology, Changsha 410004, China
2 Hunan Big Data Engineering Technology Research Center of Natural Protected Areas Landscape Resources, Changsha 410004, China
3 Department of Horticulture and Landscape, Shanghai Chenshan Botanical Garden, Shanghai 201602, China
* Correspondence: t19960222@csuft.edu.cn; Tel.: +86-156-1633-1155

Abstract: Identifying useful taxonomic indicators for classifying *Hibiscus syriacus* L. (Malvaceae) cultivars can help address challenges in their homonymy and synonymy. Moreover, analyzing which pollen traits possibly lead to their successful fruiting can serve to guide the hybridization and breeding of *H. syriacus*. For the first time, this study classified 24 cultivars of *H. syriacus* based on 24 morphological and palynological indicators assessed for flowers, leaves, and pollen grains. These indicators were a mixture of quantitative and qualitative traits, measured to contribute to the identification and classification of *H. syriacus* cultivars. The results showed that the 24 *H. syriacus* cultivars could be classified into 2–6 clusters according to different taxonomic criteria. The leading diagnostic indicators were eight quantitative and eight qualitative traits, of which two new quantitative traits—the width of the spine base (SW) and average of the pollen grain radius and spine length (D-spine)—and five new qualitative traits—the amount of pollen surface spines (O-SA), whether the petals have the red center (B-RC), whether the pollen surface ruffles strongly (B-RS), the degree of pollen surface ruffling (O-DR), and relationship between calyx and bract (O-CB)—could be used as defining traits for *H. syriacus* cultivars owing to their robust contribution to the classification. The correlations between indicators for flowers, leaves, and pollen grains were explored, which revealed that the O-SA in *H. syriacus* was strongly tied to quantitative pollen traits. Furthermore, three qualitative morphological traits—whether the stamens are heterogeneous in terms of inner petals (B-IP), O-CB, and whether the leaf lobing is strong (B-LL)—were correlated with partial quantitative pollen traits. We also found that those *H. syriacus* cultivars with micro-spines or granulate on the pollen grain surface have higher fruiting rates; additionally, pollen diameter, spine length, and spine spacing might also be potential factors influencing successful breeding. The insights gained from this study could fill a key knowledge gap concerning the taxonomic criteria suitable for distinguishing *H. syriacus* cultivars. Our findings also provide timely information on how to understand the pollination process, especially those aspects leading to pollinator selection via pollen grain features, which could influence breeding programs and outcomes.

Keywords: classification; *Hibiscus* L.; morphology and palynology; pollen morphology variation; quantitative and qualitative traits

Citation: Xiao, F.; Wang, X.; Jiang, Y.; Chen, C.; Chen, J.; Zhang, J.; Wen, Y. Combined Morphological and Palynological Classification for *Hibiscus syriacus* L. (Malvaceae): Construction of the Diagnostic Classification Framework and Implications of Pollen Morphological Variation on Fruiting. *Agronomy* **2023**, *13*, 828. https://doi.org/10.3390/agronomy13030828

Academic Editors: Guanglong Wang, Lijun Ou and Aisheng Xiong

Received: 17 February 2023
Revised: 8 March 2023
Accepted: 10 March 2023
Published: 11 March 2023

1. Introduction

Hibiscus syriacus L., a species of the genus *Hibiscus* of the Malvaceae family, is one of the traditional Chinese flowers and the national flower of South Korea [1,2]. It is noted for its diverse floral morphological characteristics, wide distribution, and adaptability [3].

Approximately 250 cultivars of *H. syriacus* are known worldwide [2]; however, suitable cultivars for horticultural landscapes are few and limited [4], and still beset with challenges of homonymy and synonymy. Therefore, a unified nomenclature and classification of *H. syriacus* cultivars is of particular importance to maximize the use of *H. syriacus* resources.

Only few studies have attempted to classify *H. syriacus* cultivars in terms of differing color systems or pollen grains [2,5]. The spiny pollen grains of the Malvaceae are beneficial for the classification of this group, mainly for *Hibiscus* species, as demonstrated for *H. pernambucensis* and *H. tiliaceus* by the study of Oliveira et al. [6]. Accordingly, for *H. syriacus* cultivars that are homochromatic and have similar phenotypic characteristics, palynological classification must be very useful. Sung et al. classified 22 *H. syriacus* cultivars into six clusters based on six pollen feature indices [5]. We hypothesized that relying solely on pollen traits for plant taxonomy is insufficient and not intuitive enough for a species whose pollen morphology is relatively uniform. Instead, it seems more sensible and acceptable to arrive at a quick classification directly gleaned from an examination of external morphology, such as flower and leaf traits; however, that might not always be compatible with palynological investigations. Taxonomic studies of this species in other aspects such as palynological or molecular classification remain surprisingly limited. In particular, a classification that integrates all aspects of morphology, such as the combination of flower, leaf, and pollen features, is not yet available. Such studies are necessary as they will allow a more intuitive and rapid classification of most *H. syriacus* cultivars and will be essential for marketability and industrial development.

Pollen morphology has a unique structure determined by the genes of each species, meaning it is reliable, quite stable, and generally unaffected by the external environment [7,8]. Accordingly, it is often used in plant taxonomic and palynological analyses for identifying and classifying genera, species, or cultivars of horticultural plants [9–11]. In general, the pollen grains with *H. syriacus* are apolar, spheroidal monads. Quantitative pollen traits were employed to analyze the morphology of *H. syriacus* pollen grains in studies by Sung et al. (pollen diameter, spine exine length, no. of spine exines, and distance between spine exines) and Zhao et al. (spine length, width of spine base, spine distance, pollen diameter, spine length/width of spine base, and pollen diameter/spine length), and both found significant differences in spine features among cultivars [3,5]. However, taking this approach alone is inadequate; other relevant numerical traits must also be investigated, such as the exact pollen shape and the pollen diameter in two directions in the two-dimensional plane (X-axis and Y-axis), which are collectively needed to inform and analyze the variation in pollen morphology among cultivars of *H. syriacus*. Quantitative morphological traits of pollen grains are widely used in plant numerical classification [11,12]. Pollen grains are also diverse in exine sculpturing. However, related qualitative traits, such as pollen ornamentation and perforation, are mostly found in textual descriptions of general pollen morphology and are rarely involved in the numerical taxonomy of pollen grains using coding methods.

The same is possible for phenotypic characteristics of *H. syriacus*. We have already performed a preliminary classification for *H. syriacus* based its quantitative and qualitative floral morphological traits [13]; however, the contribution of its qualitative indicators to the classification framework has not been evaluated yet. Accordingly, in this study, we aimed to detect classification criteria by considering diverse morphological traits of a combination of plant parts, namely flowers, leaves, and pollen grains, which could enhance the credibility of the classification and improve its efficiency. Moreover, in general, research on reliable qualitative indicators suitable for plant numerical classification is still limited. If they are quantified and coded, a better classification framework can be produced than traditional ones, which can facilitate the establishment of new taxonomic criteria, in turn [14]. In earlier work, Fen et al. demonstrated that qualitative traits contributed substantially to a robust classification within conifer pine *Cathaya argyrophylla* [14]. Hence, we sought to apply the same methodology to *H. syriacus* in the present study.

Pollen export might help to form specific floral characteristics that attract pollinators [15], thereby indirectly influencing plant fruiting and seed set rates. However, limited knowledge is available on the co-varying relationships between the pollen morphology, flower morphology, and the fruiting rate of *H. syriacus*. Insect pests might be among the drivers of diversity in seed set levels among *H. syriacus* cultivars. Additionally, the effects of pollen traits on the pollination mechanism or pollinator selection, or both, might be another reason for it, yet these have not been extensively studied. The available evidence suggests that the relationship between pollen and fertility is complex because the whole developmental mechanism is jointly influenced by many factors, such as pollen's grain size, spine length, and spacing, and even the pollinator species or their ability to carry certain amounts of pollen [16,17].

In this study, 24 morphological and palynological indicators were used together to classify 24 cultivars of *H. syriacus*, revealing parsimonious criteria for the first time. The variation in pollen morphology among cultivars was also assessed, as well as the contribution of qualitative traits to the classification. We detected possible links between pollen traits affecting the fruiting rate for *H. syriacus*. Altogether, these findings could fill a knowledge gap concerning the classification of *H. syriacus* cultivars and provide further insight into pollen features leading to the successful fertility enhancement of *H. syriacus*.

2. Materials and Methods

2.1. Plant Materials and Sources

A total of 24 *H. syriacus* cultivars were collected from Hunan, Henan, Zhejiang, and Shanghai, China, as well as from South Korea and the USA, and a resource garden was established at the Central South University of Forestry and Technology (Table 1). From selected healthy, clean, pest-free buds, and incompletely bloomed or blooming flowers, the petals were separated from the anthers, with the pollen grains obtained from the anthers using a laboratory blade. All these operations were carried out in a sterilized chamber and on sulfate paper. Five individual plants per cultivar were taken and mixed together as one composite sample. The collected pollen grains were dried under room conditions for 1 day, placed in centrifuge tubes, and immediately subjected to a scanning electron microscope, and remaining pollen materials were stored in a refrigerator at 4 °C.

Table 1. Information about the 24 *Hibiscus syriacus* cultivars.

No.	Cultivar	Source	No.	Cultivar	Source
1	Marina	US	13	Purple pillar	US
2	White chiffon	SH	14	Rubis	KR
3	Pink giant	SH	15	Chungmu	KR
4	China chiffon	SH	16	Suminokurahanagasa	KR
5	Paeoniflorus	HN	17	Pyonghwa	KR
6	Woodbridge	SH	18	Blue bird	US
7	Diana	SH	19	Akagionmamori	KR
8	Lavender	SH	20	Red heart	SH
9	Hamabo	SH	21	Qiancenghong	HEN
10	Elegantissimus	SH	22	Hongyun	HEN
11	Arang	HN	23	Huaban	HEN
12	Lavandula chiffon	SH	24	Naesarang	KR

Note: SH: Shanghai, China; HN: Hunan Province, China; HEN: Henan Province, China; KR: South Korea; US: the United States.

2.2. Scanning Electron Microscopy (SEM) and Coding the Pollen Qualitative Traits

The collected pollen grains were evenly spread on a sampling tray, then sprayed with metal (SCD 500) and positioned under a scanning electron microscope (JSM 6360-LV) operating at a voltage of 10.0 kV for observation, and electromicrographed at magnifications of 150×, 700×, 1000×, and 1600×. For each cultivar, 10 mature, well-formed pollen grains were randomly selected for measurement. In total, 240 pollen grains were measured.

A total of 12 pollen trait indicators were chosen for quantification and encoded using the scalar quantifying method (Table 2) [18], including eight quantitative traits ("N"), which are numerical and calculated in raw data form without coding. Of these, width of the spine base (SW), spine index (SL/SW), and the average pollen grain radius and spine length (D-spine) were evaluated for the first time for *H. syriacus*. Two binary traits ("B"), coded by "0" or "1", denoted negative and positive states, respectively. Two ordered multistate traits ("O") were each coded with consecutively arranged positive integers ("1", "2", "3", . . .). All four qualitative traits were new indicators (Table 2) never assessed before in *H. syriacus* cultivars, as were the quantified characteristics used to describe their degree of pollen surface ruffling and the number of spines. The pollen description terminology used in this study follows that developed by Erdtman et al. [19], Wang et al. [20], and Punt et al. [21].

Table 2. Twelve pollen traits and their codes for *Hibiscus syriacus*.

No.	Characteristic	Code Type	Code Details
1	Pollen diameter parallel to the X-axis (D_1)	N	/
2	Pollen diameter parallel to the Y-axis (D_2)	N	/
3	Pollen shape ratio (D_2/D_1)	N	/
4	Length of the spine (SL)	N	/
5	Width of the spine base (SW)	N	/
6	Spine index (SL/SW)	N	/
7	Radius of the pollen grain and spine length, averaged (D-spine)	N	/
8	Spacing between spines (S-spine)	N	/
9	Whether the pollen surface has micro-spines or granular verrucae (B-GW)	B	Yes, 1; No, 0
10	Whether the pollen surface ruffles strongly (B-RS)	B	Yes, 1; No, 0
11	Number of pollen surface spines (O-SA)	O	Few (<40), 1; Medium (40–50), 2; Many (≥50), 3
12	Degree of pollen surface ruffling (O-DR)	O	Smooth and largely unruffled, 1; Light ruffles, 2; Strong ruffles, 3

Note: D_1 refers to the diameter parallel to the X-axis in the observation view; D_2 refers to the diameter parallel to the Y-axis in the observation view; the three classes of O-SA were determined by the average range values of pollen spine numbers of 24 *H. syriacus* cultivars (30–60).

2.3. Flower and Leaf Morphological Indicators and Their Measurement

Eleven flower morphological indicators were employed in this study, consisting of seven numerical ("N") characteristics, two binary ("B"), and two ordered multistate ("O") characteristics (Table 3). These 11 floral morphological traits were used in our previous morphological classification study of 27 *H. syriacus* cultivars [13]. However, how they are each defined, and consequently their contribution to the classification, are reported and evaluated here for the first time. An indicator of the degree of leaf lobing was used as a binary trait and applied here to *H. syriacus* (Table 3). The investigation methodology had three components [2]: (1) MS: Survey of individual measurements of the subject plant and its parts. (2) VG: Observation survey with one overall observation of the subject plant and its parts. (3) VS: Observation survey with individual observations of the subject plant and its parts.

2.4. Data Analysis

The mean, maximum, minimum, and coefficient of variation (CV) of eight numerical pollen traits were analyzed for each cultivar (Table A1). Here, the pollen shape ratio (D2/D1), which corresponds to the P/E (the length of the polar axis/equatorial diameter) indicator for polar pollen grains, was classified according to the criteria proposed by Erdtman [22]: oblate spheroidal (0.89–0.99), spheroidal (1.00), prolate spheroidal (1.01–1.14), and subprolate (1.15–1.33). To compare the 24 cultivars in terms of the eight quantitative pollen traits, a multivariate analysis of variance (MANOVA) based on the Shapiro–Wilk normality test was used, followed by one-way analysis of variance (ANOVA), and with Tukey's HSD test to determine which of the 24 cultivars differed from each other.

Table 3. Twelve flower and leaf traits and their codes for *Hibiscus syriacus*.

No.	Characteristic	Investigation Method	Code Type	Code Details
1	Stalk length (ST)	MS	N	/
2	Petal length (PL)	MS	N	/
3	Petal width (PW)	MS	N	/
4	Petals index (PL/PW)	MS	N	/
5	Red center length (RC)	MS	N	/
6	Length of red center line (RCL)	MS	N	/
7	Red center index (RCL/RC)	MS	N	/
8	Whether the stamens are heterogeneous in terms of inner petals (B-IP)	VS	B	Yes, 1; No, 0
9	Whether the petals have a red center (B-RC)	VS	B	Yes, 1; No, 0
10	Relationship of the calyx with the bract (O-CB)	VS	O	Shorter, 1; Near Equal Length, 2; Beyond, 3
11	Relationship of the red center line with the red center (O-RC)	VS	O	Near Equal Length, 1; Beyond, 2; Beyond Obvious, 3
12	Whether the leaf lobing is strong (B-LL)	VG	B	Yes, 1; No, 0

Next, bivariate correlations among the 12 pollen traits were examined using Pearson's r coefficient. To explore the developmental patterns between phenotypic features of *H. syriacus*, the associations between pollen, flower, and leaf morphological indicators were investigated. We also collected data on the fruiting rate of 11 *H. syriacus* cultivars (Table A2) [2], to test for pollen morphology effects on fruiting rates based on the correlation analysis.

A classification framework based on the 24 combined morphological indicators (Tables 2 and 3) was applied to the 24 *H. syriacus* cultivars. To do this, we first standardized (STD) the raw data for each of the 24 traits to eliminate differences in their dimensions. To this STD dataset, a principal component analysis (PCA) of 24 traits was applied, which yielded new uncorrelated variables (PCs) [23–25]. Next, variables were compared in the R-type clustering analysis (between-groups linkage as the clustering method and Pearson's r as a correlation measure), and each variable was placed as a unit in a cluster. In the Q-type analysis (Ward's method as the clustering method and squared Euclidean length as the distance measure), samples were compared using uncorrelated or least correlated variables (implemented in R-type clustering) [26]. Based on the clustering results, it was possible to identify whether the qualitative indicators helped to distinguish *H. syriacus* cultivars. All calculations and analyses described above were carried out using SPSS Statistics 19.0 software.

3. Results

3.1. Floral and Leaf Morphology of 24 H. syriacus Cultivars

The *H. syriacus* cultivars were evidently morphologically diverse. Two of the 24 cultivars in this study lacked a red center (White chiffon and Diana). All seven numerical traits varied considerably among the cultivars: ST (range: 4.00~36.30 mm; mean: 12.46 mm), PL (range: 39.60~70.60 mm; mean: 49.77 mm), PW (range: 22.20~52.30 mm; mean: 36.85 mm), and PL/PW (range: 1.15~1.81; mean: 1.39). They also varied considerably among those cultivars with a red center and red center line: RC (range: 7.30~19.60 mm; mean: 10.25 mm), RCL (range: 7.40~37.90 mm; mean: 14.51 mm), and RCL/RC (range: 1.00~2.30 mm; mean: 1.28 mm). In addition, significant differences were found among cultivars with respect to the presence or absence of inner petals (B-IP), the relationship of the calyx with the bract (O-CB), the relationship of the red center line with the red center (O-RC), and the degree of leaf lobing (B-LL) (Figure 1).

Figure 1. The description and codes for five qualitative traits of *Hibiscus syriacus* flower and leaf parts, for which the yellow arrow indicates the typical characteristic. Whether the stamens are heterogeneous in terms of inner petals (B-IP): (**a-1**,**a-2**) stamens not differentiated into inner petals (No, 0); (**b-1**,**b-2**) stamens differentiated into inner petals (Yes, 0). Whether the petals have the red center (B-RC): (**c-1**,**c-2**) petals without the red center (No, 0); (**d-1**,**d-2**) petals with the red center (Yes, 0). Relationship of the calyx with the bract (O-CB): (**e-1**) calyx shorter than bract (Shorter, 1); (**e-2**) calyx almost equal to bract (Near equal length, 2); (**e-3**) calyx longer than bract (Beyond, 3). Relationship of the red center line with the red center (O-RC): (**f-1**) red center line is almost equal to red center (Near equal length, 1); (**f-2**) red center line is longer than red center (Beyond, 2); (**f-3**) red center line is significantly longer than red center (Beyond obviously, 3). Whether leaf lobing is strong (B-LL): (**g-1**,**g-2**,**g-3**) low degree of leaf lobing (No, 0); (**g-4**,**g-5**,**g-6**) high degree of leaf lobing (Yes, 1).

3.2. Overall Pollen Morphology of *H. syriacus*

The electromicrographs obtained via SEM were used to observe and illustrate the pollen morphology of *H. syriacus* samples (Figure 2). Most *H. syriacus* pollen grains were subspherical monads with a symmetrical structure, having a spiny pollen surface with blunt spine ends. The germination pores were panpori (following the pollen terminology of Erdtman et al. and Wang et al. [19,20]). The pollen surface appeared smooth or ruffled, with varying degrees of ruffling, and the ornamentation was mostly striate, cerebroid, or a mixture of striate and cerebroid. All studied samples differed markedly in terms of pollen spines, while their pollen shapes were less variable.

Figure 2. *Hibiscus syriacus* pollen grains electromicrographed by SEM. (**a**,**b**) An overall full view of pollen grains as spheroidal monads with spines. (**c**) The germinal porus (panpori) on the pollen surface. (**d**) A pollen grain with a smooth surface (pollen surface ornamentation is inconspicuous). (**e**) A pollen grain with a ruffled surface (pollen surface with distinctive cerebroid ornamentation). (**f**) A pollen grain with micro-spines or granulate on its surface.

3.2.1. Quantitative Traits of *H. syriacus* Pollen Grains

Details of the quantitative pollen traits of *H. syriacus* are presented in Table A1 and Figure 3. For polar pollen grains, the equatorial diameter (E) is commonly used to evaluate species' pollen size [27,28]. Here, pollen diameter parallel to the X-axis (D_1) is equivalent to this indicator. For the 24 cultivars of *H. syriacus*, the D_1 and pollen diameter parallel to the Y-axis (D_2) had mean values of 124.19 μm and 124.85 μm, respectively, ranging from 111.33 to 144.34 μm and 112.39 to 146.03 μm. The D_1 (90.19–160.95 μm) and D_2 (90.90–158.22 μm) values for the 240 measured pollen grains varied more significantly (Figure 3a,b). The exact pollen shape (D_2/D_1) of *H. syriacus* was assessed here for the first time, revealing less variation in this characteristic. For the 24 cultivars of *H. syriacus*, the pollen shape (D_2/D_1) was frequently oblate spheroidal (38%) and prolate spheroidal (46%) with a mean value of 1.01, ranging from 0.95 to 1.07. For the 240 measured pollen grains, the pollen shape likewise mostly presented as oblate spheroidal (36%) and prolate spheroidal (54%). The D_2/D_1 reached its maximum in the cultivar White chiffon (1.18), and its minimum in the cultivar Woodbridge (0.73) (Figure 3c).

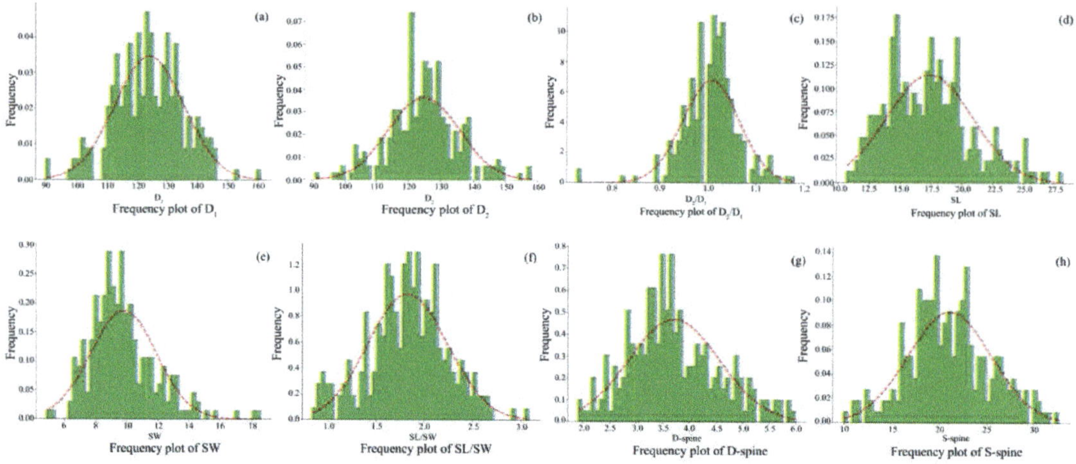

Figure 3. Histograms of eight quantitative traits of *Hibiscus syriacus* based on 240 measured pollen grains. Shown is the normalized frequency of D_1, D_2, D_2/D_1, SL, SW, SL/SW, D-spine, and S-spine for each sample of each *H. syriacus* cultivar. (**a**) The D_1 ranged from 90.19 to 160.95 μm, peaking at 120~130 μm. (**b**) The D_2 ranged from 90.90 to 158.22 μm, peaking at 120~130 μm. (**c**) The D_2/D_1 ranged from 0.73 to 1.18, peaking at 1.0~1.05. (**d**) The SL ranged from 10.73 to 28.24 μm, peaking at 17.00~17.50 μm. (**e**) The range value of SW ranged from 6.91 to 18.61 μm, peaking at 9.00~10.00 μm. (**f**) The SL/SW ranged from 0.84 to 3.09, peaking at 1.50~2.00. (**g**) The D-spine ranged from 1.92 to 6.00 μm, peaking at 3.50~4.00 μm. (**h**) The S-spine ranged from 9.98 to 32.75 μm, peaking at 20~25 μm.

Length of the spine (SL) was 17.40 μm, on average, and varied widely among the 24 *H. syriacus* cultivars (13.42~25.04 μm), especially across the 240 measured samples (10.73–28.24 μm) (Figure 3d). The width of the spine base (SW) is described here for the first time. This trait measured 9.80 μm on average, ranging from 8.25 to 15.00 μm among the 24 cultivars, while showing a near four-fold difference across the 240 samples (4.91~18.61 μm) (Figure 3e). The spine index (SL/SW) was 1.83, on average, ranging from 1.06 to 2.29 among the 24 cultivars, while spanning from 0.84 to 3.09 among the 240 studied pollen grains (Figure 3f).

The average of pollen grain radius and spine length (D-spine) was first used to describe the pollen characteristic for *H. syriacus*. D-spine values ranged from 2.22 to 5.38, with a mean of 3.67; for the 240 pollen grains of *H. syriacus*, the D-spine attained its maximum in the cultivar Woodbridge (6.00) and its minimum in the cultivar Naesarang (1.92) (Figure 3g). The spacing between spines (S-spine) was also determined in this study, having a mean value of 21.14 μm (12.85~27.37 μm) among the 24 cultivars. Across the 240 pollen grains, S-spine was largest in the cultivar Elegantissimus (32.75 μm) and smallest in the cultivar Pink giant (9.98 μm) (Figure 3h).

3.2.2. Qualitative Traits of *H. syriacus* Pollen

Four new qualitative traits were employed to describe the characteristics of the pollen surface for *H. syriacus* (Figure 4). We found relatively few cultivars with micro-spines or granulate verrucae, these amounting to 20.83% of the samples (Figure 4(a-1,a-2,b-1,b-2)). The spine numbers varied considerably between cultivars (30~54) (Figure 4c,(d-1,d-2),e), being greatest in the cultivar Pink giant. We divided this trait into three classes; most of the cultivars (70.83%) featured a lower number of spines. In addition, a significant difference in pollen surface ruffling was found among the cultivars, with ruffled surfaces predominating. Two qualitative indicators were used to code their ruffling features (Figure 4f,g,(h-1,h-2)).

We used the apparent degree of surface ornamentation to describe the roughness or smoothness of the pollen grain surface.

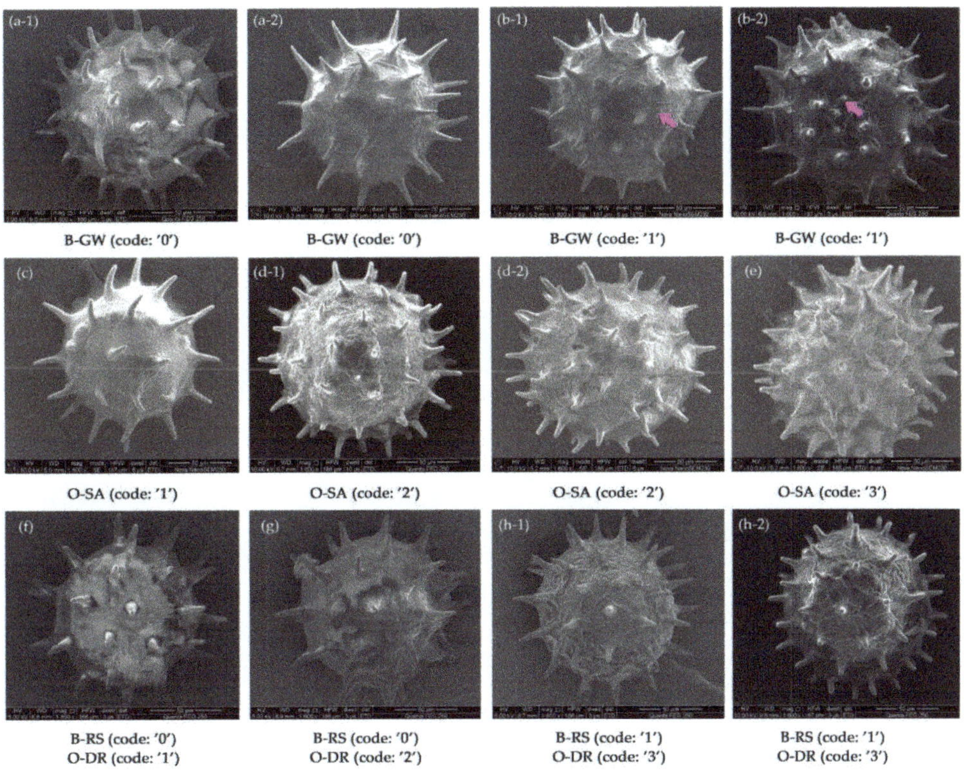

Figure 4. The description and codes for four qualitative traits of *Hibiscus syriacus* pollen grains, where the red arrow indicates the typical characteristic. Whether the pollen surface has granular verrucae (B-GW): (**a-1**,**a-2**) pollen exine is smooth or ruffled without micro-spines or granular verrucae (No, 0); (**b-1**,**b-2**) pollen exine has micro-spines or granular verrucae (Yes, 0). Number of pollen surface spines (O-SA): (**c**) pollen surface spines less than forty in number (<40, 1); (**d-1**,**d-2**) pollen surface spines varying between forty and fifty (40~50, 2); (**e**) pollen surface spines greater than or equal to fifty in number (≥50, 3). Whether the pollen surface ruffles strongly (B-RS): (**f**,**g**) pollen surface not strongly ruffled, smooth, or slightly ruffled (pollen surface ornamentation is inconspicuous or with a small number of ornaments) (No, 0); (**h-1**,**h-2**) pollen surface strongly ruffled (pollen surface with distinctive ornamentation) (Yes, 1). Degree of pollen surface ruffling (O-DR): (**f**) pollen surface smooth and free of ruffles (pollen surface ornamentation is inconspicuous) (smooth and largely unruffled, 1); (**g**) pollen surface is slightly ruffled (pollen surface with a small number of ornaments) (light ruffles, 2); (**h-1**,**h-2**) pollen surface strongly ruffled with striate or a mixture of striate and cerebroid (pollen surface with distinctive ornamentation) (strong ruffles, 3).

3.3. *Variation and Correlation of Pollen Morphology between the 24 H. syriacus Cultivars*

The range, mean, and cv for eight quantitative traits revealed pronounced differences among the 24 cultivars of *H. syriacus* (Table A1). For example, Chungmu displayed a high level of variation in the traits P and S-spine, while the cultivars Diana, Woodbridge, White chiffon, Qiancenghong, Rubis, and Pyonghwa showed a high level of variation in the traits D_1, D_2/D_1, SL, SW, SL/SW, and D-spine, respectively. Eight quantitative traits were simultaneously tested for a difference among cultivars. This MANOVA result revealed a significant difference among 24 cultivars (Wilk's $\lambda = 0.01$, $F = 7.318$, $p < 0.01$). The follow-up ANOVA results for the eight quantitative traits were as follows: D_1 (F = 8.632), D_2 (F = 8.434), D_2/D_1 (F = 2.913), SL (F = 20.768), SW (F = 12.669), SL/SW (F = 12.228), D-spine (F = 16.262), and S-spine (F = 7.152), which demonstrated there was significant variation among the studied cultivars, as all the eight traits exhibited a high level of statistical significance ($p < 0.01$). According to our post hoc study (Turkey's HSD test) of the 24 cultivars, Paeoniflorus and Naesarang were distinct from the other cultivars by a separate subset of trait A. Other separate subsets were: Pink giant and Woodbridge for trait SL/SW, Pink giant for trait S-spine, and Woodbridge for trait D-spine.

Correlations between pollen quantitative and qualitative traits of *H. syriacus* were also determined. We found more correlations that were significant between quantitative traits than qualitative traits (Figure 5). The trait D-spine was correlated with all six quantitative traits except for SW, which had positive correlations with traits D_1 (0.64, $p < 0.01$) and D_2 (0.46, $p < 0.05$) and negative correlations with traits D_2/D_1 (−0.43, $p < 0.05$), SL (−0.91, $p < 0.01$), SL/SW (−0.82, $p < 0.01$), and S-spine (−0.54, $p < 0.01$). Meanwhile, D-spine showed a significant positive correlation with only one qualitative trait (O-SA, 0.59, $p < 0.01$). Trait P/E had no significant correlation with any traits except for D-spine. Among the qualitative traits, O-SA was correlated with all the quantitative traits except for P/E. In addition, O-DR exhibited a significant positive correlation with B-RS, with both indicators describing the surface ruffling characteristics of *H. syriacus* pollen grains.

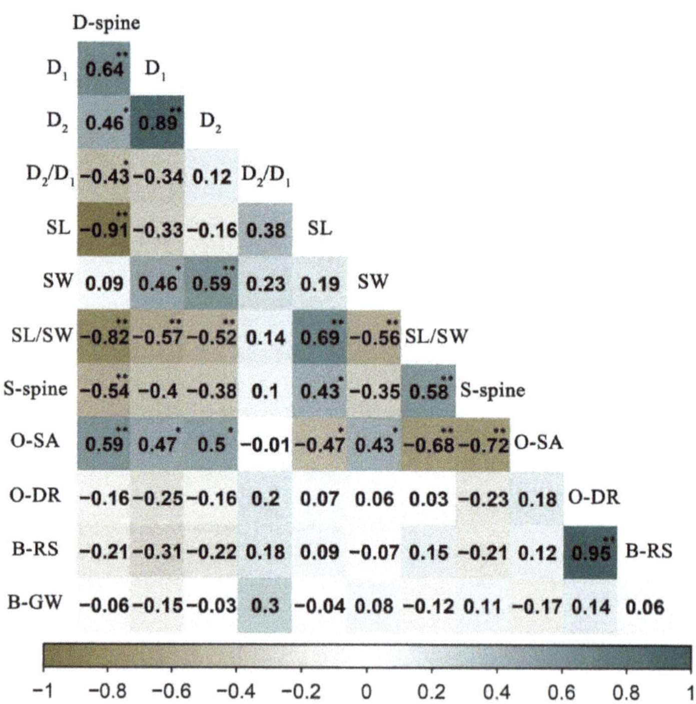

Figure 5. Heatmap of correlations between eight quantitative and four qualitative traits of pollen grains of *Hibiscus syriacus*. The degree of a correlation is indicated by its shaded coloring, with positive or negative numbers in the small square boxes indicating positive or negative correlations between row and column traits, respectively. * $p < 0.05$, ** $p < 0.01$.

3.4. Correlations among Pollen, Flower, and Leaf Morphological Characteristics of *H. syriacus* Cultivars

We tested for correlations between quantitative pollen traits and morphological traits of flowers and leaves of *H. syriacus* (Table 4). The traits D_2, D_2/D_1, SW, and S-spine did not correlate with any flower or leaf traits. Trait E exhibited a significant positive correlation with O-CB (0.550, $p < 0.01$). Trait SL showed positive correlation with both ST (0.496, $p < 0.05$) and B-IP (0.434, $p < 0.05$), but negative correlation with PL (−0.410, $p < 0.05$) and B-LL (−0.478, $p < 0.05$). Trait SL/SM was negatively correlated with PL (−0.565, $p < 0.01$) and O-CB (−0.485, $p < 0.05$), yet positively correlated with B-IP (0.587, $p < 0.01$). Finally, D-spine was positively correlated with three other traits: PL (0.483, $p < 0.05$), O-CB (0.561, $p < 0.01$), and B-LL (0.506, $p < 0.05$), while negatively correlated with B-IP (−0.495, $p < 0.05$).

Table 4. Correlation matrix of pollen vis-à-vis morphological traits of flower and leaf in *Hibiscus syriacus*.

Morphological Traits		Pollen Quantitative Traits							
		D_1	D_2	D_2/D_1	SL	SW	SL/SW	S-Spine	D-Spine
ST	Flower traits	−0.008	0.111	0.247	**0.496** *	0.137	0.265	0.336	−0.387
PL		0.279	0.171	−0.233	**−0.410** *	0.330	**−0.565** **	−0.383	**0.483** *
PW		0.179	0.081	−0.197	−0.080	0.324	−0.292	−0.257	0.173
PL/PW		−0.020	0.002	0.029	−0.342	−0.213	−0.120	0.025	0.246
RC		0.006	0.012	−0.128	0.060	0.379	−0.210	0.049	0.037
RCL		−0.089	−0.151	−0.115	0.094	0.309	−0.149	0.063	−0.053
RCL/RC		−0.033	−0.152	−0.256	0.121	0.077	0.060	−0.014	−0.072
B-IP		−0.301	−0.198	0.237	**0.434** *	−0.302	**0.587** **	0.204	**−0.495** *
B-RC		0.086	0.018	−0.190	0.148	0.105	0.089	−0.004	−0.068
O-CB		**0.550** **	0.381	−0.376	−0.403	0.219	**−0.485** *	−0.194	**0.561** **
O-RC		−0.136	−0.182	−0.038	−0.164	−0.001	−0.177	−0.094	0.123
B-LL	Leaf trait	0.223	0.062	−0.338	**−0.478** *	−0.239	−0.220	−0.110	**0.506** *

Note: Pearson's linear correlation coefficient values between eight pollen quantitative traits and flower and leaf traits. Corresponding details can be found in Tables 2 and 3. The positive or negative numbers in bold indicate positive or negative correlations between row and column traits, respectively. * $p < 0.05$, ** $p < 0.01$.

We also tested for correlations between the 12 pollen traits (Table 2) and the fruiting rates based on data from 11 cultivars of *H. syriacus* (Table A1). These results showed that only B-GW was positively correlated with fruiting rate (0.678, $p < 0.05$). This was an encouraging finding and demonstrated that pollen surface spines and granulates were intrinsically related to pollination.

3.5. Clustering Analysis of 24 H. syriacus Cultivars Based on the 24 Combined Morphological Traits

The PCA results for the 24 *H. syriacus* morphological traits showed a total contribution of 84.55%, in which seven principal components (PCs) were derived. The crucial traits of PC1 (25.81%) were D_2 and D_1; likewise, for PC2 (15.02%), they were SL and D-spine; for PC3 (11.89%), they were RC, RCL, RCL/RC, and B-RC; for PC4 (10.92%), it was PW; for PC5 (10.56%), they were B-RS and O-DR; for PC6 (6.07%), it was O-CB; and for PC7 (4.27%), it was B-GW. PC1, PC2, PC5, and PC7 were indicators related to pollen morphological characteristics, while PC3, PC4, and PC6 were related to flower morphological characteristics.

R-type clustering was used to convey the rationality of indicator selection (Figure 6a). The pairs of traits O-DR and B-RS, RC and RCL, PL and PW, and D_1 and D_2 were clustered at close distances, indicating an equal contribution to the classification of *H. syriacus*. In order to better interpret and compare the effects of traits on the classification, we retained these indicators and used them in a follow-up Q-type clustering analysis.

The results obtained from Q-type clustering revealed that *H. syriacus* cultivars could be divided into 2–6 clusters based on different indicators (Figure 6b):

(1) The 24 cultivars could be divided into two clusters at the grade bond line L1 (D = 20.00) based on the main indicators (D_2, D_1, SW, O-SA, D-spine, and SL). The first cluster included two cultivars (Pink giant and Woodbridge), which presented larger values for the traits D_2, D_1, SW, O-SA, and D-spine, but a smaller value for trait SL. The second cluster comprised the remaining 22 cultivars.

(2) The 24 cultivars could be divided into three clusters at the grade bond line L2 (D = 18.43). The first cluster still consisted of Pink giant and Woodbridge. The remaining 22 cultivars were further classified into two more clusters based on their flower traits (B-RC, RCL, RC, RCL/RC). The second cluster harbored two cultivars (White chiffon and Diana), whose petals lack a red center. The remaining 20 cultivars with the red center constituted the third cluster.

(3) The 24 cultivars could be divided into four clusters at the grade bond line L3 (D = 15.00). In the first cluster was Pink giant and Woodbridge, and in the second cluster was White chiffon and Diana. Then, the other 20 cultivars were further classified into two clusters based on the SL, SW, PW, D-spine, D_2, D_1, O-SA, and O-CB traits. The third cluster had seven cultivars: Pyonghwa, Huaban, Paeoniflorus, Narsarang, Purple pillar, Elegantissimus, and Arang; the fourth cluster entailed the remaining 15 cultivars. The cultivars forming the third cluster presented larger values for SL (17.22~25.04 μm), SW (8.33~12.00 μm), and PW (32.20~50.40 mm), and smaller values for D-spine (2.22~3.32), D_2 (112.39~129.51 μm), and D_1 (111.33~127.13 μm) when compared with those of the fourth cluster. Moreover, the cultivars of the third cluster all had a small number of spines (O-SA, code: 1), and their calyx length was shorter than that of the bract (O-CB, code: 1).

(4) The 24 cultivars were classified into five clusters based on the grade bond line L4 (D = 10.00)—with the same first three clusters for L4 as for L3. The other 15 cultivars were further split into two clusters based on two pollen grain traits (B-RS and O-DR). Two cultivars (Marina and Qiancenghong) were included in the fourth cluster as their pollen grains' surface was weakly and slightly ruffled. The fifth cluster included the remaining 13 cultivars, all of which exhibited strong ruffles.

(5) The 24 cultivars were divided into six clusters based on the grade bond line L5 (D = 7.50). The clustering results at L5 were the same as those at L4, except for the latter's fourth cluster. However, based on two pollen traits (B-RS and O-DR), the fourth cluster at L4 was further divided into two clusters, of which Elegantissimus and Arang grouped together as the fifth cluster as their pollen grains' surface was weakly ruffled. By contrast, the remaining five cultivars were strongly ruffled and thus formed the sixth cluster.

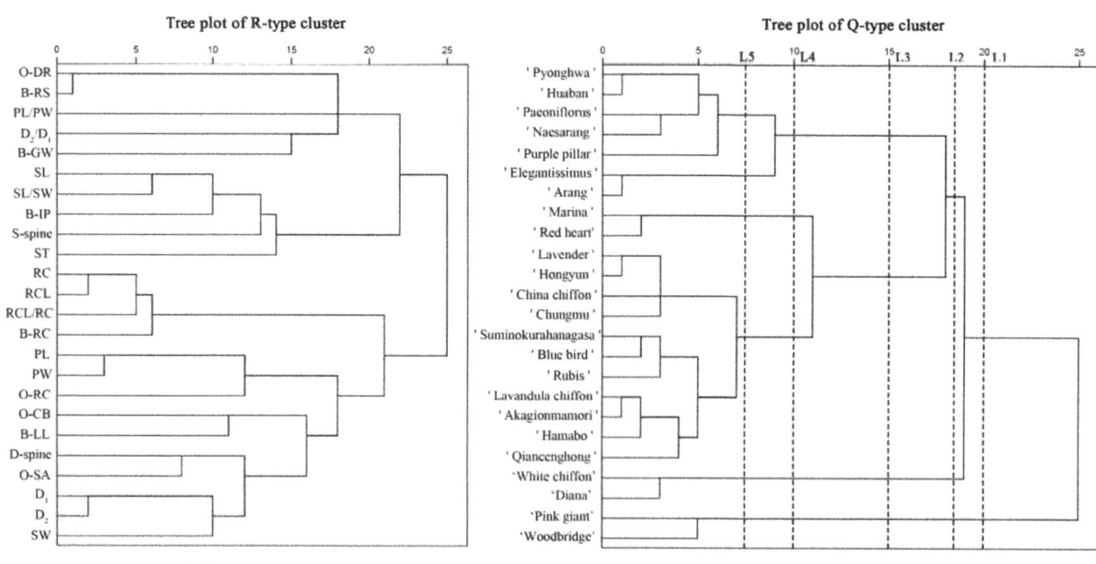

Figure 6. Tree plot of R-type cluster and Q-type cluster of *Hibiscus syriacus*. (**a**) R-type conducted on 24 combined morphological traits based on between-groups linkage method. (**b**) Q-type conducted on *Hibiscus syriacus* 24 cultivars based on ward method.

4. Discussion

4.1. Pollen Morphological Variation of H. syriacus Cultivars and Novel Diagnostic Pollen Traits

Pollen variation diversity can contribute to taxonomic and phylogenetic analyses within the Malvaceae family, whose species can be identified by their spine characteristics [3,29–31]. The pollen of Malvaceae plants is large, being 50~242 µm in diameter, with spines and panpori on the pollen surface, while *Hibiscus* pollen generally varies in size, from 118 to 252.5 µm in diameter [20]. Based on the Palynological Database, *H. syriacus* pollen grains were recorded as a spheroidal shape of large size (>100 µm) [32], and according to Sung et al. and Zhao et al. [3,5], the *H. syriacus* pollen grain diameters ranged from 111.33 to 144.34 µm and 111.65 to 148.98 µm, respectively, while both found high variability in its spine features. In contrast, two indicators of pollen diameter were introduced in this study, D_1 and D_2, with ranges from 111.33 to 144.34 µm and 112.39 to 146.03 µm, respectively, while the ratio of the two indicators (D_2/D_1) described the pollen grain shape. We observed little variation in the pollen shape (D_2/D_1), with that being principally oblate spheroidal (38%) and prolate spheroidal (46%). Four quantitative traits (D_2, D_1, SW, SL, and D-spine) and three new qualitative traits (O-SA, O-DR, and B-RS) examined here could thus be considered as useful diagnostic pollen traits for the classification of *H. syriacus* cultivars.

The trait D–spine takes pollen diameter parallel to the X-axis (D_1) and spine length (SL) into account. Andrade et al. applied this index to distinguish mature pollen grains of *H. rosa-sinensis*, and it was sufficiently accurate to distinguish the species from others in the same family [30]. Here, we employed the D-spine trait for the first time to detect variation between *H. syriacus* cultivars, finding that it made a prominent contribution to the diagnosis and classification of the 24 cultivars, as it showed significant variation among them (range: 2.22~5.38). We suggest the D-spine trait may be an essential pollen trait that can assist in distinguishing species or cultivars of *Hibiscus*. Furthermore, the SW trait, as a new quantitative trait, differed markedly between *H. syriacus* cultivars. At the same time, SW exhibited a positive correlation with D_2, D_1, and O-SA, indicating that cultivars with larger pollen sizes or more spines on their pollen surfaces tend to be accompanied by a broader spine base. The spine index (SL/SW) was also applied here for the first time; however, a weaker contribution to the classification was found for it.

The contribution of qualitative indicators to plant classification can be easily overlooked when they are not quantified. A few qualitative indicators are typically used to describe pollen morphology, usually the pollen surface germinal colpus or porus, and exine ornamentation [20,33]. Spine abundance is often used as a numerical trait for description, as done for *H. syriacus* and *H. rosa-sinensis* [3,30]. Meo et al. found that the number of spine rows between colpi was also the taxonomically important characteristic of *Parthenium hysterophorus* [34]; however, it is weak in its ability to distinguish between plant species or cultivars. Instead, we employed a qualitative trait (O-SA) here to describe the spine abundance, which played a significant role in the classification, with its use as the primary basis for classification at L1 and L3. Furthermore, because the trait O-SA correlated with most quantitative pollen traits, we could assume that pollen grains with large sizes, short spines, or wide spine bases generally also have a higher number of spines. In addition, we employed two new qualitative indicators to describe pollen surface ruffling in two ways. The first is the trait B-RS, for which a smooth or slightly ruffled surface was defined as not strongly ruffled. Most cultivars had a ruffled pollen surface, the five exceptions being Marina, Woodbridge, Elegantissimus, Arang, and Red heart. The second is O-DR, which has three levels of pollen surface ruffling (smooth, lightly ruffled, and extremely ruffled). This trait could further help to distinguish between smooth and slightly ruffled as key features, so that the five cultivars mentioned above could be further clustered into two clusters (one for the first three and one for the last two).

4.2. Combining Morphological Indicators Helps to Better Distinguish Cultivars of H. syriacus and the Contribution of Qualitative Indicators to Clustering

From the results for the Q-type clustering of *H. syriacus* cultivars, the main criteria classification was based on pollen traits at the L1 grade line: D_2, D_1, SW, D-spine, SL, and O-SA. At the L2 grade line, the traits related to the red center of petals (B-RC, RC, RCL, and RCL/RC) enabled us to further classify the remaining 22 cultivars: in this way, two cultivars without the red center were identified (White chiffon and Diana). Prior evidence suggested these (B-RC, RC, RCL, and RCL/RC) indicators could effectively distinguish *H. syriacus* cultivars as PC3 (11.89%) in a classification based on floral morphological traits alone [13]. The present study's results lend further support to using these traits as an independent basis for *H. syriacus* cultivars in a combined classification, in that they were uncorrelated with pollen indicators. At grade line L3, those cultivars with a red center could be further classified into two clusters based on their pollen (SL, SW, D-spine, D_2, D_1, and O-SA) and floral traits (PW and O-CB). Given that O-CB was found positively correlated with E, we may presume those *H. syriacus* cultivars with smaller pollen grains are usually accompanied by a calyx shorter than the bract. Knowledge of this correlation can enable horticulturists to promptly identify *H. syriacus* cultivars, but admittedly more samples are still needed to support this finding. In addition, at grade lines L4 and L5, the qualitative pollen traits related to ruffle features (B-RS and O-DR) were the main classification criteria elucidated. Although the R-type cluster results indicated these two traits are closely related, we kept both since O-DR provided a more in-depth distinction of results generated via B–RS. Lastly, a contribution of the leaf morphological trait (B-LL) to *H. syriacus* classification was not found.

Overall, the results of this study demonstrate that a number of *H. syriacus* cultivars are distinguishable using only quantitative traits. Accordingly, coding and quantifying quantitative traits could contribute to the identification of a greater number of cultivars. Similar results and patterns were reported for *Cathaya argyrophylla* [14]. We propose that the integrated use of morphological indicators of species could generate a broader taxonomic basis.

4.3. Effects of H. syriacus Pollen Traits on Fruiting

Pollen morphological characteristics can influence pollination and breeding, as demonstrated by studies by Mccallum et al., Mendoza et al., and Xia et al., as reported for *Ipomoea purpurea*, *Orius laevigatus*, and *Ottelia acuminata* [27,35,36]. Pollen, seed, and fruit characteristics often have positive correlations [37–39]. The mechanisms by which pollen spines affect fruiting success are complex and challenging to elucidate; they may be related to the spine distribution pattern, spine length, spine density, or even the space distance between spines [17,40]. It would be helpful if we could provide breeding guidance based on the relationships between pollen morphology and pollination or fruit set. We found that the trait B-GW is positively correlated with fecundity (0.678, $p < 0.05$), which indicates that high fruiting rates occur in cultivars that have micro-spine or granular verrucae on their pollen surface; e.g., Blue bird (26.2%) and Red heart (26.0%). Of the 24 studied cultivars, just five harbored this feature (Arang, Rubis, Suminokurahanagasa, Akagionmamori, and Qiancenghong).

Pollen grains with spines and granulate verrucae may adhere to the long hairs of bees for transport, as was previously found for an *Pavonia* sp. (Malvaceae) [41]. However, either pollen size or echinate exine structure alone was not an excellent factor for pollen collectability [40] since the pollination mechanism is complex and related to pollen-collecting bee species, with different genera of bees showing divergence in their collecting behavior [40,42–44]. Fruit set is likely also affected by insect pests [45,46], and whether the incidence of insect pests correlates with pollen morphology or a specific substance warrants further investigation.

Furthermore, floral pollinators' behaviors reflect a selection of pollen traits. Evidence from a study by Lynn et al. of *Taraxacum ceratophorum* revealed that bumblebee pollinators were prone to picking up pollen grains within a narrow distribution of spine distance [17], indicating that a certain spine distance can favor the likelihood of pollination (i.e., a trait selected because it enhances plant fitness vis-à-vis pollinator community). In this respect, *H. syriacus*, a species pollinated chiefly by bumblebees [47], probably has the same selection pattern as described above, and we observed that those cultivars with high fruiting rates also featured a smaller value for spine spacing (S-spine) (Blue bird, 20.78 μm; Red heart, 20.86 μm) than the average for all 24 cultivars (21.14 μm). In addition to pollinators' selection mechanism, which concerns pollen size, spines, and spacing, a pollinator's body size also determines the amount of pollen it can carry, as does its degree of hairiness [16,48]. A model showed that the poor interaction bond between pollenkitt-free spines and pollenkitt-covered exine might weaken the compacting within pollen storage organs; this pattern was prominent on large pollen grains as they reduced the contact surface and thus affected pollen collection [49]. Spine length might also influence pollen adherence to a pollinator's body, but though proven, it is not a significant trait for *Taraxacum ceratophorum* during its pollen pickup [17]. In this study, we obtained an interesting finding that those cultivars with high fruiting rates mainly had smaller values for both E and spine length (SL), e.g., Blue bird and Red heart, which had high fruit sets, whereas their D_1 was 123.36 or 124.52 μm, respectively. By contrast, cultivars with higher D_1 values, namely Pink giant (141.09 μm) and Woodbridge (144.34 μm), had very low fruit sets, at 4.4% and 1.5%, respectively. The traits S-spine and SL did not directly correlate with fruiting rate in our correlation analysis, but they are worth discussing, as effective explanations for this phenomenon are still limited. In this respect, according to the mechanical-defense hypothesis, bumblebees do not collect pollen with bent spines [49]. We observed that some *H. syriacus* cultivars produce pollen with bent spines, but the limited data available in this study prevented us from verifying this hypothesis. Therefore, more *H. syriacus* samples are needed for further testing, to better discern and interpret the relationships between the pollination mechanism and breeding system of *H. syriacus*.

5. Conclusions

Significant variation among studied *H. syriacus* cultivars was demonstrated, especially in their pollen spine features. The derived classification scheme based on flowers and pollen morphological indicators let us classify 24 *H. syriacus* cultivars into 2–6 clusters. The main diagnostic quantitative traits are D_2, D_1, SW, SL, D-spine, RC, RCL, and RCL/RC, while the main diagnostic qualitative traits are O-SA, B-RC, B-RS, O-DR, and O-CB. Among all of those, two new quantitative traits (SW and D-spine) and five new qualitative traits (O-SA, B-RC, B-RS, O-DR, and O-CB) made a robust contribution to the classification of *H. syriacus* cultivars. The number of pollen spines (O-SA) of *H. syriacus* is strongly correlated with its quantitative pollen traits, and three floral (B-IP and O-CB) and leaf (B-LL) phenotypic traits are correlated with certain quantitative pollen traits. The trait B-GW is correlated with fruiting rate, and pollen diameter parallel to the X-axis (D_1), spine length (SL), and spine spacing (S-spine) might all be potential factors that lead to successful breeding in *H. syriacus*.

Author Contributions: Conceptualization, X.W. and F.X.; methodology, F.X.; software, F.X., J.C. and J.Z.; formal analysis, F.X.; investigation, F.X. and Y.J.; resources, X.W. and Y.W; writing—original draft preparation, F.X.; writing—review and editing, X.W.; funding acquisition, X.W., Y.W. and C.C. All authors have read and agreed to the published version of the manuscript.

Funding: This research was funded by the Key Discipline of the State Forestry Administration (LinRenFa [2016] No. 21), "Double First-Class" Cultivation Discipline of Hunan Province (XiangJiaoTong (2018) No. 469), and Scientific Research Project of Hunan Provincial Department of Education (21B0224).

Institutional Review Board Statement: The study did not require ethical approval.

Informed Consent Statement: The study did not involve humans.

Data Availability Statement: Data available from the author.

Conflicts of Interest: The authors declare no conflict of interest.

Appendix A

Table A1. The minimal, maximal, mean values and coefficient of variation (cv) for eight quantitative pollen traits of *Hibiscus syriacus*.

Samples	D_1			D_2			P/E			SL		
	Mean	Range	cv	Mean	Range	cv	Mean	Range	cv	Mean	Range	cv
1	130.70	110.89–145.87	7.67%	124.04	104.11–131.79	6.47%	0.95	0.82–1.13	8.75%	15.07	13.32–17.91	8.94%
2	123.03	110.35–136.40	9.52%	131.06	120.92–139.25	5.04%	1.07	0.93–1.18	7.50%	16.66	13.14–22.76	17.09%
3	141.09	132.97–145.58	2.45%	146.03	132.60–158.22	6.32%	1.03	0.97–1.13	4.91%	15.65	13.25–17.80	10.45%
4	127.95	122.82–134.50	3.31%	129.78	123.55–135.09	2.88%	1.01	1.00–1.04	1.06%	16.81	12.48–20.26	14.83%
5	127.13	119.37–133.47	3.60%	129.30	110.46–138.07	6.04%	1.02	0.85–1.06	6.31%	23.70	20.15–27.10	8.82%
6	144.34	133.46–160.95	5.64%	139.12	112.37–148.10	9.58%	0.97	0.73–1.05	13.09%	13.42	12.09–15.49	8.25%
7	120.83	98.91–135.14	13.77%	117.76	103.68–128.76	9.53%	0.98	0.93–1.05	4.77%	15.31	13.84–17.55	8.02%
8	123.44	112.96–140.53	8.66%	121.44	116.96–126.63	3.09%	0.99	0.90–1.04	5.43%	15.55	12.73–19.05	12.33%
9	126.29	119.89–129.83	2.88%	125.44	120.65–129.62	2.11%	0.99	0.97–1.03	1.90%	19.56	16.64–22.36	9.50%
10	125.19	105.58–133.40	9.44%	129.51	105.82–145.29	10.70%	1.03	0.99–1.10	3.71%	20.15	15.08–23.40	11.81%
11	119.08	90.19–131.66	10.38%	123.24	105.56–138.39	9.84%	1.04	0.99–1.17	5.21%	19.73	17.73–21.13	4.98%
12	118.66	112.55–125.56	4.22%	118.32	104.18–127.02	5.91%	1.00	0.92–1.07	5.27%	16.73	14.02–19.65	11.38%
13	114.59	104.39–128.85	5.31%	119.76	111.69–129.57	4.43%	1.05	0.97–1.15	5.04%	19.33	14.68–23.21	13.45%
14	123.16	113.81–135.99	5.95%	126.94	119.62–134.17	4.04%	1.03	0.96–1.10	4.17%	14.45	11.50–18.30	15.58%
15	122.76	90.72–137.44	12.08%	119.16	90.9–135.96	12.10%	0.97	0.90–1.05	4.50%	14.20	11.32–17.95	14.25%
16	116.54	109.08–127.62	5.93%	119.18	111.99–127.62	4.35%	1.02	0.92–1.12	5.35%	15.06	12.14–19.45	16.80%
17	117.64	108.77–128.85	4.93%	116.95	113.16–121.18	1.99%	1.00	0.91–1.06	4.53%	20.24	13.58–24.71	15.39%
18	121.06	114.61–130.31	3.80%	123.36	113.45–138.66	4.15%	1.02	0.95–1.10	5.40%	16.13	12.71–19.12	13.25%
19	126.71	102.54–144.72	8.79%	126.80	103.63–131.22	7.64%	1.00	0.95–1.04	2.75%	18.54	17.05–22.53	8.84%
20	125.33	117.65–138.44	5.55%	124.52	111.99–137.85	5.34%	0.99	0.94–1.04	3.75%	16.08	10.73–19.42	15.35%
21	138.20	132.71–143.62	2.54%	134.37	131.24–138.45	1.80%	0.97	0.94–1.00	2.00%	18.47	16.41–20.07	4.99%
22	121.18	103.63–125.44	5.26%	121.79	99.99–138.45	7.65%	1.00	0.96–1.11	4.22%	14.54	12.00–19.55	16.41%
23	114.31	102.90–123.26	5.79%	112.39	102.9–118.68	5.38%	0.98	0.93–1.02	3.45%	17.22	14.38–18.61	8.40%
24	111.33	98.55–120.09	7.61%	116.22	97.27–126.29	9.04%	1.04	0.99–1.10	3.58%	25.04	22.89–28.24	6.78%

Samples	SW			SL/SW			D–spine			S–spine		
	Mean	Range	cv	Mean	Range	cv	Mean	Range	cv	Mean	Range	cv
1	8.59	6.61–10.15	13.85%	1.79	1.31–2.29	18.30%	4.36	3.50–5.01	10.76%	20.36	16.26–23.51	11.20%
2	9.01	7.13–10.67	12.35%	1.85	1.53–2.37	11.96%	3.77	2.93–5.19	17.18%	22.09	16.52–27.74	16.03%
3	15.00	10.99–18.61	16.54%	1.06	0.90–1.37	16.21%	4.56	3.78–5.43	11.45%	12.85	9.98–19.73	20.57%
4	8.37	6.73–9.86	10.84%	2.01	1.78–2.42	8.71%	3.88	3.12–4.99	15.57%	19.19	14.80–22.74	12.66%
5	10.43	8.99–12.14	10.53%	2.29	1.97–2.61	10.20%	2.70	2.23–3.24	10.59%	20.49	16.14–26.44	16.78%
6	12.73	11.54–14.36	6.43%	1.06	0.84–1.28	11.54%	5.40	4.72–6.00	7.56%	18.75	16.20–23.38	12.21%
7	9.47	8.13–11.56	11.07%	1.63	1.36–1.83	9.18%	3.96	3.33–4.83	13.62%	20.28	14.57–27.85	18.96%
8	8.96	6.94–10.51	12.32%	1.75	1.57–2.11	11.10%	4.00	3.62–4.62	8.81%	18.98	15.15–22.57	15.51%
9	9.93	8.12–12.70	17.00%	2.02	1.38–2.63	18.93%	3.26	2.82–3.76	10.69%	25.63	20.22–31.54	14.08%
10	9.66	8.73–11.28	8.16%	2.09	1.64–2.36	9.77%	3.15	2.48–4.38	17.25%	27.37	20.57–32.75	13.79%
11	10.66	9.32–11.72	7.34%	1.86	1.60–2.12	8.49%	3.03	2.23–3.46	12.69%	24.14	15.73–30.56	19.24%
12	9.18	6.57–11.33	14.99%	1.84	1.40–2.22	12.72%	3.59	2.93–4.26	11.79%	18.04	11.51–21.48	16.09%
13	12.00	9.19–14.48	13.54%	1.66	1.01–2.24	23.53%	3.01	2.41–3.56	12.51%	23.74	16.28–30.04	20.59%
14	9.64	6.50–12.36	19.02%	1.56	0.93–2.42	28.34%	4.36	3.17–5.65	17.97%	18.57	14.37–26.65	21.80%
15	9.24	7.28–12.56	15.69%	1.56	1.21–1.88	15.62%	4.44	2.86–5.85	22.93%	22.70	11.66–29.09	22.96%
16	8.25	7.27–10.35	10.88%	1.84	1.43–2.32	16.82%	3.97	3.01–5.26	18.22%	21.90	17.93–28.00	17.12%
17	9.19	7.25–11.59	15.83%	2.22	1.87–2.72	13.02%	3.00	2.20–4.74	23.33%	24.65	18.59–31.32	15.44%
18	8.74	5.31–11.31	22.37%	1.91	1.40–2.51	20.75%	3.81	3.08–4.75	13.89%	20.78	17.90–28.74	17.10%
19	8.50	7.02–9.84	9.72%	2.19	1.93–2.53	8.75%	3.44	2.68–3.87	11.56%	23.45	20.04–27.43	10.08%
20	8.48	4.91–10.04	17.15%	1.91	1.70–2.19	8.55%	3.99	3.26–5.74	17.63%	20.86	18.13–23.08	8.60%
21	10.88	6.92–14.54	22.89%	1.78	1.38–2.64	23.71%	3.75	3.46–4.30	5.91%	20.92	14.03–27.63	21.24%
22	8.43	6.57–10.94	16.13%	1.73	1.50–2.05	10.94%	4.26	3.20–5.13	16.02%	20.23	15.00–24.85	14.00%
23	8.34	6.50–10.60	17.78%	2.11	1.67–2.58	14.32%	3.34	3.00–4.06	8.87%	20.40	18.57–23.90	7.57%
24	11.59	9.15–14.28	17.70%	2.23	1.76–3.09	21.53%	2.23	1.92–2.48	8.93%	21.11	14.46–28.10	17.76%

Table A2. The fruiting rates of 11 *Hibiscus syriacus* cultivars.

No.	Cultivar	Fruiting Rate (%)
1	'Pink giant'	4.4
2	'Woodbridge'	1.5
3	'Elegantissimus'	0.1
4	'Arang'	15.8
5	'Rubis'	13.2
6	'Chungmu'	0.1
7	'Suminokurahanagasa'	2.3
8	'Blue bird'	26.2
9	'Akagionmamori'	0.1
10	'Red heart'	26.0
11	'Naesarang'	1.0

References

1. Liu, Z.; Gao, L. Ancient *Hibiscus syriacus* culture of China. *Agric. Archaeol.* **2010**, *4*, 228–231.
2. National Institute of Forest Science. *Illustration of Hibiscus syriacus L. cultivars*; Seoul Press: Seoul, Republic of Korea, 2014.
3. Bae, S.H.; Younis, A.; Hwang, Y.J.; Lim, K.B. Various Pollen Morphology in *Hibiscus syriacus*. *Flower Res. J.* **2015**, *23*, 125–130. [CrossRef]
4. He, L. Comparision in Cytology and Molecular Biology of Simple and Double Flower on *Hibiscus syriacus*. Ph.D. Thesis, Beijing Forestry University, Beijing, China, 2006.
5. Zhao, Y.; Feng, Q.; Tian, L.; Zhang, J.; Wang, X.; Liu, Y.; Liu, D. Pollen morphology and numerical taxonomy of 22 *Hibiscus syriacus*. *Guihaia* **2021**, *41*, 103–113. [CrossRef]
6. Oliveira, P.P.; dos Santos, F.d.A.R. Morfologia polínica de Hibiscus pernambucensis Arruda e *Hibiscus tiliaceus* L. (Malvaceae). *Acta Biol. Leopoldensia* **2004**, *26*, 203–211.
7. EI, N.S. Pollen morphology of Egyptian Malvaceae: An assessment of taxonomic value. *Turk. J. Bot.* **2004**, *28*, 227–240.
8. Yuan, T.; Wang, L. Pollen Morphology of Several Tree Peony Wild Species and Discussion on Its Evolution and Taxonomy. *J. Beijing For. Univ.* **1999**, *21*, 18–21.
9. Palazzesi, L.; Pujana, R.R.; Burrieza, H.P.; Steinhardt, A.P. Pollen grain morphology of selected allergenic species native to Southern South America. *J. Torrey Botanic. Soc.* **2007**, *134*, 527–533. [CrossRef]
10. Korszun, S.; Klimko, M. Microsporangia and pollen morphology of Ginkgo biloba cultivars. *Dendrobiology* **2014**, *71*, 83–92. [CrossRef]
11. Ma, S.L.; Lu, Y.M. Classification and phylogenetic analysis of Chinese hawthorn assessed by plant and pollen morphology. *Genet. Mol. Res. Gmr* **2016**, *15*, gmr.15038739. [CrossRef] [PubMed]
12. Ma, Z.H.; Glc, B.; Zhang, D.X. Pollen morphology of Callicarpa L. (Lamiaceae) from China and its systematic implications. *Plant Syst. Evol.* **2016**, *302*, 67–88. [CrossRef]
13. Xiao, F.; Wang, X.H.; Wang, Y.Q.; Liu, C.L.; Xie, L.S.; Ren, X.Y. Numerical Classification and Principle Component Analysis of 27 Hibiscus Cultivars. *J. Cent. South Univ. For. Technol.* **2019**, *39*, 59–64.
14. Xiao, F.; She, Y.; She, J.; Wang, Y.; Wu, F.; Xie, P.; Chen, Q. Intraspecific Pollen Morphology Variation and Its Responses to Environmental Factors of Wild *Cathaya argyrophylla* Chun Et Kuang Endemic to China. *Forests* **2022**, *13*, 651. [CrossRef]
15. Minnaar, C.; Anderson, B.; de Jager, M.L.; Karron, J.D. Plant–pollinator interactions along the pathway to paternity. *Ann. Bot.* **2018**, *123*, 225–245. [CrossRef] [PubMed]
16. Phillips, B.B.; Williams, A.; Osborne, J.L.; Shaw, R.F. Shared traits make flies and bees effective pollinators of oilseed rape (*Brassica napus* L.). *Basic Appl. Ecol.* **2018**, *32*, 66–76. [CrossRef]
17. Lynn, A.; Piotter, E.; Harrison, E.; Galen, C. Sexual and natural selection on pollen morphology in Taraxacum. *Am. J. Bot.* **2020**, *107*, 364–374. [CrossRef]
18. Xu, K.X. *Numerical Taxonomy*; Science Publishing: Beijing, China, 1994.
19. Erdtman, G.; Vishnu-Mittre. On Terminology in Pollen and Spore Morphology. *Grana Palynol.* **1958**, *1*, 6–9. [CrossRef]
20. Wang, F.X.; Qian, N.F.; Zhang, Y.L.; Yang, H.Q. *Pollen Flora of China*, 2nd ed.; Science Publishing: Beijing, China, 1995; pp. 5–13.
21. Punt, W.; Hoen, P.P.; Blackmore, S.; Nilsson, S.; Le Thomas, A. Glossary of pollen and spore terminology. *Rev. Palaeobot. Palynol.* **2007**, *143*, 1–81. [CrossRef]
22. Erdtman, G. The Acetolysis Method. A Revised Description. *Sven. Bot. Tidskr.* **1960**, *54*, 561–564.
23. Abdi, H.; Williams, L.J. Principal Component Analysis. *WIREs Comput. Stat.* **2010**, *2*, 433–459. [CrossRef]
24. Martinović, S.; Vlahović, M.; Gajić-Kvaščev, M.; Vuksanović, M.; Glišić, D.; Volkov-Husović, T. Principal component analysis of morphological descriptors for monitoring surface defects induced by thermal shock. *J. Eur. Ceram. Soc.* **2021**, *41*, 426–429. [CrossRef]
25. Fukunaga, K. *Introduction to Statistical Pattern Recognition*; Academic Press: Cambridge, MA, USA, 1990. [CrossRef]
26. Parks, J.M. Part 2: Applications of Multivariate Statistics in Geology | | Cluster Analysis Applied to Multivariate Geologic Problems. *J. Geol.* **1966**, *74*, 703–715. [CrossRef]
27. Mccallum, B.; Chang, S.M. Pollen competition in style: Effects of pollen size on siring success in the hermaphroditic common morning glory, *Ipomoea purpurea*. *Am. J. Bot.* **2016**, *103*, 466–468. [CrossRef] [PubMed]
28. Xiao, F.; She, Y.; She, J.; Zhang, J.; Zhang, X.; Luo, C. Assessing habitat suitability and selecting optimal habitats for relict tree Cathaya argyrophylla in Hunan, China: Integrating pollen size, environmental factors, and niche modeling for conservation. *Ecol. Indic.* **2022**, *145*, 109669. [CrossRef]
29. EI, N.S.; Sawady, N. Pollen morphology of Malvaceae and its taxonomic significance in Yemen. *Flora Medit* **2008**, *18*, 431–439.
30. Andrade, K.; Guerra, S.; Debut, A. Fullerene-Based Symmetry in Hibiscus rosa-sinensis Pollen. *PLoS ONE* **2014**, *9*, e102123. [CrossRef]
31. Shaheen, N.; Khan, M.A.; Hayat, M.Q.; Yasmin, G. Pollen morphology of 14 species of Abutilon and Hibiscus of the family Malvaceae (sensu stricto). *J. Med. Plants Res.* **2009**, *3*, 921–929.
32. Halbritter, H.; Heigl, H. *Hibiscus syriacus*. In *PalDat—A Palynological Database*. Available online: https://www.paldat.org/pub/Hibiscus_syriacus/304207 (accessed on 7 March 2023).
33. Hao, M.; Cao, F.; Zhang, W.; Wang, G. Variation of pollen morphology among male trees of Ginkgo. *J. For. Eng.* **2006**, *20*, 53–55.

34. Meo, A.A.; Khan, M.A. Pollen morphology of invasive species *Parthenium hysterophorus* L. (Heliantheae—Asteracea) from Islamabad and Rawalpindi, Pakistan. *J. Agric.* **2005**, *21*, 227–230.
35. Mendoza, J.E.; Balanza, V.; Cifuentes, D.; Bielza, P. Selection for larger body size in *Orius laevigatus*: Intraspecific variability and effects on reproductive parameters. *Biol. Control* **2020**, *148*, 104310. [CrossRef]
36. Xia, J.; Lu, J.; Wang, Z.; Hao, B.; Wang, H.; Liu, G. Pollen limitation and Allee effect related to population size and sex ratio in the endangered *Ottelia acuminata* (Hydrocharitaceae): Implications for conservation and reintroduction. *Plant Biol.* **2013**, *15*, 376–383. [CrossRef]
37. Satil, F.; Kaya, A.; Ünal, M. Fruit, seed and pollen morphology of Chorispora DC. Species (Brassicaceae) of Turkey. *Bangladesh J. Bot.* **2018**, *47*, 459–466. [CrossRef]
38. Doi, K.; Inoue, R.; Iwasaki, N. Seed weight mediates effects of pollen on berry weight, ripening, and anthocyanin content in highbush blueberry. *Sci. Hortic.* **2021**, *288*, 110313. [CrossRef]
39. Maryam; Jaskani, M.J.; Ahmad, S.; Awan, F.S. Metaxenial Effects on Morphological Attributes in Date Palm cvs. Hillawi and Khadrawy. *Pak. J. Agric. Sci.* **2015**, *52*, 387–393.
40. Konzmann, S.; Koethe, S.; Lunau, K. Pollen grain morphology is not exclusively responsible for pollen collectability in bumble bees. *Sci. Rep.* **2019**, *9*, 4705. [CrossRef] [PubMed]
41. Schlindwein, C.; Pick, R.A.; Martins, C.F. Evaluation of oligolecty in the Brazilian bee *Ptilothrix plumata* (Hymenoptera, Apidae, Emphorini). *Apidologie* **2009**, *40*, 106–116. [CrossRef]
42. Michener, C.D. The bees of the world. In *Systematic Entomology*; J.H.U. Press: Baltimore, MD, USA, 2010; Volume 26, pp. 255–256.
43. Aguilar, C.; Azucena, M.; Parra-Tabla, V. Importance of conserving alternative pollinators: Assessing the pollination efficiency of the squash bee, *Peponapis limitaris* in *Cucurbita moschata* (Cucurbitaceae). *J. Insect Conserv.* **2000**, *4*, 201–208. [CrossRef]
44. Ballantyne, G.A.; Baldock, K.; Willmer, P.G. Constructing more informative plant-pollinator networks: Visitation and pollen deposition networks in a heathland plant community. *Proc. R. Soc. Biol. Sci.* **2015**, *282*, 20151130. [CrossRef]
45. Jung, J.-K.; Kim, M.; Lee, C.Y.; Jang, B.-J.; Kim, D.; Kwon, H.Y.; Park, Y. Comparison of Insect Pest Communities on 30 Cultivars of *Hibiscus syriacus*. *J. Korean Soc. For. Sci.* **2021**, *110*, 116–127.
46. Kim, Y.; Cho, Y.; Kang, Y.K.; Choi, M.; Nam, S.H. A study of the major insect pest communities associated with *Hibiscus syriacus* (Columniferae, Malvaceae). *J. Ecol. Environ.* **2013**, *36*, 125–129. [CrossRef]
47. You, X.; Zhang, D.; Chi, J.; Wu, H. ComparatiVe study on three *Hibiscus syriacus* cultivars' characteristics of nectar secretion, and pollen presentation and entomophilous pollination. *Acta Agric. Shanghai* **2009**, *25*, 81–86.
48. Stavert, J.R.; Linan-Cembrano, G.; Beggs, J.R.; Howlett, B.G.; Pattemore, D.E.; Bartomeus, I. Hairiness: The missing link between pollinators and pollination. *Peerj* **2016**, *4*, e2779. [CrossRef] [PubMed]
49. Lunau, K.; Piorek, V.; Krohn, O.; Pacini, E. Just spines—Mechanical defense of malvaceous pollen against collection by corbiculate bees. *Apidologie* **2015**, *46*, 144–149. [CrossRef]

agronomy

MDPI

Article

Genome-Wide Analysis and Expression of *MYC* Family Genes in Tomato and the Functional Identification of *slmyc1* in Response to Salt and Drought Stress

Yang Feng, Senlin Zeng, Jinping Yan, Kunzhi Li and Huini Xu *

Faculty of Life Science and Technology, Kunming University of Science and Technology, Jingming South Street, Kunming 650224, China
* Correspondence: xuhn@kust.edu.cn

Abstract: Myelocytomatosis (*MYC*) transcription factors are crucial mediators of the jasmonate signaling pathway, which mediates the growth and developmental processes of plants. However, the function of *MYC* genes in tomato, *Solanum lycopersicum* (*SlMYC*), remains poorly understood. In this study, we have identified 14 non-redundant *SlMYC* genes across the genome of tomatoes. Six of the twelve chromosomes included these genes, and four syntenic pairs of *SlMYC* were identified. According to the results of phylogenetic analysis, 14 *SlMYC* genes were clustered into classes I, II, III, and IV, and their functional domains were predicted. The *SlMYC* upstream promoter region contained a variety of light-, stress-, and hormone-response regulatory elements. The expression of the 14 *SlMYC* genes differed significantly across organs. *SlMYCs* primarily showed an upregulation trend after methyl jasmonate (MeJA) treatment. In contrast, after treatment with sodium chloride (NaCl), *SlMYCs* showed a trend of downregulation. However, there were differences in the expression patterns of *SlMYCs* after mannitol treatment. Using clustered regularly interspaced short palindromic repeats/Cas 9 (CRISPR/Cas 9) technology, the loss-of-function of *SlMYC1* (*slmyc1*) was obtained. The *slmyc1* tomato plants demonstrated reduced resistance to NaCl and mannitol stress compared to wild-type plants due to their shorter root length and higher reactive oxygen species (ROS) content. In brief, this study provides valuable information about the taxonomy of the *SlMYC* genes in tomato. It establishes a foundation for future research on the mechanism by which *SlMYC* influences plant development and stress response.

Keywords: tomato; *MYC*; transcription factor; jasmonate; salt stress; drought stress

Citation: Feng, Y.; Zeng, S.; Yan, J.; Li, K.; Xu, H. Genome-Wide Analysis and Expression of *MYC* Family Genes in Tomato and the Functional Identification of *slmyc1* in Response to Salt and Drought Stress. *Agronomy* **2023**, *13*, 757. https://doi.org/10.3390/agronomy13030757

Academic Editors: Guanglong Wang, Lijun Ou and Aisheng Xiong

Received: 11 January 2023
Revised: 4 February 2023
Accepted: 5 February 2023
Published: 5 March 2023

1. Introduction

The basic helix–loop–helix (bHLH) family is the second largest transcription factor family in plants. Myelocytomatosis (*MYC*), a subfamily of the bHLH family, has an N-terminal bHLH_MYC_N region and a C-terminal bHLH region. The bHLH region is a dimerization domain composed of a helix of hydrophobic amino acid residues connected by distinct loops [1]. It is a DNA-binding domain that enables the attachment of HLH proteins to an E-box [2,3]. In addition, bHLH domains allow for the development of homodimeric or heterodimeric complexes and protein–protein interactions [2]. *MYC* transcription factors are a core regulator in the jasmonic acid signaling pathway [4]. As oxylipin phytohormones, jasmonates (JAs) play a crucial role in plant growth, development, and stress response by reprogramming gene expression across the genome [5–7]. JA zinc-finger inflorescence meristem (ZIM) domain (JAZ) proteins act as JA signaling pathway repressors.

The *MYC* gene plays a role in several physiological phases of plants, particularly in the growth and development of distinct species [8]. In rice (*Oryza sativa* L.), *OsMYC2* is expressed in all plant tissues; in particular, a higher level of expression is evident in spikelets and floral organs [9]. *OsMYC2* regulates rice spikelet development through interactions with OsJAZ1 and the activation of downstream genes [10]. *MYC* genes have been extensively studied in *Arabidopsis* sp. They regulate leaf senescence, root development, and stamen development and play a role in seed production and development, seed protein accumulation, and chlorophyll degradation [11–15]. In addition, studies have demonstrated that overexpression of *MYC3* and *MYC4* in *Arabidopsis* sp. results in increased anthocyanin accumulation [16]. *MYC2*, *MYC3*, and *MYC4* could regulate the biosynthesis of glucosinolates [17]. In grasses, *MYC1* of barley and wheat regulates the biosynthesis of anthocyanin in the pericarp [18,19]. *MYC* genes also play an essential role in the response to abiotic stress. In *Arabidopsis* sp., overexpressed *AtMYC2* could enhance salt tolerance [20], and *AtMYC67* and *AtMYC70* could enhance cold tolerance [21]. In addition, MYCs activate the JA signaling pathway by binding to the G-box sequence of the downstream gene promoter [22–26]. *AtMYC2* plays an important role in the JA-induced activation of defense genes [27]. In addition, the *MYC1* gene in tomato (*SlMYC1*) could affect the relationship between JA and gibberellin hormone signal transduction; it could also affect pathogen defense [28]. *SlMYC1* could play a crucial role in the morphological development of type VI trichomes as well as in the division and proliferation of glandular cells [29,30]. *MYC1* and *MYC2* have a dual role in the regulation of constitutive and stress-inducible specialized metabolism in tomato [31].

Until presently, the *MYC* gene family has been identified in multiple species such as 26 *TaMYC* of *Triticum aestivum*, 7 *OsMYC* of *Oryza sativa*, 7 *BdMYC* of *Brachypodium distachyon*, and 8 *ZmMYC* of *Zea mays* [8]. However, the bioinformatics analysis of the *SlMYC* family has not been documented. Using bioinformatics methods, 14 members of the tomato *MYC* gene family were identified. Their phylogenetic relationships, protein-conserved motifs, chromosome structure analysis, and localization were then thoroughly examined. Finally, the expression profiles of *SlMYC* genes were determined using real-time quantitative PCR in various organs and under stress treatment with methyl jasmonate (MeJA), sodium chloride (NaCl), and mannitol. Using clustered regularly interspaced short palindromic repeats/Cas 9 (CRISPR/Cas 9), the mutant of *SlMYC1* was obtained, and reduced salt and mannitol resistance was observed. This study established a foundation for further research into the role of the tomato *MYC* gene under abiotic stress.

2. Materials and Methods

2.1. Plant Materials and Treatments

Tomato seeds were soaked in warm water at 55 °C for 1 to 2 h. The seeds were placed in a Petri dish covered with two layers of wet filter paper and germinated in an incubator (BIOBASE, China) at 28 °C. Following 80% germination, the seeds were planted on perlite, covered with perlite, and sprayed with tomato nutrient solution, including $Ca(NO_3)_2 \cdot 4H_2O$ 590 $mg \cdot L^{-1}$, KNO_3 404 $mg \cdot L^{-1}$, KH_2PO_4 136 $mg \cdot L^{-1}$, $MgSO_4 \cdot 7H_2O$ 246 $mg \cdot L^{-1}$, EDTA·Na_2-Fe 40 $mg \cdot L^{-1}$, H_3BO_3 2.86 $mg \cdot L^{-1}$, $MnSO_4 \cdot 4H_2O$ 2.13 $mg \cdot L^{-1}$, $ZnSO_4 \cdot 7H_2O$ 0.22 $mg \cdot L^{-1}$, $CuSO_4 \cdot 5H_2O$ 0.08 $mg \cdot L^{-1}$, $(NH_4)_6Mo_7O_{24} \cdot 4H_2O$ 0.02 $mg \cdot L^{-1}$. The tomato seedlings were transferred to a 4 L square pot for hydroponic growth once they had produced a true leaf. The roots, stems, leaves, flowers, immature fruit, and mature fruits were collected for tissue analysis during the growth stage. When the tomato seedlings had three leaves, they were treated for 3 h, 6 h, and 12 h with 150 mM NaCl, 100 μM MeJA, or 100 mM mannitol (Sangon Biotech., Shanghai, China), and their leaves were collected. Meanwhile, another set of plants that had not been exposed to any stress was utilized as a control.

2.2. Identification and Characterization of the Tomato MYC Transcription Factors

The hidden Markov model (HMM) of the bHLH-MYC_N domain (PF14215) and HLH domain (PF00010) and the characteristics of *MYCs* were retrieved from the Pfam website. A Venn diagram was created using TBtools (v1.112) [32]. The tomato genome files (SL3.0) were downloaded from the database of the National Center for Biotechnology Information (NCBI) (https://www.ncbi.nlm.nih.gov/) (accessed on 20 September 2021.), and *MYC* protein sequences were obtained (with an E-value cutoff of 0.005) from the tomato genome employing BLAST and hmm search functions; thereafter, the bHLH-MYC_N and HLH domains were further determined using Pfam analysis. Based on the database accession, the number of proteins and the nucleotide and genomic sequences of each *TaMYC*, *NtMYC*, and *AtMYC* gene copy were validated [8]. The theoretical predicted isoelectric point (PI), molecular weight (Da), and grand average of hydropathy (GRAVY) of the *SlMYC* protein were examined by the online analysis software Expasy (http://www.expasy.ch/tools/protparam.html) (accessed on 21 September 2021).

2.3. Protein Sequence, Phylogenesis, Gene Structure, Mapping MYC Genes, and Subcellular Localization of SlMYC Analyses

The structural and functional domains of *SlMYC* were analyzed. The coding sequence of *SlMYC* was obtained from the transcriptome (https://www.ncbi.nlm.nih.gov/refseq) (accessed on 22 September 2021.). Using Pfam (http://pfam.xfam.org/) (accessed on 24 September 2021.) and HMMER (https://www.ebi.ac.uk/Tools/hmmer/) (accessed on 24 September 2021.), the protein domain of *SlMYC* was predicted. The protein sequences of bHLH-MYC_ N and HLH domains were determined using the multiple sequence alignment of cluster omega and visualization of jalview software (2.11.1.4). The tree was visualized using the iTol online website (https://itol.embl.de/tree/) (accessed on 28 September 2021.), which was built using adjacency joining (NJ) and 1000 repeat bootstrap methods. The phylogenetic tree and conserved protein domain of *SlMYC* were visualized using the MEGAX program (1.0.0.0) and TBtools (v1.112), respectively. The exon and intron structure, as well as the conserved protein domain of the *MYC* gene, were determined with gene structure display server 2.0 (http://mg2c.iask.in/mg2c_v2.1/) (accessed on 5 October 2021.) using information from the genome database. The location of *SlMYC* on the tomato chromosome and the synteny relationship were determined through TBtools software (v1.112). Subcellular localization prediction was examined through the online analysis Cell-PLoc 2.0 (http://www.csbio.sjtu.edu.cn/bioinf/Cell-PLoc-2/) (accessed on 8 March 2022).

2.4. Analysis of cis-Regulatory Elements of the SlMYC

To analyze cis-elements in the *SlMYC* promoters, 2000 base pair (bp) sequences upstream of the transcription start sites of the *SlMYC* genes were obtained from the genomic sequence of tomato, and cis-acting elements were predicted in these sequences employing PlantCARE (http://bioinformatics.psb.ugent.be/webtools/plantcare/html/) (accessed on 15 February 2022).

2.5. Gene Expression Analysis

Using the Eastep^{TM} Universal RNA Extraction Kit (Promega, Shanghai, China), total RNA was extracted from leaves of various treatments (Promega, Shanghai, China). All of the quantitative real-time PCR (qRT-PCR) primers were designed using the NCBI online program Primer-BLAST (https://www.ncbi.nlm.nih.gov/tools/primer-blast/) (accessed on 2 April 2022). qRT-PCR was performed in three technical repetitions with complementary DNAs (cDNAs) synthesized from three biological replicates of leaves. The relative expression of specific genes was quantified using the $2^{-\Delta\Delta Ct}$ method, and the *SlActin* gene was used as a constitutive internal control. These primer sequences are listed in Supplemental Table S1.

2.6. Plasmid Construction, Plant Transformation, and Mutant Analysis

The CRISPR-Cas9 technique was performed in accordance with an earlier study [33]. *SlMYC1* mutant primers were designed using the CRISPR-Cas9 website (http://crispr.hzau.edu.cn/cgi-bin/CRISPR2/CRISPR) (accessed on 5 November 2019.). The primer sequences are listed in Supplemental Table S2. After that, pHEE401E-pME6DT1DT2 was used as a template for PCR amplification, and the amplified band was 600–700 bp in size. The PCR product was mixed with the final vector pHEE401E to cut and ligate (15 cycles of 37 °C for 5 min, 10 °C for 5 min, 20 °C for 5 min, and 37 °C for 5 min). The plasmid was then transformed into "Alisa Craig" tomato using *Agrobacterium* LBA4404-mediated transformation. The transgenic tomato was selected using hygromycin resistance.

In order to identify transgenic tomatoes, we designed primers (forward primer: 5′-ATGACGGACTATAGATTA-3′; reverse primer: 5′-TTTCATCGCGATTCAGCA-3′). Tomato genomic DNA was extracted using the cetyltrimethylammonium bromide (CTAB) method [34]. Genomic DNA was employed as the template for PCR amplification. The PCR products were electrophoresed to confirm their sizes and sequenced.

2.7. Salt and Mannitol Treatments to slmyc1 Seeds

The *slmyc1* seeds were soaked in warm water for 1 to 2 h at 55 °C. The soaked seeds were subsequently placed in Petri dishes covered with two layers of filter paper and allowed to germinate in an incubator at 28 °C. The germinated *slmyc1* and Ailsa Craig (AC) seeds were subjected to nutrient solution added to 100 mM NaCl or 100 mM mannitol (Sangon Biotech., Shanghai, China) treatment for 96 h. A tomato nutrient solution was employed as a control (CK). As the last step, the root length was measured, and the contents of reactive oxygen species (ROS) in root tips were examined with 2,7-dichlorodihydrofluorescein diacetate (H_2 DCF-DA) [35]. The root tip of the tomato seedlings was cut into approximately 0.5 cm–1 cm with scissors, rinsed with pure water, and placed in an EP tube containing 2 μM H_2DCFDA (Beyotime Biotech., China) for 30 min in the dark. The root tip was then immersed three times in a 20 mM HEPES-KOH (pH 7.8) buffer solution (Beyotime Biotech., Shanghai, China) for 45 min. The rinsed sample was photographed under a fluorescence microscope.

2.8. Statistical Analysis

All data represent the average of three biological replicates ± standard deviation (SD). One-way analysis of variance (ANOVA) of the *t*-test was performed on the data. (p-values < 0.05 are summarized with one asterisk, and p-values < 0.01 are summarized with two asterisks).

3. Results

3.1. Identification and Characterization of SlMYC Genes

SL3.0 tomato genome annotations for both bHLH-MYC_N and HLH domains were scanned. Approximately 199 HLH domain-containing genes and 32 bHLH-MYC_N domain-containing genes were identified, and 17 candidates of *SlMYC* were found in both groups (Figure S1). We retrieved 14 *SlMYC* genes upon removing the redundant genes. The *SlMYC* genes were assigned their respective names and chromosomal locations (*SlMYC1-SlMYC12*) (Table 1). The coding sequence length of the *SlMYC* genes ranged between 1182 (*SlMYC2-like-1*) and 2070 (*SlMYC4*) bp. The length of predicted proteins ranged between 393 (*SlMYC2-like-2*) and 689 (*SlMYC2*) amino acids. The molecular weight ranged from 45.11 (*SlMYC2-like-2*) to 75.69 (*SlMYC12*) kDa, while the theoretical PI ranged from 5.10 (*SlMYC12*) to 8.4 (*SlMYC11*). It was expected that all *SlMYCs* would remain in the nucleus, implying that they could act as transcription factors in regulating plant-specific processes. All values of GRAVY were <0, indicating that all *SlMYC* proteins are hydrophilic.

3.2. Chromosomal Localization and Synteny Relationship of SlMYC Genes

The chromosomal location of each *SlMYC* gene was determined according to the genomic sequence of the tomato. A total of 14 *SlMYC* genes were located on 6 of the 12 chromosomes (Figure 1). The maximum number of *SlMYC* genes was found in chromosome 8, which comprised six *SlMYC* genes. Two *SlMYC* genes were found in chromosomes 1, 5, and 10. A single *SlMYC* gene was located in chromosomes 6 and 9. Five syntenic pairs of *SlMYC* were identified and linked by colored lines.

Table 1. Characteristics of the MYC gene family in *Solanum lycopersicum*. The bioinformatics online software ExPASy (https://web.expasy.org/protparam/) (accessed on 21 September 2021). was used to predict the physical and chemical properties of 14 tomato MYC family genes at the chromosomal positions and their corresponding protein sizes, molecular weights, and isoelectric and subcellular localization prediction points.

Gene	Accession Number	Chr. Location	No. of Amino Acids (aa)	Length of Gene (bp)	Length of CDS (bp)	Molecular Weight (Da)	Predicted Isoelectric Point (PI)	Subcellular Localization Prediction	GRAVY
MYC1	NM_001301178.2	Chromosome 8: 33,463–35,295	630	2174	1893	69,346.51	5.78	Nucleus	−0.564
MYC2	NM_001324483.1	Chromosome 8: 60,990,466–60,992,535	689	2485	2070	75,041.11	5.51	Nucleus	−0.598
MYC2-like-1	XM_004244608.3	Chromosome 8: 2,993,511–2,996,747	452	1556	1359	51,296.03	6.47	Nucleus	−0.431
MYC2-like-2	XM_004248047.4	Chromosome 10: 3,248,348–3,250,830	393	1608	1182	45,112.04	5.85	Nucleus	−0.454
MYC3-like	XM_004248044.4	Chromosome 10: 3,273,458–3,274,813	451	1769	1356	50,463.80	5.98	Nucleus	−0.435
MYC4	XM_004229943.4	Chromosome 1: 87,017,635–87,020,664	605	3030	1818	67,261.91	6.83	Nucleus	−0.528
MYC5	XM_004229974.4	Chromosome 1: 87,325,345–87,327,252	450	1665	1353	49,496.80	5.44	Nucleus	−0.348
MYC6	XM_010322267.3	Chromosome 5: 240,028–243,178	482	2143	1449	54,013.27	5.84	Nucleus	−0.394
MYC7	XM_010323205.3	Chromosome 5: 61,560,052–61,562,617	579	2556	1740	64,271.00	7.59	Nucleus	−0.425
MYC8	XM_004242307.4	Chromosome 6: 49,306,848–49,309,630	477	2182	1434	52,944.89	6.59	Nucleus	−0.472
MYC9	XM_019215077.2	Chromosome 8: 51,962,518–51,966,949	602	2432	1809	68,432.78	5.61	Nucleus	−0.874
MYC10	NM_001347001.1	Chromosome 8: 64,341,490–64,346,297	626	2353	1881	70,053.83	5.49	Nucleus	−0.546
MYC11	XM_004246037.4	Chromosome 8: 65,832,927–65,835,049	413	2123	1242	46,922.74	8.41	Nucleus	−0.234
MYC12	XM_010328066.3	Chromosome 9: 63,177,650–63,185,525	686	2425	2061	75,694.23	5.10	Nucleus	−0.407

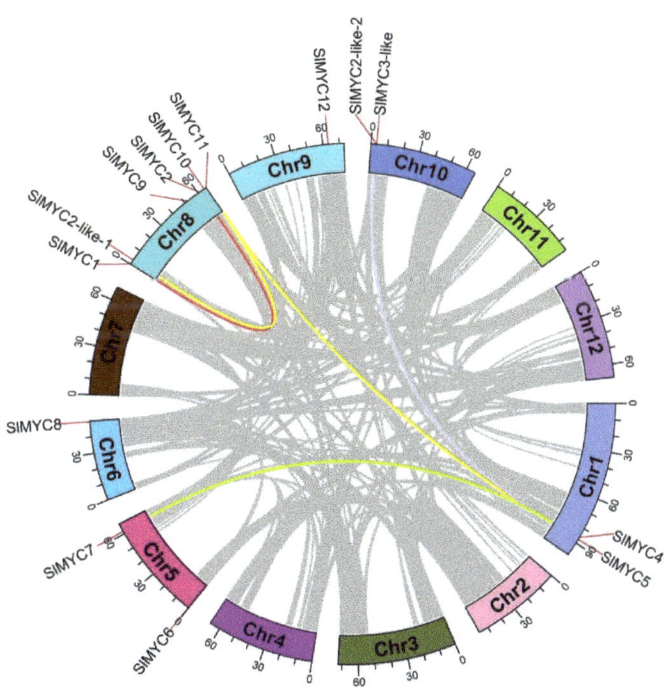

Figure 1. Genome distribution and synteny relationship of *SlMYC* genes in *Solanum lycopersicum*. Different chromosomes were used different colors. Colored lines represent the syntenic pairs of *SlMYC*.

3.3. Phylogenetic Analysis of SlMYC Proteins

All of the *SlMYCs* were expected to remain in the nucleus, implying that they could act as transcription factors in the regulation of plant-specific processes. The *SlMYC* proteins were categorized into four groups based on the expression of conserved motifs (Figure 2), such as class I (*SlMYC4, SlMYC7,* and *SlMYC8*), class II (*SlMYC1, SlMYC2, SlMYC2-like-1, SlMYC2-like-2, SlMYC3-like, SlMYC5,* and *SlMYC11*), class III (*SlMYC10* and *SlMYC12*), and class IV (*SlMYC6* and *SlMYC9*).

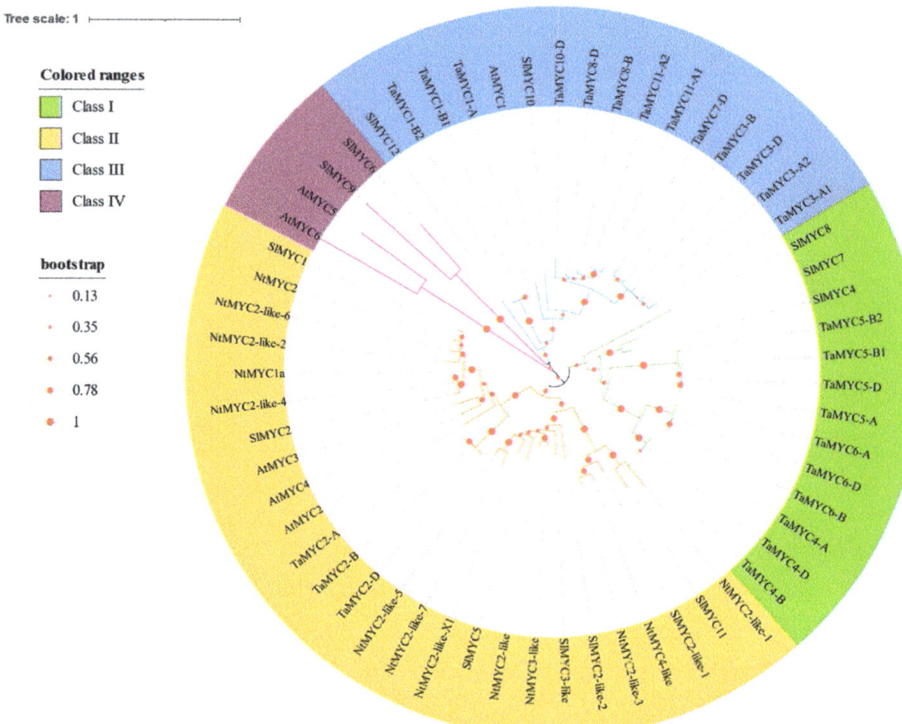

Figure 2. Phylogenetic tree of *MYC* proteins. The phylogenetic tree was carried out based on protein alignments. *SlMYC* of *Solanum lycopersicum*, *TaMYC* of *Triticum aestivum*, *NtMYC* of *Nicotiana tabacum*, and *AtMYC* of *Arabidopsis* were used. The neighbor-joining method is used with 1000 bootstrap trials.

3.4. Conserved Motifs and SlMYC Protein Sequence Alignment

The top seven potential conserved motifs in *SlMYCs* from diverse classes were identified using the MEME software. *SlMYC* proteins belonging to the same class shared similar motifs. Class I (*SlMYC1, SlMYC2, SlMYC2-like-1, SlMYC2-like-2, SlMYC3-like, SlMYC5,* and *SlMYC11*) exhibited five to seven motifs. Class II (*SlMYC4, SlMYC7,* and *SlMYC8*) exhibited six to seven motifs. Class III (*SlMYC10* and *SlMYC12*) had five motifs. In comparison, Class IV (*SlMYC6* and *SlMYC9*) had the lowest number of motifs with three to four motifs, indicating that proteins on the same branches might have substantially similar functions (Figure 3A). Motifs three, four, five, and seven corresponded to the bHLH-MYC_N domains. In contrast, motif one corresponded to the bHLH domain (Figure 3A,B). The bHLH_MYC_N and bHLH domains are located at the N- and C-termini of the *MYC* family, respectively (Figure 3C). An extremely conserved bHLH domain was predicted for all 14 *SlMYC* proteins. The highly conserved bHLH domain, which is associated with DNA binding, consists of approximately 50 amino acids.

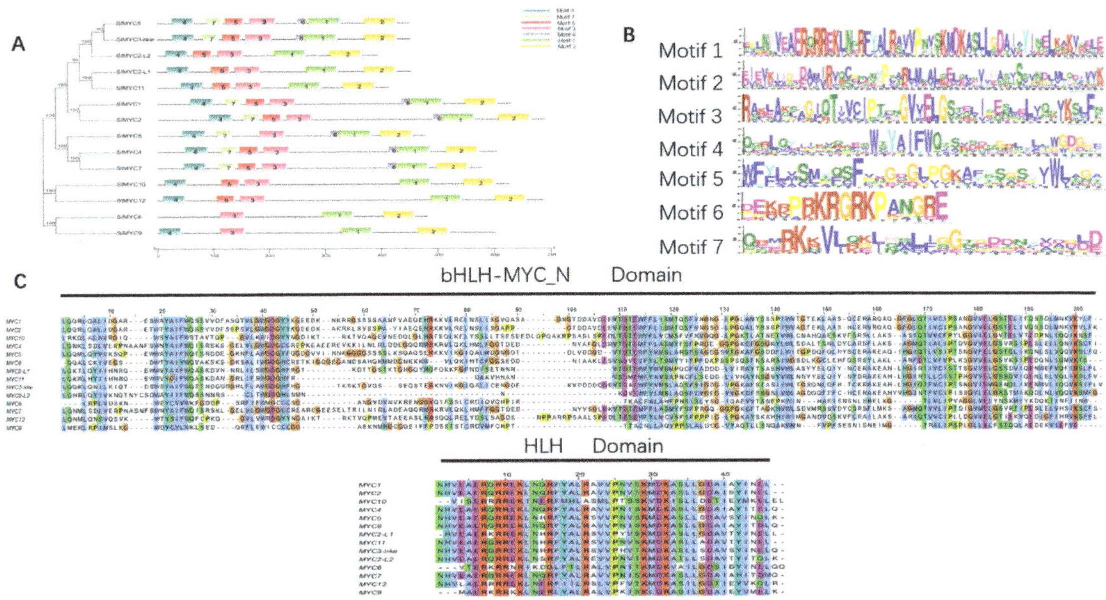

Figure 3. Phylogenetic and conserved motifs of *MYC* proteins in tomato. (**A**) Phylogenetic and conserved motif distribution of 14 tomato *MYCs*. The tree is rooted. (**B**) Logo sequences for 14 *SlMYC* conserved motifs. (**C**) Multiple alignments of *MYC* and two conserved domains (bHLH_MYC_N and bHLH) of tomato *MYC* protein. The conservative amino acids were used different colours.

3.5. Analysis of Exon–Intron Structure and cis-Regulatory Elements of SlMYC Genes

The full lengths of the *SlMYC* genes ranged between 1556 (*SlMYC2-like-1*) and 3030 bp (*SlMYC4*). Various exon–intron locations were compared to gain insight into potential mechanisms of structural diversity found in *SlMYC*. The number of introns varied in this study from 0 to 12. *SlMYC12* exhibited a maximum of 12 introns; however, *SlMYC1*, *SlMYC2*, *SlMYC3-like*, *SlMYC4*, *SlMYC7*, *SlMYC8*, and *SlMYC11* were found to exist without introns in their structure (Figure 4A). Furthermore, *SlMYC6*, *SlMYC9*, and *SlMYC10* contained seven introns, the *SlMYC2-like-2* gene contained two introns, while the *SlMYC2-like-1* and *SlMYC5* genes contained one intron (Figure 4A).

The upstream promoter region (2000 bp) of the *SlMYC* genes was extracted from the tomato genome to identify cis-regulatory components. In *SlMYC* promoter regions, four hormone-responsive regulatory elements were identified, including abscisic-acid-responsive elements (ABRE), a TGACG motif, a TGA element, and a TATC box, which were correlated with abscisic acid, MeJA, auxin, gibberellin, and salicylic acid responses (Figure 4B). The promoter region of *SlMYC* copies contained three light-responsive regulatory elements, including circadian, a TCT motif, and an MSL recognition element (MRE). Moreover, other regulatory elements (such as CAT box, TC-rich repeats, a CCAAT-box, a myosin-binding subunit (MBS), and a TCA element) associated with meristem expression, defense, and stress response as well as myeloblastosis (MYB) binding site involved in drought-inducibility and salicylic acid response were identified (Figure 4B).

3.6. Expression Profiles of SlMYC in Various Organs of Tomato Plants

Using qRT-PCR, the expression levels of the 14 *SlMYC* genes were examined in the roots, stems, leaves, immature fruit, and mature fruit of tomato plants. *SlMYC6* and *SlMYC12* were expressed in the roots, and *SlMYC11* was expressed in the stem. *SlMYC5*, *SlMYC7*, and *SlMYC10* were extremely exhibited in the leaves. *SlMYC2-like-1*, *SlMYC3-like*, *SlMYC8*, and *SlMYC9* were extremely expressed in the flowers. *SlMYC1* and *SlMYC2* were extremely expressed in the immature fruit. *SlMYC4* was exhibited in the mature fruit (Figure 5). Among these 14 *SlMYC* genes, the expression of *SlMYC2-like-2* and *SlMYC11* was not detected in any of the organs of tomato. According to this finding, the *SlMYC* gene is expressed to varying degrees in each of these organs, with leaves and flowers exhibiting the highest expression levels.

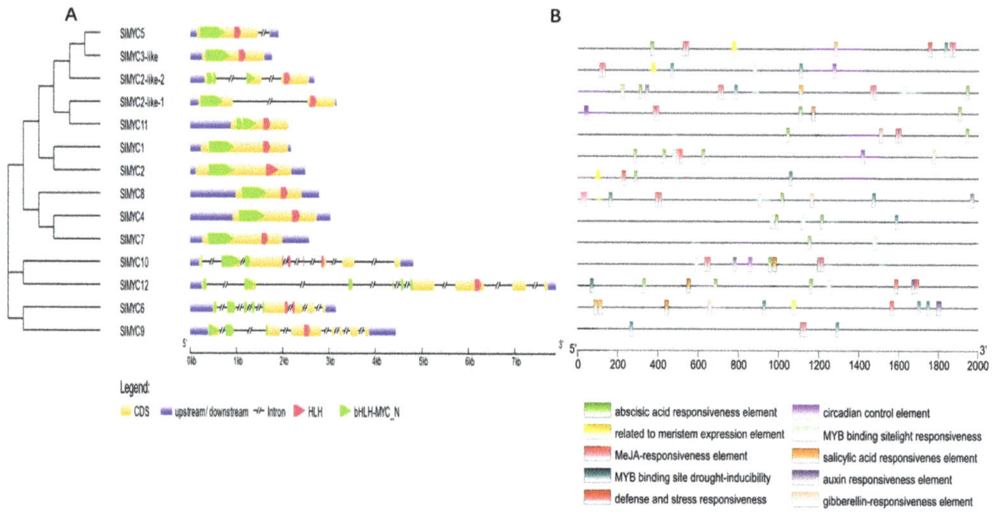

Figure 4. Analysis of exon–intron structures and cis-regulatory elements of *SlMYC* genes. (**A**) Exon–intron organization of *MYC* genes in *Solanum lycopersicum* (Sl) grouped according to their gene orthology. Blue and yellow bars indicate untranslated (UTR) regions and exons. Black interrupted lines indicate introns. The tree is rooted. (**B**) The number and composition of cis-acting regulatory elements in the promotor region of *SlMYC* genes. The 2000 base pair promoter region of each gene copy is displayed. Different colorful shapes show different elements.

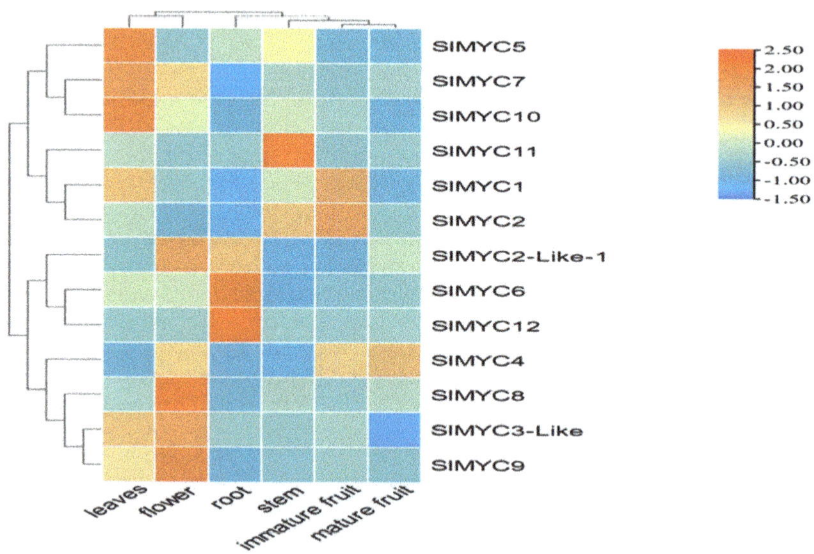

Figure 5. Response analysis of *SlMYC* genes in the roots, stems, leaves, flowers, immature fruit, and mature fruit by qRT−PCR. RNA samples were extracted from different tissues of tomato.

3.7. Expression Patterns of SlMYC under MeJA, NaCl, and Mannitol Stress

To better understand the role of *SlMYCs* in the response to abiotic stress, tomato seedlings were subjected to MeJA, NaCl, and mannitol treatment. As illustrated in Figure 6, *SlMYC2-like-1*, *SlMYC6*, *SlMYC9*, and *SlMYC12* exhibited a similar expression pattern and increased significantly by 14.0-, 41.0-, 1.9-, and 19.0-fold in comparison to the control following 3 h of MeJA treatment. Subsequently, the expression level was progressively decreased, and *SlMYC2-like-1*, *SlMYC9*, and *SlMYC12* reduced to the same level as the control at 6 h and 12 h. At 6 and 12 h, *SlMYC6* was expressed at 2.0- and 9.5-fold higher levels compared to the control. In contrast, *SlMYC5* expression was distinctly downregulated by 68.2% at 3 h compared to the control. At 6 h, following MeJA treatment, *SlMYC1*, *SlMYC8*, and *SlMYC10* exhibited a significant upregulation of 2.1-, 5.0-, and 2.7-fold, respectively. *SlMYC7* was gradually increased at 3 h, 6 h, and 12 h following MeJA treatment, and the expression peaked at 6 h. Following MeJA treatment, the expression of *SlMYC2*, *SlMYC3-like*, and *SlMYC4* did not differ substantially from the control (Figure 6).

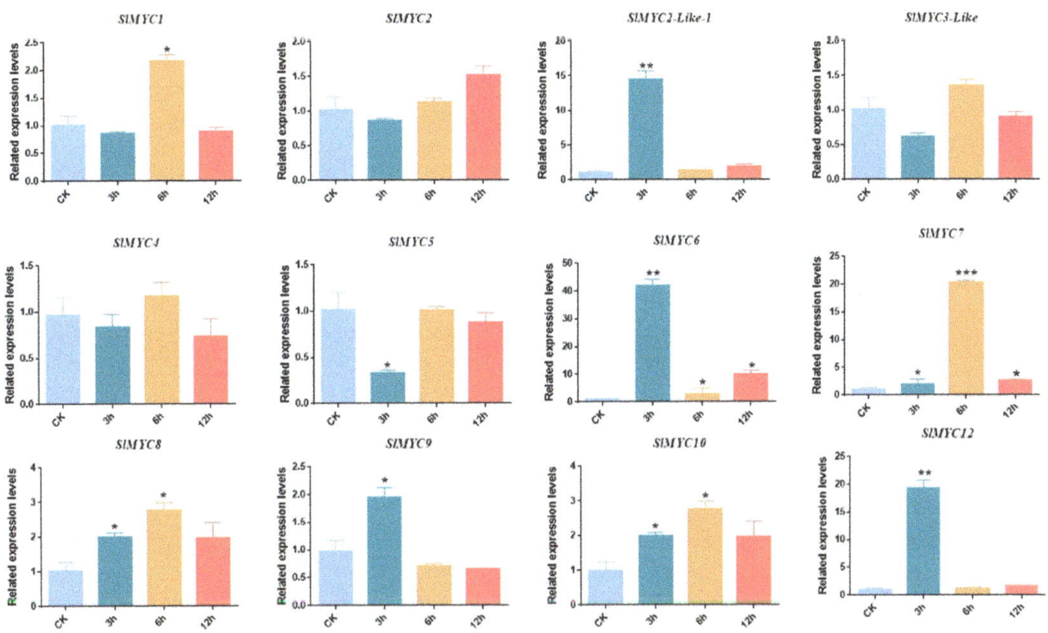

Figure 6. Response analysis of *SlMYC* genes in leaves under MeJA treatment by qRT-PCR. The expression patterns of *SlMYC* in response to 100 μM MeJA treatment for 3 h, 6 h, 12 h. Data represent the means of three biological replicates ±SD. Different colors were used to indicate different treatment times (*p*-values less than 0.05 are summarized with one asterisk, *p*-values less than 0.01 are summarized with two asterisks and *p*-values less than 0.0003 are summarized with three asterisk).

Following 3 h of NaCl treatment, *SlMYC1* and *SlMYC10* showed 4.1- and 2.0-fold upregulation, respectively, in comparison to the control. After that, the expression level gradually decreased with time, and at 12 h after NaCl treatment, *SlMYC10* was downregulated by 65.0% compared to the control. However, although sharing a similar expression pattern, *SlMYC2-like-1*, *SlMYC3-like*, *SlMYC4*, and *SlMYC12* were significantly downregulated following exposure to NaCl. *SlMYC5* and *SlMYC6* progressively decreased following NaCl treatment and were downregulated significantly by 82.0% and 95.0% at 6 h and 85.1% and 80.2% at 12 h, respectively. Following NaCl treatment, the expression of *SlMYC2*, *SlMYC7*, *SlMYC8*, and *SlMYC9* did not differ substantially from the control (Figure 7).

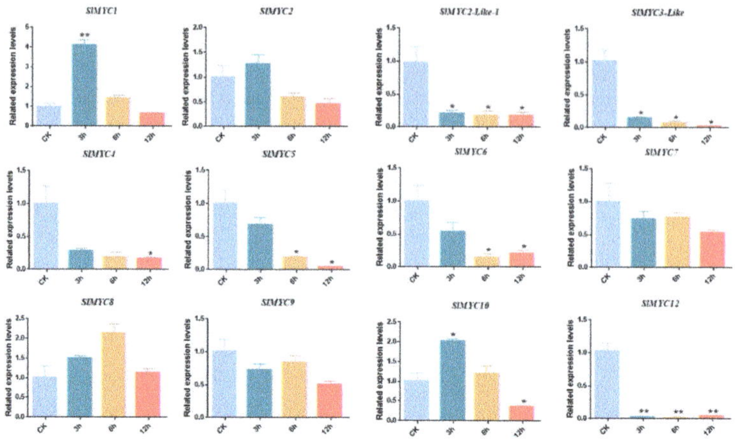

Figure 7. Response analysis of *SlMYC* genes in tomato leaves under NaCl stress by qRT-PCR. The expression patterns of *SlMYC* in response to 150 mM NaCl treatment for 3 h, 6 h, and 12 h. Data represent the means of three biological replicates ±SD. Different colors were used to indicate different treatment times (*p*-values less than 0.05 are summarized with one asterisk, and *p*-values less than 0.01 are summarized with two asterisks).

Finally, the expression of *SlMYC* genes under 100 mM mannitol treatment was examined. At 3 h, 6 h, and 12 h following mannitol treatment, the expression levels of *SlMYC1*, *SlMYC3-like*, *SlMYC4*, *SlMYC7*, and *SlMYC8* were lower. *SlMYC2-like-1*, *SlMYC6*, and *SlMYC12* exhibited a similar expression pattern, and their expression gradually increased over the treatment period and peaked at 12 h. The expression of *SlMYC5* was significantly downregulated by 78.5% and 68.4% 3 and 6 h after mannitol treatment, respectively. Following 6 h of mannitol treatment, the expression of *SlMYC10* was 2.1-fold higher than in the control. The expression of *SlMYC2* and *SlMYC9* did not differ substantially from the control group following mannitol treatment (Figure 8).

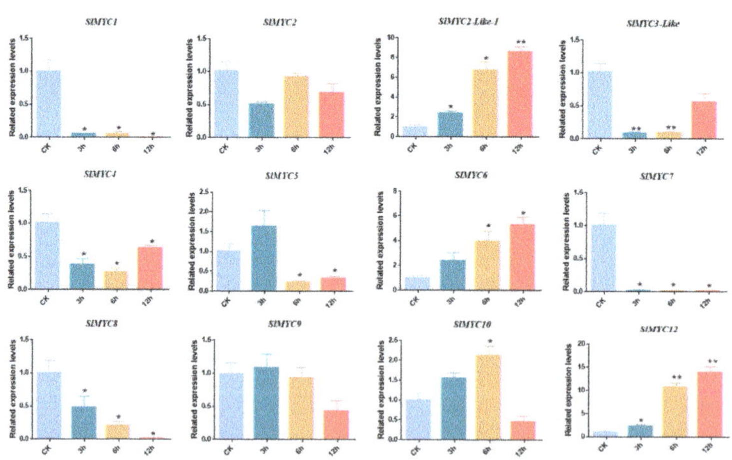

Figure 8. Response analysis of *SlMYC* genes in tomato leaves under mannitol by qRT-PCR. The expression patterns of *SlMYC* in response to 100 mM mannitol treatment for 3 h, 6 h, and 12 h. Data represent the means of three biological replicates ±SD. Different colors were used to indicate different treatment times (*p*-values less than 0.05 are summarized with one asterisk, and *p*-values less than 0.01 are summarized with two asterisks).

3.8. CRISPR/Cas9-Mediated Mutagenesis in slmyc1

Our previous transcriptome data of tomatoes under low nitrogen stress showed that the transcript of *SlMYC1* was significantly changed (data not shown). To further understand the function of *SlMYC1* under abiotic stress, the *SlMYC1* was mutated in the Alisa Craig tomato using the CRISPR/Cas9 technique. In the exons encoding the function domain of *SlMYC1*, two target sites, target 1 and target 2, were designed. A hygromycin-resistant transgenic line (T0 generation) was created and grown. A pair of primers was designed to amplify the sequence including target 1 and target 2 in the genomic DNA. The size of the PCR product of the T0 generation was 435 bp, according to agarose gel electrophoresis of the PCR products with the genomic DNA containing the two target sequences, whereas that of the wild-type (WT) was 1895 bp (Figure 9A). In addition, DNA sequence analysis revealed that 1461 bp was deleted in the mutation (Figure 9B).

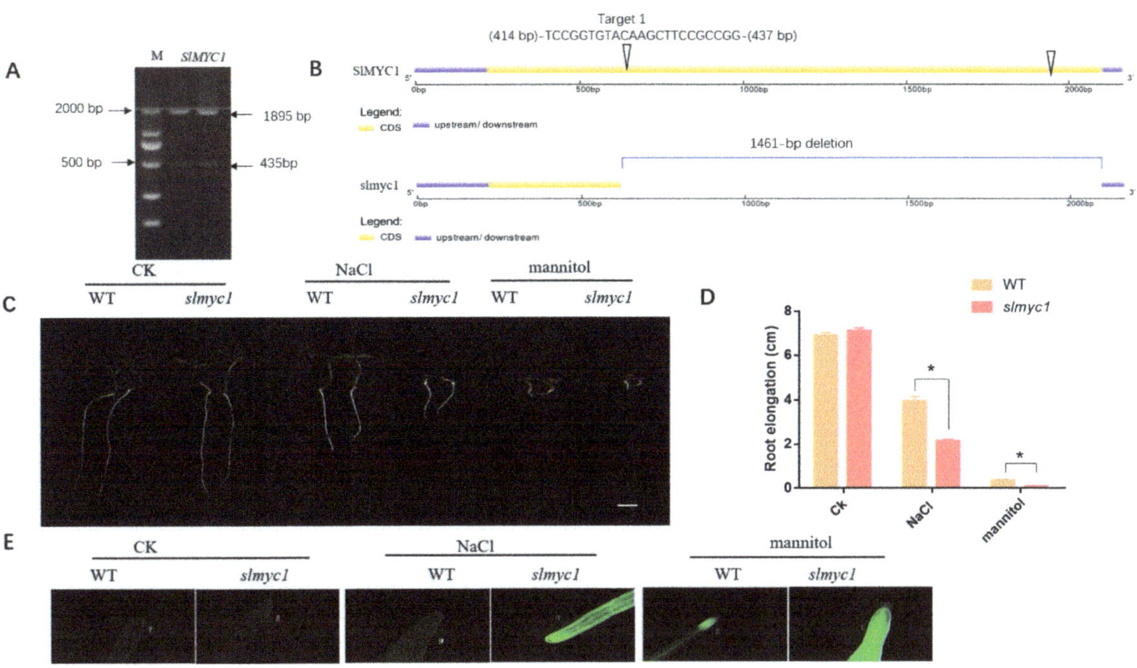

Figure 9. CRISPR/Cas9-mediated mutation of the *SlMYC* gene and the response of *slmyc1* to NaCl and mannitol stress. (**A**) The size of DNA fragments amplified by PCR with genomic DNA. M: DNA marker. (**B**) Schematic diagram of the sgRNA target sites on the *SlMYC1* gene and the sequencing results of the sequence near the mutation site of the *SlMYC1* gene. M: marker. (**C**) Growth of the *slmyc1* mutant and wild-type (WT) seeds after 96 h of NaCl and mannitol treatment. The tomato nutrient solution treatment was employed as a control (CK). (**D**) Statistical analysis of the root length of the *slmyc1* mutant and the WT at 96 h after NaCl and mannitol treatment (p-values less than 0.05 are summarized with one asterisk). (**E**) Determination of ROS content in root tips of the *slmyc1* mutant after 96 h of NaCl and mannitol treatment.

3.9. slmyc1 Had Decreased Tolerance to NaCl and Mannitol Stress

The effects of NaCl and mannitol on the growth of *slmyc1* were analyzed. We observed that the seed germination of *slmyc1* was not affected under normal conditions (Figure 9C). However, following 96 h of 100 mM NaCl treatment, the growth of *slmyc1* was substantially inhibited, and the root elongation decreased by 50.1% in comparison to the control (Figure 9D). Following 96 h of treatment with 100 mM mannitol, both the WT and *slmyc1* exhibited significant root length inhibition, with *slmyc1* exhibiting more inhibition compared to WT (Figure 9D). In addition, following NaCl and mannitol treatment, ROS fluorescence in the root tips increased in both WT and *slmyc1*; however, *slmyc1* had more ROS in comparison to the WT (Figure 9E). These results indicated that the *slmyc1* mutant significantly decreased the resistance to NaCl and mannitol stress and *SlMYC1* might function as a positive regulator under salt and dehydration stress.

4. Discussion

In this study, 14 members of the *MYC* gene family were isolated and identified to better understand the function of *MYC* genes in tomato plants (Table 1). Phylogenetic analysis separated these *SlMYCs* into four subgroups (Figure 2). Cis-acting elements may function as an important molecular switch to regulate gene expression under stressful conditions [36]. Four hormone regulatory elements, four light response regulatory elements, and one emergency stress response element were identified in our study (Figure 4). Besides *SlMYC3-like*, *SlMYC4*, *SlMYC6*, and *SlMYC9*, other members of the *MYC* family contain varying quantities of hormone-responsive elements, particularly ABA (ABRE). As a "stress hormone", ABA could regulate many of the growth and development processes of plants, including seed germination, the senescence of tissues and organs, and stress tolerance [37]. Studies have demonstrated that *MYC2* is associated with the ABA signaling pathway in *Arabidopsis* sp. [38]. Fourteen *SlMYC* genes have different types and numbers of cis-acting elements in their respective promoter regions, suggesting that these genes might be regulated differently and respond to different stresses.

As a subfamily of the bHLH family, *SlMYC* possesses a highly conserved domain HLH at the C-terminus (Figure 3C), which binds to the G-box cis-element in the JA-responsive promoter [39,40]. Phylogenetic tree analysis revealed that *SlMYC6* and *SlMYC9* have more independent evolutionary trends than other members of the *MYC* family (Figure 1). According to gene structure analysis, the majority of *SlMYC* family members have fewer introns (71%). A significant proportion has no introns (70%) (Figure 3). This variation in gene structure could be due to alternative splicing. Genes with fewer introns could evolve quickly through rapid replication or reverse transcription, whereas genes with more introns could evolve and acquire new functions [41]. Sequence logos of conserved amino acid residues (protein motif) exhibit an extremely conserved pattern of *SlMYC* family members at the C- and N- termini (Figure 3), indicating that *SlMYC* is evolutionarily conserved.

Earlier studies have demonstrated that *MYC* transcription factors play a crucial role in the growth and development of plants [14,42,43]. For example, *MYC* proteins could regulate plant seed germination, root elongation, leaf senescence, and flower development [44]. In one study, *OsMYC* genes were significantly expressed in stems [44]. In another study, *BdMYC* genes were exhibited substantially in inflorescences [44]. In an additional study, *TaMYC* genes were significantly expressed in leaves [8]. In our study, *SlMYC* gene expression varied to different degrees in all the organs of tomato plants and was expressed primarily in the leaves and flowers (Figure 5).

The expression profiles of *MYC* genes indicates their potential roles in the response to abiotic stress. For example, *FaMYC2* was significantly increased at 15 min, 30 min, and 6 h following MeJA treatment [45]. JA could induce the expression of *AtMYC* [23]. In our study, the expression of most *SlMYC* genes was upregulated following MeJA treatment (Figure 6). These results indicate that the *SlMYC* gene is involved in the JA signaling pathway. Salinity stress induced the expression of *TaMyc-B1* and *TaMyc-A2*; however, *TaMyc-A1* was suppressed [46]. In our experiment, *SlMYC* primarily exhibited a declining trend following NaCl treatment (Figure 7). *ZmMYC* was significantly upregulated under drought stress [47]. *TcMYC* in *Taxus* sp. was upregulated as a result of drought and excessive salinity stress [48]. However, in this study, *SlMYC* exhibited various expression patterns under mannitol treatment (Figure 8). *SlMYC2-like-1*, *SlMYC6*, and *SlMYC12* exhibited a similar expression pattern, suggesting that they might have similar functions in the response to abiotic stress. In our study, we used the CRISPR/Cas 9 technique to knock out the function of *SlMYC1*. In comparison to WT, the *slmyc1* had a shorter root length and higher ROS after NaCl and mannitol treatments (Figure 9). Our findings indicate that *slmyc1* reduces NaCl and mannitol stress tolerance and has beneficial effects in NaCl and mannitol treatments.

5. Conclusions

In conclusion, this study offers a genome-wide analysis of *MYC* genes in tomatoes. In total, 14 *SlMYCs* were identified and divided into four classes. The expression patterns of *SlMYCs* provide information about their various physiological functions in tomato growth and development, particularly in MeJA, NaCl, and mannitol regulation. Moreover, the *slmyc1* mutant exhibited decreased tolerance to NaCl and mannitol stress.

Supplementary Materials: The following supporting information can be downloaded at: https://www.mdpi.com/article/10.3390/agronomy13030757/s1. Figure S1: Identification of the SlMYC gene family in tomato; Table S1: Specific primers of qRT-PCR amplification of MYC genes. Table S2: Specific primers of CRISPR-Cas9 amplification of SlMYC1 gene.

Author Contributions: Y.F. and S.Z. performed the experiments and analyzed the data. Y.F. wrote the manuscript. J.Y. and K.L. helped in the revision of the manuscript. H.X. designed the research and revised the manuscript. All authors have read and agreed to the published version of the manuscript.

Funding: This research was funded by the National Natural Science Foundation of China (grant no. 32260753) and the Yunnan Ten Thousand Talents Plan: Young and Elite Talents Project.

Data Availability Statement: The data presented in this study are available on request from the corresponding author.

Conflicts of Interest: The authors declare no competing interest.

References

1. Ferré-D'Amaré, A.R.; Pognonec, P.; Roeder, R.G.; Burley, S.K.J.E.J. Structure and function of the b/HLH/Z domain of USF. *EMBO J.* **1994**, *13*, 180–189. [CrossRef] [PubMed]
2. Massari, M.E.; Murre, C. Helix-Loop-Helix Proteins: Regulators of Transcription in Eucaryotic Organisms. *Mol. Cell. Biol.* **2000**, *20*, 429–440. [CrossRef] [PubMed]
3. Atchley, W.R.; Terhalle, W.; Dress, A. Positional Dependence, Cliques, and Predictive Motifs in the bHLH Protein Domain. *J. Mol. Evol.* **1999**, *48*, 501–516. [CrossRef] [PubMed]
4. Penuelas, M.; Monte, I.; Schweizer, F.; Vallat, A.; Reymond, P.; Garcia-Casado, G.; Franco-Zorrilla, J.M.; Solano, R. Jasmonate-Related MYC Transcription Factors Are Functionally Conserved in Marchantia polymorpha. *Plant Cell* **2019**, *31*, 2491–2509. [CrossRef]
5. Reinbothe, C.; Springer, A.; Samol, I.; Reinbothe, S. Plant oxylipins: Role of jasmonic acid during programmed cell death, defence and leaf senescence. *FEBS J.* **2009**, *276*, 4666–4681. [CrossRef]
6. Zheng, Y. Roles of jasmonate signalling in plant inflorescence and flower development. *Curr. Opin. Plant Biol.* **2015**, *27*, 44–51.
7. Kazan, K. Diverse roles of jasmonates and ethylene in abiotic stress tolerance. *Trends Plant Sci.* **2015**, *20*, 219–229. [CrossRef]
8. Chen, S.; Zhao, H.; Luo, T.; Liu, Y.; Li, H. Characteristics and Expression Pattern of MYC Genes in Triticum aestivum, Oryza sativa, and Brachypodium distachyon. *Plants* **2019**, *8*, 274. [CrossRef]
9. Cai, Q.; Yuan, Z.; Chen, M.; Yin, C.; Luo, Z.; Zhao, X.; Liang, W.; Hu, J.; Zhang, D.J.N.C. Jasmonic acid regulates spikelet development in rice. *Nat. Commun.* **2014**, *5*, 3476. [CrossRef]
10. Uji, Y.; Taniguchi, S.; Tamaoki, D.; Shishido, H.; Akimitsu, K.; Gomi, K.J.P.; Physiology, C. Overexpression ofOsMYC2Results in the Up-Regulation of Early JA-Rresponsive Genes and Bacterial Blight Resistance in Rice. *Plant Cell Physiol.* **2016**, *57*, 1814–1827. [CrossRef]
11. Qi, T.; Wang, J.; Huang, H.; Liu, B.; Xie, D.J.P.C. Regulation of Jasmonate-Induced Leaf Senescence by Antagonism between bHLH Subgroup IIIe and IIId Factors in Arabidopsis. *Plant Cell* **2015**, *27*, 1634–1649. [CrossRef] [PubMed]
12. Debora, G.; Aurore, C.; Acosta, I.F.; Jonas, G.; Laurens, P.; Alain, G.; René, D.; Esteban, A.; Farmer, E.E.; Hao, Y.J.P.G. Multilayered Organization of Jasmonate Signalling in the Regulation of Root Growth. *PLoS Genet.* **2015**, *11*, e1005300.
13. Qi, T.; Huang, H.; Song, S.; Xie, D.J.P.C. Regulation of Jasmonate-Mediated Stamen Development and Seed Production by a bHLH-MYB Complex in Arabidopsis. *Plant Cell* **2015**, *27*, 1620–1633. [CrossRef] [PubMed]
14. Gao, C.; Qi, S.; Liu, K.; Li, D.; Jin, C.; Li, Z.; Huang, G.; Hai, J.; Zhang, M.; Chen, M.; et al. MYC2, MYC3, and MYC4 function redundantly in seed storage protein accumulation in Arabidopsis. *Plant Physiol. Biochem.* **2016**, *108*, 63–70. [CrossRef]
15. Zhu, X.; Chen, J.; Xie, Z.; Gao, J.; Ren, G.; Gao, S.; Zhou, X.; Kuai, B. Jasmonic acid promotes degreening via MYC2/3/4- and ANAC019/055/072-mediated regulation of major chlorophyll catabolic genes. *Plant J.* **2015**, *84*, 597–610. [CrossRef]

16. Niu, Y.; Figueroa, P.; Browse, J. Characterization of JAZ-interacting bHLH transcription factors that regulate jasmonate responses in Arabidopsis. *J. Exp. Bot.* **2011**, *62*, 2143–2154. [CrossRef]
17. Schweizer, F.; Fernández-Calvo, P.; Zander, M.; Diez-Diaz, M.; Fonseca, S.; Glauser, G.; Lewsey, M.G.; Ecker, J.R.; Reymond, S.; Reymond, P. Arabidopsis Basic Helix-Loop-Helix Transcription Factors MYC2, MYC3, and MYC4 Regulate Glucosinolate Biosynthesis, Insect Performance, and Feeding Behavior. *Plant Cell* **2013**, *25*, 3117–3132. [CrossRef]
18. Yuan, Z.; Xinyuan, X.; Shiming, L.; Wenjie, C.; Bo, Z.; Dengcai, L.; Baolong, L.; Daowen, W.; Huaigang, Z. Allelic Variation and Transcriptional Isoforms of Wheat TaMYC1 Gene Regulating Anthocyanin Synthesis in Pericarp. *Front. Plant Sci.* **2017**, *8*, 1645.
19. Yu, S.O.; Hans-Peter, M.; Kukoeva, T.V.; Andreas, B.; Khlestkina, E.K. Regulation of the Flavonoid Biosynthesis Pathway Genes in Purple and Black Grains of Hordeum vulgare. *PLoS ONE* **2016**, *11*, e0163782.
20. Abe, H.; Yamaguchi-Shinozaki, K.; Urao, T.; Iwasaki, T.; Hosokawa, D.; Shinozaki, K. Role of Arabidopsis MYC and MYB Homologs in Drought and Abscisic Acid-Regulated Gene Expression. *Plant Cell* **1997**, *9*, 1859–1868.
21. Yamaguchi-Shinozaki, S. Gene networks involved in drought stress response and tolerance. *J. Exp. Bot.* **2007**, *58*, 221. [CrossRef] [PubMed]
22. Shyu, C.; Figueroa, P.; Depew, C.L.; Cooke, T.F.; Sheard, L.B.; Moreno, J.E.; Katsir, L.; Zheng, N.; Browse, J.; Howe, G.A. JAZ8 lacks a canonical degron and has an EAR motif that mediates transcriptional repression of jasmonate responses in Arabidopsis. *Plant Cell* **2012**, *24*, 536–550. [CrossRef] [PubMed]
23. Pauwels, L.; Barbero, G.F.; Geerinck, J.; Tilleman, S.; Grunewald, W.; Perez, A.C.; Chico, J.M.; Bossche, R.V.; Sewell, J.; Gil, E.; et al. NINJA connects the co-repressor TOPLESS to jasmonate signalling. *Nature* **2010**, *464*, 788–791. [CrossRef] [PubMed]
24. Ke, J.; Ma, H.; Gu, X.; Thelen, A.; Brunzelle, J.S.; Li, J.; Xu, H.E.; Melcher, K.J.S.A. Structural basis for recognition of diverse transcriptional repressors by the TOPLESS family of corepressors. *Sci. Adv.* **2015**, *1*, e1500107. [CrossRef] [PubMed]
25. Du, M.; Chen, Q.; Zhai, Q.; Fang, M.; Li, C.; Shen, J.; Lu, Y.; Deng, L.; Wang, Q.; Liu, Y. MYC2 Regulates the Termination of Jasmonate Signaling via an Autoregulatory Negative Feedback Loop. *Plant Cell* **2019**, *31*, 106–127.
26. Chini, A.; Fonseca, S.; Fernandez, G.; Adie, B.; Chico, J.M.; Lorenzo, O.; Garcia-Casado, G.; Lopez-Vidriero, I.; Lozano, F.M.; Ponce, M.R.; et al. The JAZ family of repressors is the missing link in jasmonate signalling. *Nature* **2007**, *448*, 666–671. [CrossRef]
27. Boter, M.; Ruiz-Rivero, O.; Abdeen, A.; Prat, S. Conserved MYC transcription factors play a key role in jasmonate signaling both in tomato and Arabidopsis. *Gene Dev.* **2004**, *18*, 1577–1591. [CrossRef]
28. Panda, S.; Jozwiak, A.; Sonawane, P.D.; Szymanski, J.; Kazachkova, Y.; Vainer, A.; Kilambi, H.V.; Almekias-Siegl, E.; Dikaya, V.; Bocobza, S.; et al. Steroidal alkaloids defence metabolism and plant growth are modulated by the joint action of gibberellin and jasmonate signalling. *New Phytol.* **2022**, *233*, 1220–1237. [CrossRef]
29. Hua, B.; Chang, J.; Wu, M.L.; Xu, Z.J.; Zhang, F.Y.; Yang, M.N.; Xu, H.M.; Wang, L.J.; Chen, X.Y.; Wu, S. Mediation of JA signalling in glandular trichomes by thewoolly/SlMYC1regulatory module improves pest resistance in tomato. *Plant Biotechnol. J.* **2021**, *19*, 375–393. [CrossRef]
30. Xu, J.S.; van Herwijnen, Z.O.; Drager, D.B.; Sui, C.; Haring, M.A.; Schuurink, R.C. SlMYC1 Regulates Type VI Glandular Trichome Formation and Terpene Biosynthesis in Tomato Glandular Cells. *Plant Cell* **2018**, *30*, 2988–3005. [CrossRef]
31. Swinnen, G.; De Meyer, M.; Pollier, J.; Molina-Hidalgo, F.J.; Ceulemans, E.; Venegas-Molina, J.; De Milde, L.; Fernandez-Calvo, P.; Ron, M.; Pauwels, L.; et al. The basic helix-loop-helix transcription factors MYC1 and MYC2 have a dual role in the regulation of constitutive and stress-inducible specialized metabolism in tomato. *New Phytol.* **2022**, *236*, 911–928. [CrossRef] [PubMed]
32. Chen, C.; Rui, X.; Hao, C.; He, Y. TBtools, a Toolkit for Biologists integrating various HTS-data handling tools with a user-friendly interface. *bioRxiv* **2018**, *10*, 289660.
33. Kim, W.-N.; Kim, H.-J.; Chung, Y.-S.; Kim, H.-U. Construction of Multiple Guide RNAs in CRISPR/Cas9 Vector Using Stepwise or Simultaneous Golden Gate Cloning: Case Study for Targeting the FAD2 and FATB Multigene in Soybean. *Plants* **2021**, *10*, 2542. [CrossRef] [PubMed]
34. Porebski, S.; Bailey, L.G.; Baum, B.R. Modification of a CTAB DNA extraction protocol for plants containing high polysaccharide and polyphenol components. *Plant Mol. Biol. Report.* **1997**, *15*, 8–15. [CrossRef]
35. Qi, Q.; Yanyan, D.; Yuanlin, L.; Kunzhi, L.; Huini, X.; Xudong, S. Overexpression of SlMDHAR in transgenic tobacco increased salt stress tolerance involving S-nitrosylation regulation. *Plant Sci.* **2020**, *299*, 110609. [CrossRef]
36. Nakashima, K.; Ito, Y.; Yamaguchi-Shinozaki, K.J.P.P. Transcriptional regulatory networks in response to abiotic stresses in Arabidopsis and grasses. *Plant Physiol.* **2009**, *149*, 88. [CrossRef] [PubMed]
37. Wang, Y.; Qiao, L.; Bai, J.; Wang, P.; Duan, W.; Yuan, S.; Yuan, G.; Zhang, F.; Zhang, L.; Zhao, C. Genome-wide characterization of JASMONATE-ZIM DOMAIN transcription repressors in wheat (*Triticum aestivum* L.). *BMC Genom.* **2017**, *18*, 152. [CrossRef] [PubMed]
38. Abe, H.; Urao, T.; Ito, T.; Seki, M.; Yamaguchi-Shinozaki, S.K.J.P.C. Arabidopsis AtMYC2 (bHLH) and AtMYB2 (MYB) Function as Transcriptional Activators in Abscisic Acid Signaling. *Plant Cell* **2003**, *15*, 63–78. [CrossRef]
39. Fernández-Calvo, P.; Chini, A.; Fernández-Barbero, G.; Chico, J.M.; Gimenez-Ibanez, S.; Geerinck, J.; Eeckhout, D.; Schweizer, F.; Godoy, M.; Franco-Zorrilla, J. The Arabidopsis bHLH Transcription Factors MYC3 and MYC4 Are Targets of JAZ Repressors and Act Additively with MYC2 in the Activation of Jasmonate Responses. *Plant Cell* **2011**, *23*, 701–715. [CrossRef]
40. Figueroa, B. The Arabidopsis JAZ2 promoter contains a G-box and thymidine-rich module that are necessary and sufficient for jasmonate-dependent activation by MYC transcription factors and repression by JAZ proteins. In Proceedings of the Society for Advancement of Hispanics/Chicanos & Native Americans in Science National Conference, Kansas, MO, USA, 28–30 October 2021.

41. Iwamoto, M.; Maekawa, M.; Saito, A.; Higo, H.; Higo, K. Evolutionary relationship of plant catalase genes inferred from exon-intron structures: Isozyme divergence after the separation of monocots and dicots. *Theor. Appl. Genet.* **1998**, *97*, 9–19. [CrossRef]
42. Kazan, K.; Manners, J.M. MYC2: The master in action. *Mol. Plant* **2013**, *6*, 686–703. [CrossRef] [PubMed]
43. Song, S.; Huang, H.; Wang, J.; Liu, B.; Qi, T.; Xie, D.J.P.; Physiology, C. MYC5 Is Involved in Jasmonate-regulated Plant Growth, Leaf Senescence and Defense Responses. *Plant Cell Physiol.* **2017**, *58*, 1752. [CrossRef] [PubMed]
44. Bai, J.F.; Wang, Y.K.; Guo, L.P.; Guo, X.M.; Guo, H.Y.; Yuan, S.H.; Duan, W.J.; Liu, Z.; Zhao, C.P.; Zhang, F.T.; et al. Genomic identification and characterization of MYC family genes in wheat (*Triticum aestivum* L.). *BMC Genom.* **2019**, *20*, 1032. [CrossRef] [PubMed]
45. Adrián, G.-B.; Figueroa, N.E.; Figueroa, P.M.; Figueroa, C.R.; Blazquez, M.A.J.P.O. Jasmonate signalling pathway in strawberry: Genome-wide identification, molecular characterization and expression of JAZs and MYCs during fruit development and ripening. *PLoS ONE* **2018**, *13*, e0197118.
46. Strygina, K.V.; Khlestkina, E.K. Myc-like transcriptional factors in wheat: Structural and functional organization of the subfamily I members. *BMC Plant Biol.* **2019**, *19*, 50. [CrossRef]
47. Wei, K.; Chen, H. Comparative functional genomics analysis of bHLH gene family in rice, maize and wheat. *BMC Plant Biol.* **2018**, *18*, 309. [CrossRef]
48. Yanfang, Y.; Kaikai, Z.; Liying, Y.; Xing, L.; Ying, W.; Hongwei, L.; Qiang, L.; Duanfen, C.; Deyou, Q. Identification and characterization of MYC transcription factors in *Taxus* sp. *Gene* **2018**, *675*, 1–8. [CrossRef]

agronomy MDPI

Perspective

More or Less: Recent Advances in Lignin Accumulation and Regulation in Horticultural Crops

Guang-Long Wang [1,*], Jia-Qi Wu [1], Yang-Yang Chen [1], Yu-Jie Xu [1], Cheng-Ling Zhou [1], Zhen-Zhu Hu [1], Xu-Qin Ren [1] and Ai-Sheng Xiong [2,*]

1 School of Life Science and Food Engineering, Huaiyin Institute of Technology, Huai'an 223003, China; wujiaqi011015@163.com (J.-Q.W.); 17834830201@163.com (Y.-Y.C.); 13236206281@163.com (Y.-J.X.); zhoucling@hyit.edu.cn (C.-L.Z.); huzhenzhu1001@hyit.edu.cn (Z.-Z.H.); jsrxq@hyit.edu.cn (X.-Q.R.)
2 State Key Laboratory of Crop Genetics and Germplasm Enhancement, College of Horticulture, Nanjing Agricultural University, Nanjing 210095, China
* Correspondence: guanglongwang@hyit.edu.cn (G.-L.W.); xiongaisheng@njau.edu.cn (A.-S.X.)

Abstract: Lignin is an important secondary metabolite that maintains the mechanical strength of horticultural plants and enhances their ability to respond to external environmental changes such as biotic and abiotic stresses. However, excessive accumulation of lignin can lead to lignification of horticultural products, reducing their taste quality and nutritional value. Therefore, the lignin content of horticultural products needs to be controlled at a reasonable level, and studying and regulating lignin metabolism is very meaningful work. This article focuses on the synthesis, accumulation, and regulation of lignin in horticultural crops in recent years, provides a systematic analysis of its molecular mechanism and application prospects, and sheds insights into the directions that need further research in the future. This article provides an important basis for the regulation of lignin accumulation and lignification in horticultural crops and proposes new ideas for improving the quality of horticultural crops.

Keywords: lignin; horticultural crops; accumulation; regulation; quality

Citation: Wang, G.-L.; Wu, J.-Q.; Chen, Y.-Y.; Xu, Y.-J.; Zhou, C.-L.; Hu, Z.-Z.; Ren, X.-Q.; Xiong, A.-S. More or Less: Recent Advances in Lignin Accumulation and Regulation in Horticultural Crops. *Agronomy* **2023**, *13*, 2819. https://doi.org/10.3390/agronomy13112819

Academic Editor: Steven R. Larson

Received: 27 October 2023
Revised: 10 November 2023
Accepted: 14 November 2023
Published: 15 November 2023

1. Introduction

Horticultural plants generally refer to crops with high economic value that are used for human viewing or consumption, mainly including fruit trees, vegetables, flowers, tea, edible fungi, and medicinal plants. The relationship between horticultural plants and humans is extremely close, and we cannot do without these plants in our daily life. Moreover, with the progress of human civilization, many new horticultural plants have been continuously domesticated and cultivated by humans.

On the one hand, lignin can confer on horticultural crops high rigidity, providing plants with resistance to the external environment [1,2]. However, for many horticultural plants, excessive accumulation of lignin can lead to a decrease in the taste, texture, and general quality of horticultural products. Therefore, understanding the accumulation and regulatory mechanisms of lignin in horticultural plants is crucial for improving their resistance and quality.

2. Lignin

Lignin is an important component of the cell wall in vascular plants such as ferns, gymnosperms, and angiosperms. It is mainly deposited in the secondary cell wall of transport tissues, mechanical tissues, and protective tissues [3]. It plays a key role in the evolution of terrestrial vascular plants and has very important biological functions. Lignin is very abundant in nature, second only to cellulose, and its use as a binder and various additives in the chemical industry accounts for its primary economic value [4]. In woody plants, lignin accounts for 25% of the composition, making it the second most abundant organic matter in the world. As the main component of the plant cell wall, lignin is cross-linked with cellulose, hemicellulose, pectin, and other substances [5]. On the one hand, the plant skeleton formed by this cross-linked lignin carbohydrate complex improves the mechanical strength and hardness of plant tissue, increases the penetration ability of cell wall pathogens, and improves the plant's ability to resist biotic and abiotic stresses [6]. On the other hand, it has a compressive effect and its hydrophobicity makes plant cells less permeable, facilitating the long-distance transportation of water, minerals, and organic matter within the plant body [7].

3. Lignin Biosynthesis and Transcriptional Regulation

3.1. Lignin Biosynthesis

Lignin is synthesized through the phenylpropane pathway and lignin-specific pathway [8]. Phenylalanine goes through deamination, hydroxylation, methylation, and reduction reactions to form three main monomers, coumarin, coniferol, and sinapinol, which are subsequently polymerized to form p-hydroxyphenyl lignin (H lignin), guaiacyl lignin (G lignin), and syringal lignin (S lignin), respectively [9]. The key enzymes involved in this pathway consist of phenylalanine ammonia lyase (PAL), cinnamate 4-hydroxylase (C4H), 4-coumarate-CoA ligase (4CL), cinnamoyl-CoA reductase (CCR), cinnamyl alcohol dehydrogenase (CAD), hydroxycinnamoyl-CoA shikimate/quinate hydroxycinnamoyl transferase (HCT), *p*-coumaroyl coumaroyl shikimate/quinate 3′-hydroxylase (C3′H), caffeoyl-CoA *O*-methyltransferase (CCoAOMT), ferulate 5-hydroxylase (F5H), caffeic acid *O*-methyltransferase (COMT), peroxidase (PER), and laccase (LAC) [10].

3.2. Transcriptional Regulation

MYB (v-myb avian myeloblastosis viral oncogene homolog), the main regulatory factor for lignin synthesis, directly induces the expression of most genes. MYB58, MYB63, and MYB85 in Arabidopsis can directly bind to AC elements in the promoter regions of PAL, C4H, 4CL, C3′H, CCoAOMT, CCR, and CAD [11]. And NAC transcription factors are mostly located upstream of the lignin transcription regulation network, indirectly regulating lignin biosynthesis by binding to different MYB transcription factors [12]. For instance, an SG2-type R2R3-MYB transcription factor from chrysanthemum, CmMYB15-like, directly interacts with the AC cis-element in the promoter region of Cm4CL2 and results in lignin accumulation and cell wall thickening [13]. PbrMYB169 from pear is evidently integrated with the AC elements [ACC(T/A)ACC] in the promoter of lignin genes *C3H1*, *CCR1*, *CCOMT2*, *CAD*, *4CL1*, *4CL2*, *HCT2*, and *LAC18* to regulate lignin deposition and cell wall thickness [14]. PbrMYB24 can also bind to AC elements and MYB binding sites (MBSs) in the promoter of lignin biosynthesis genes to activate their transcription [15]. In apple, transgenic plants overexpressing *MdMYB46* presented enhanced secondary cell wall and lignin deposition by directly triggering the promoter of lignin biosynthesis genes [16]. Overexpression of PnMYB2 from *Panax notoginsengin* resulted in increased lignin accumulation and evidently reinforced thickness of primary and secondary cell walls compared with wild-type plants. Additionally, PnMYB2 could directly interact with the promoter of *PnCCoAOMT1*, a key lignin biosynthesis gene [17]. In pomelo (*Citrus grandis*), CgNAC043 is located upstream of CgMYB46 and can simultaneously bind to the promoter of the lignin biosynthesis genes *CgCCoAOMT* and *CgC3H*, thereby modulating juice sac granulation [18]. In apple, MdSND1, an NAC transcription factor related to secondary wall development, can activate the expression of MdMYB46/83 to regulate lignin biosynthesis and accumulation [19]. An NAC STONE CELL PROMOTING FACTOR essential for stone cell formation in pear, namely PbrNSC, can bind to the promoters and induce the expression of *PbrMYB169*, *Pbr4CL4*, and *PbrLAC4* [20].

Further studies indicate that bHLH, bZIP, SPL, and other transcription factors are also involved in the transcription regulation of lignin. The atypical bHLH transcription factor CmHLB (HLH PROTEIN INVOLVED IN LIGNIN BIOSYNTHESIS) from chrysanthemum has been confirmed to interact with chrysanthemum KNOTTED ARABIDOPSIS

THALIANA7 (CmKNAT7) and has a positive role in stem mechanical strength, cell wall thickness, and lignin content [21]. EjAGL15, a senescence-specific MADS-box gene in loquat, triggered lignin-biosynthesis-related genes to positively modulate the variation in lignin content [22]. CcBLH6, a bell-like homeodomain-containing (BLH) transcription factor from *Camellia chekiangoleosa*, was consistent with the unique lignification pattern observed during fruit development, and transgenic experiments confirmed the role of CcBLH6 in the control of fruit lignification [23]. The grapevine transcription factor VvWRKY2 showed an ability to bind to the promoter of the *VvC4H* gene to regulate lignification in response to biotic or abiotic stresses [24]. The LIM transcription factor CaβLIM1a from chickpea (*Cicer arietinum*) is a transcriptional regulator of CaPAL1, the gatekeeping enzyme of the phenylpropanoid pathway. Its interaction with the nuclear effector PEXEL-like Effector Candidate 25 (ArPEC25) from *Ascochyta rabiei* would interfere with its DNA-binding ability and result in altered lignin production [25]. An ethylene response factor in apple, MdERF114, can directly interact with the GCC-box in the promoter of MdPEROXIDASE63 (MdPRX63) and activate its transcription, leading to increased lignin accumulation and conferring root resistance to *Fusarium solani* [26]. The role of EjHSF3, a heat shock factor from loquat, in regulating fruit lignification was realized by activating the promoters of the downstream lignin biosynthesis genes and interacting with the regulator, EjAP2-1 [27]. Overexpression of tomato SlHB8 belonging to the homeodomain-leucine zipper class III transcription factor gene family decreased stem diameter, xylem width, and number of xylem cell layers, accompanied by reduced lignin production [28].

4. Lignin Accumulation

4.1. Growth Developmental Stages

Lignin accumulation is altered in response to developmental stages and can impact plant and product organ development. In pomelo, juice sac granulation and lignin content increased after anthesis with the development of pomelo fruit, and the lignin content near the core was higher than that further away from the core [29]. Similarly, the lignin content showed a continuous increase in winter jujube pericarp during pigmentation [30]. During the growth of *Camellia oleifera* shell, lignin content increased continuously, leading to a thickened stone cell wall [31]. By contrast, lignin content in storage root gradually decreased during the developmental stages in two sweet potato cultivars [32]. The concentrations of total lignin and its precursors displayed marked decreases during ginger tissue maturation; however, the lignin component was dramatically altered with syringyl lignin mainly enriched in mature rhizomes, contributing to ginger lignification [33]. Lignin content exhibited a decrease in pepper fruits at different stages during growth and maturation, following the same pattern as the lignin precursors [34]. In carrot roots, lignin was mostly deposited in xylem vessels, and with the development of carrot taproots, lignin content continuously decreased, accompanied by a reduced transcript abundance of the lignin biosynthesis genes [35]. Indeed, plant development is a dynamic process that includes changes in structure, phenotype, and internal metabolic substances. Lignin is a secondary metabolite, and its content in plants is largely influenced by other metabolites. In addition, the pattern of lignin content changes in different organs or tissues within the same plant may differ at different developmental stages.

4.2. Postharvest Storage

The change in postharvest fruit quality is a complex biological process, and texture is an important aspect of fruit quality which directly affects the storage life, product merchantability, and market competitiveness of horticultural products. Lignification is a common occurrence in fruits and vegetables such as pear, loquat, mangosteen, kiwi and zucchini, and improper storage can lead to a decline in fruit quality and severely limit storage time. Under normal circumstances, postharvest fruits and vegetables will generally experience the process of lignification strengthening. Increased activities of PAL and PER promoted lignin synthesis and accumulation during postharvest pumelo storage, which in turn induced sucrose degradation and an energy deficit [36]. With the extension of shelf life, broccoli florets underwent apparent increases in lignin accumulation, and this effect was reversed by exogenous diacetyl [37]. During cold storage processes, lignification was gradually enhanced in the epidermis of the shoots in water bamboo with the reinforcement of firmness [38]. Compared to the control group, the water bamboo shoot samples subjected to melatonin displayed enhanced lignification during storage [39]. With increasing storage time, two types of snow pea underwent enhanced lignin accumulation, and small snow peas were more susceptible to lignification than large sweet broad peas [40]. Lignin accumulation in samples in both modified polyethylene packages and normal polyethylene packaging gradually increased during storage [41]. Over the period of storage, enhanced respiration may require a significant amount of energy consumption, leading to accelerated senescence and the generation of reactive oxygen species, as well as the strengthened lignification process of horticultural products.

5. Lignin Regulation by External Stimuli

5.1. Environmental Changes

5.1.1. Temperature

Studies have demonstrated that the enhanced accumulation and distribution of lignin within plant tissues can confer resistance to temperature fluctuations. In Chinese cabbage, the lignin content in the samples exposed to low temperatures was apparently higher than that of the control, resulting in the assumption that lignin adjustments are a defense mechanism or response to low temperatures [42]. Similarly, low-temperature-induced enriched accumulation of lignin biosynthesis was seen in postharvest banana and Lei bamboo shoots [43,44]. Just like those under low-temperature conditions, bamboo shoots also underwent lignification when exposed to high temperatures [45]. The primary monolignol contents were substantially increased after heat stress in coffee leaves [46]. These findings again favor the theory that lignin is a polymer that defends against environmental temperature changes.

5.1.2. Drought

Drought stress is one of the most important environmental stresses adversely affecting plant growth and quality. Lignin deposition in plant cells is of great significance for drought adaptability and resistance. It was reported that drought stress in tea plants resulted in a significant increase in lignin content [47]. In tomato, a higher proportion of lignin in the cell walls of leaves was observed due to repeated drought cycles [48]. Drought induced H_2O_2, ABA, and JA accumulation and promoted the expression of CAD genes, contributing to increased lignin synthesis in melon stems [49,50]. Water stress activated the phenylalanine ammonia-lyase genes as well as the primary and secondary metabolism of postharvest carrots, accelerating the lignification process [51].

5.1.3. Salinity

Soil salinization is becoming increasingly severe and is profoundly affecting various aspects of crop growth and development. Plants can adapt to salinity and the resulting injury by modifying the accumulation and distribution of the main cell wall components, especially lignin, to maintain water balance and reduce the transport of harmful ions into the plant. The levels of total and individual phenolic acids in multi-leaf lettuce, such as lignin, gradually increased with increasing concentrations of salt [52]. Similarly, when subjected to salt stress, the monolignol content in coffee leaves was substantially increased as compared to the control [53]. Salt stress might enhance xylem development in tomato root, accompanied by an increased number of lignified cells in vascular bundles as well as a greater deposition of lignin [54]. Increased lignin accumulation under salt stress may enhance the cell-to-cell pathway for water transport and confer reinforced selectivity and decreased ion uptake to protect plant cells exposed to salt stress.

5.1.4. Heavy Metals

Due to the rapid development of industrialization, various heavy metals are constantly added to soil and water, which has become a major global concern as this leads to the loss of agricultural productivity. Heavy metals induce the production of phenolic secondary metabolic pathways and enhance the lignin deposition within cell walls. Heavy metals can damage the cell integrity of the roots, and in severe cases, may lead to cell death in the roots. In the presence of copper (Cu) exposure, lignin deposition of the cell wall in mustard roots was significantly greater than in stress-free control plants [55]. Cu intensively increased PAL activity and lignin accumulation in a dose-dependent manner in copper-treated *Matricaria chamomilla*, revealing the establishment of a barrier against metal intrusion [56]. The lignin content in lettuce roots treated with cadmium increased by 18% compared with control group [57], whereas increased Zn exposure in growth media evidently reduced the lignin content in the leaves of watermelon seedlings [58].

5.1.5. Pest and Disease Invasion

Lignin accumulation enhances the hardness and reinforcement of cell walls, provides a solid and strong barrier, and is recognized as the first line of defense against biotic infection. Furthermore, lignin deposition can diffuse pathogens by restricting the access of fungal enzymes and toxins into the host plants and preventing the movement of nutrients from hosts to invaders [2]. Lignification in pear results from plant resistance against diseases caused by *Alternaria alternata* and *Botryosphaeria dothidea*, and the PcMYB44-mediated PcmiR397-PcLACs module may be involved in the process of defense-induced lignification [59]. Leaf-spot-infected leaves had increased transcription of lignin biosynthesis genes and enhanced lignin accumulation in an important medicinal plant, *Withania somnifera* [60]. Exogenous carvacrol enhanced lignin accumulation and delayed cell wall degradation in postharvest pummelo fruit, conferring disease resistance on *Diaporthe citri* [61]. Alterations in the expression of caffeoyl shikimate esterase (CsCSE1), a key novel enzyme in lignin biosynthesis in cucumber, significantly affected lignin accumulation and plant resistance to *Podosphaera xanthii* [62]. Enhanced total lignin deposition in kiwifruit stems can induce disease resistance against kiwifruit canker [63]. Early production of lignin might be a key mechanism in repressing the spread of the red leaf blotch disease caused by the fungus *Polystigma amygdalinum* within the host leaf tissues in almond [64]. Comparison of resistant and susceptible pigeon pea genotypes demonstrated that the genotype resistant against *Fusarium wilt* possessed higher lignin contents [65]. These findings indicate that lignin accumulation and distribution can be altered to react to pest and disease invasion.

5.1.6. Carbon Dioxide (CO_2)

With the aggravation of climate change, atmospheric CO_2 levels may rise significantly by the end of the 21st century. Elevated CO_2 levels could greatly impact plant growth and development, including changes in phenotype, anatomy, and secondary metabolite accumulation. In the presence of enriched CO_2 levels, there was an upward trend in lignin concentration in celery leaves, and the transcription of most lignin metabolic genes changed in response to elevated CO_2 levels [66]. Similarly, the lignin content in carrot roots under elevated CO_2 conditions was apparently higher than that in the control group [67]. Pea plants grown in an atmosphere enriched with CO_2 showed improved cell wall fortification due to enhanced lignin accumulation, thus efficiently coping with parasite penetration [68]. By contrast, under cold storage, higher CO_2 levels may reduce the lignification process through the inhibition of PAL activity and lignin deposition [69].

5.1.7. Nitric Oxide (NO)

NO acts as a messenger, taking part in various aspects of processes during plant growth and development. Wax apple fruit treated with NO had a lower total lignin content, indicating that NO treatment can alleviate the process of lignification and senescence of wax apple fruit during storage [70]. Cold stress induced increased lignin accumulation and promoted lignification in okra pods and water bamboo shoots, which was reversed with the application of exogenous NO [71,72]. The abundance of mRNA in lignin biosynthesis genes was decreased in postharvest carrots exposed to NO donor sodium nitroprusside (SNP), leading to lower lignin deposition and a lesser degree of taproot lignification [73]. When exposed to the NO scavenger cPTIO, no detectable changes in total lignin content could be detected in sunflower roots, whereas the ratio of G/S lignin was increased, indicating that NO may impact lignin composition [74]. SNP elevated the accumulation of lignin to suppress anthracnose decay in postharvest mango fruit [75]. Similarly, the application of SNP improved the activities of PAL, C4H, CAD, and PER and increased the concentrations of lignin and its monomers, enhancing the hardness of the healing tissue in muskmelon [76].

5.1.8. Melatonin

Melatonin is a kind of biomolecule found in almost all living kingdoms and can regulate many life processes and responses to abiotic and biotic stresses. Recently, melatonin has been demonstrated to be involved in lignin regulation in horticultural crops. Melatonin strengthened the degree of lignification in tea leaves by adjusting the expression of genes involved in lignin synthesis pathway [77]. The melatonin content within plant stems is consistent with lignin content and stem strength in different *Paeonia lactiflora* cultivars; furthermore, the application of melatonin significantly reinforced stem strength by enhancing lignin production [78]. In the process of preserving horticultural crops postharvest, melatonin can also regulate lignin production to modify senescence, freshness, and disease resistance. Postharvest blueberries exposed to melatonin treatment showed stimulated accumulation of lignin in the fruits and increased disease resistance [79]. However, melatonin has the opposite effect on lignification manipulation during the postharvest storage of asparagus. In postharvest asparagus, the application of melatonin relieved the senescence process by restricting PAL and PER activities and decreasing lignin production, thereby retarding the increase in firmness [80].

5.1.9. Other Bioactive Regulators

Kiwifruit is prone to lignification induced by chilling stress, and this process can be alleviated by hydrogen-rich water [81]. A chitosan coating suppressed postharvest juice sac granulation and lignification in pummelo fruit by inhibiting the activities and transcript levels of lignin synthesis genes [82]. Cucumber seeds primed with exogenous chitosan induced enhanced resistance against cucumber powdery mildew disease in cucumber seedlings, which was well correlated with lignin production [83]. Exogenous γ-aminobutyric acid promoted the expression of lignin-related genes and increased the accumulation of lignin, thereby conferring on apple seedlings resistance to long-term drought stress [84]. Glucose acts not only as a structural substance but also a signaling molecule, playing an important regulatory role in plant growth and development. The application of exogenous glucose markedly improved the accumulation of lignin in pear calli [85]. Exogenous silicon reinforced stem strength by enhancing lignin accumulation, particularly G lignin and S lignin, and secondary cell wall thickness in herbaceous peony [86].

5.2. Hormonal Stimuli

5.2.1. Gibberellin

It is reported that gibberellin can greatly impact lignification in horticultural crops by regulating lignin accumulation and secondary growth. Gibberellin can induce lignin accumulation and enlarged secondary xylem in carrot root, accompanied by increased transcript levels of lignin-related genes [87,88]. Similar results were achieved for sweet potato roots exposed to gibberellin [89]. In table grape, gibberellic acid promotes the expression of the lignin-related genes and results in alterations in cell wall composition and pedicel structure, leading to an increase in berry drop [90]. However, gibberellin was demonstrated to have an opposite effect on lignification and secondary growth by directly regulating the activity of a basic peroxidase isoenzyme, an enzyme involved in lignin biosynthesis [91]. These results indicate that gibberellin may have a different impact on lignin accumulation depending on the plant species, organ type, and developmental stage.

5.2.2. Brassinosteroid

In garlic, brassinosteroid-mediated lignin accumulation may be largely responsible for adaption to salt stress [92]. BR triggered the transcription of *PAL*, *4CL*, and *CAD* genes and lignin-related enzyme activities and activated the synthesis of lignin precursors, thereby accelerating the wound healing of potato tubers [93]. The expression of *PAL*, *4CL*, *CCR*, and *CCoAOMT* was significantly increased after exogenous brassinosteroid application in watermelon. Furthermore, compared with Zn stress alone, brassinosteroid treatment evidently enhanced the activities of PAL, 4CL, POD, and lignin content, indicating a potential role of brassinosteroid-mediated lignin accumulation in heavy metal tolerance [94]. Similarly, 24-epibrassinolide treatment alleviated the adverse effects induced by salinity stress and *Phytophthora melonis* damping-off in cucumber seedlings by influencing cell wall reassembly and lignin-related genes [95,96]. Brassinosteroid is also involved in methane-mediated adventitious root development in marigold, altering the levels of lignin and other cell wall components [97]. Therefore, it can be concluded that the role of brassinosteroid in abiotic and biotic stress resistance may be related to its participation in lignin metabolism.

5.2.3. Ethylene

The application of ethephon, an ethylene-releasing compound, increased lignin content and delayed stem bending of snapdragon cultivars, whereas plants exposed to silver thiosulfate (an ethylene action inhibitor) exhibited a higher bending rate compared with the control [98,99]. Fruits of *Fragaria chiloensis* subjected to ethephon at the large green developmental stage displayed stimulated lignin biosynthesis and increased mRNA abundance of *FcPOD27* [100]. Exogenous ethylene resulted in the accumulation of cellulose and lignin in tomato, indicating that ethylene treatment also brought about changes in cell wall composition [101]. Ethylene significantly stimulated the expression of lignin-related genes and strengthened lignification in common beans during storage, whereas 1-methylcyclopropene, an ethylene inhibitor, had the opposite effect [102]. By contrast, 1-methylcyclopropene treatment accelerated lignin accumulation and delayed cellulose degradation during papaya fruit ripening [103].

5.2.4. Auxin

Exogenous treatment with 200 μM NAA, a synthetic auxin, decreased lignin accumulation in pear fruit stone cells and also dramatically reduced the transcript level of *PbrNSC* for lignin and cellulose biosynthesis [104]. In cucumber, high auxin levels resulting from *amino acid permease 2* mutation accumulated in the roots, accompanied by increased concentrations of lignin [105]. Auxin-like 2,4-dichlorophenoxyacetic acid can slow senescence in postharvest citrus by modifying lignin-biosynthesis-related genes and lignin contents in fruit peels [106]. Similar results were also observed in pummel exposed to indole-3-acetic acid (IAA) [107].

5.2.5. Other Hormones

Increased abscisic acid (ABA) and jasmonic acid (JA) levels induced by water stress positively regulated the expression levels and activity of CmCAD and lignin accumulation in melon stems [49]. ABA treatment triggered CgMYB58 expression and regulated the transcription of downstream lignin-related genes, affecting juice sac granulation and lignin accumulation in pummel [107]. Lignin concentrations in kiwifruit plants treated with methyl jasmonate (MJ) or salicylic acid (SA) were reduced by at least 20% as compared to the control group [108]. Loquat fruits are vulnerable to chilling stress, flesh firmness, and lignin accumulation resulting from postharvest storage at low temperatures. Pretreatment with MeJA can relieve this low-temperature-induced lignification [109]. By contrast, exogenous MeJA can enhance the activity of PAL, 4CL, CAD, and PER enzymes and upregulate the expression of related genes and lignin accumulation to inhibit fungal decay in kiwifruit [110].

6. Lignin Regulation at the Molecular Level

6.1. Different Layers Contributing to Lignin Accumulation

6.1.1. mRNA and Abundance Alterations Identified by Transcriptome

Transcriptome remains a common method of identifying genes and their expression patterns according to different conditions and requirements. Based on the RNA seq data generated from collected pulp samples, a co-expression network of structural genes and MYB, NAC, and WRKY family transcription factors for lignin synthesis in pear was constructed based on weighted gene co-expression network analysis (WGCNA) [111]. In response to citrus blight stress, genes involved in lignin biosynthesis were upregulated, leading to the hypothesis that citrus blight may lead to root lignification [112]. With the aid of de novo transcriptome sequencing, 66 generated unigenes and 10 candidate genes were found to be involved in lignin biosynthesis in radish [113]. Differential expression of lignin-related genes was observed in cucumber inbred lines differing in downy mildew resistance, leading to altered lignin accumulation in the two lines [114]. Shading treatments induced reduced lignin accumulation in asparagus, and 37 differentially expressed genes were found to be related to lignin metabolism via transcriptome profiling [115].

6.1.2. MicroRNA (miRNA)

miRNAs are a type of non-coding RNA with regulatory functions in various plant processes, being approximately 20–25 nucleotides in length. Blocking *IbmiR319a* in transgenic sweet potato plants resulted in increased brittleness and reduced lignin content [116]. Transient overexpression of *miR7125* or repression of its target, CCR, in apple fruit decreased lignin deposition under light-induced conditions [117]. In chickpea, overexpressing *CamiR397* decreased the transcript levels of its targets, *LAC4* and *LAC17L* and lignin accumulation in the root xylem and reduced xylem wall thickness [118]. A similar role of miR397 was observed during stone cell development in pear fruit [119]. With small RNA sequencing, *Bm-miR172c-5p* from *Bacopa monnieri* was found to cleave the *F5H* gene involved in the lignin biosynthesis pathway; furthermore, transgenic plants harboring *Bm-miR172c-5p* showed suppressed *F5H* expression and reduced lignification [120]. With the extension of Cd or alkali treatment, the lignin content in potato roots gradually increased; miR4243-x and the novel miRNA novel-m3483-5p were demonstrated to interfere with shikimate O-hydroxycinnamoyltransferase and cinnamic alcohol dehydrogenase (CAD), respectively, in the lignin biosynthesis pathway [121,122]. In summary, most recent studies discovered that miRNA mainly acts on structural genes in lignin synthesis, but there are a few reports on lignin-related regulatory factors controlled by miRNA.

6.1.3. Protein

Proteomics, as a key technology for exploring various life changes, has become one of the core components of life science research in the post-genomic era due to its ability to explain the molecular mechanisms of various physiological processes at the protein level through dynamic protein expression differences. Key enzymes involved in the lignin pathway, such as PAL, POD, 4CL, and CCoAMT1, were significantly induced in BcGs1-triggered disease resistance in tomato leaves [123]. The regulatory network for the biosynthesis of lignin, a principal component of stone cells, was reconstructed with generated pear proteome data [124]. The differential resistance to the parasitic weed *Orobanche cumana* in sunflower cultivars is partly attributed to altered defense-related proteins involved in the biosynthesis of lignin [125]. With the isobaric tag for relative and absolute quantitation (iTRAQ), PER47, LAC4, and β-glucosidase 15 (BGLU15) were considered to ultimately affect lignin biosynthesis and stone cell formation in pear pollinated with different varieties of pollen [126]. In the presence of UV-B stress or *Botrytis cinerea* infection, most of the proteins involved in the lignin pathway were substantially increased in *Morus alba*, indicating their significant role in the responses to abiotic and biotic stress [127].

6.1.4. Metabolic

Metabolomics is a method that allows for studying the internal metabolites and their changes in organisms. Its main research subjects include lipids, sugars, alkaloids, flavonoids, amino acids, and other substances. Qualitative and quantitative analysis of metabolites in organisms can reveal the types, quantities, and alterations of metabolites under specific conditions. In peach, lignin composition and accumulation affect fruit texture and quality, and metabolite and transcriptome analysis revealed that peach cultivars with different levels of lignin accumulation may be attributed to differential expression of *Pp4CL2*, *Pp4CL3*, and *PpCOMT2* [128]. Many metabolites derived from the phenylpropanoid pathway contributing to cell wall formation and the lignin production of stone cells were observed near pear fruit core [129]. As a result of Cu exposure, *Citrus grandis* leaves accumulated more lignin-related metabolites to cope with the stress induced by Cu [130].

6.2. Molecular Modules and Networks within and beyond NAC-MYB Layers

To date, most molecular modules and networks for lignin regulation have been found to be coupled with NAC-MYB layers. That is, MYB transcription factors directly bind to the AC elements in the promoters of lignin biosynthesis genes, with NAC transcription factors upstream of MYBs. For instance, CsMYB15 directly interacted with the Cs4CL2 promoter and induced its transcription, thereby leading to high lignin accumulation in citrus [131]. In recent years, with the deepening of research on lignin regulation, it has been found that there are other mechanisms that regulate lignin metabolism. EjERF39 could directly bind to the DRE element in the promoter of Ej4CL and activate its expression level, ultimately promoting low-temperature-induced lignification in postharvest loquat fruit [132]. Grapevine VlbZIP30 directly interacted with the G-box cis-element in the promoter of the lignin biosynthetic gene *VvPRX N1* to regulate lignin deposition [133].

7. Application Fields

7.1. Abiotic Resistance

Abiotic stresses refer to the negative impact of abiotic factors on organisms in a specific environment, including physical and chemical factors in the environment, such as drought, salt stress, and extreme temperature. With the deterioration of environmental conditions, abiotic stresses pose important challenges to the growth and development of horticultural crops, as well as their yield and quality. Some research has found that the modulation of lignin levels and deposition can effectively cope with or adapt to abiotic stresses. Overexpression of a caffeoyl-CoA O-methyltransferase gene, *CCoAOMT*, from *Paeonia ostia* resulted in markedly higher lignin levels and conferred drought stress tolerance on transgenic plants [134]. Transgenic plants hosting *AgNAC1*, an NAC transcription factor from celery, modified lignin contents and improved salt tolerance [135]. MdMYB46 could induce lignin deposition and confer salt and osmotic stress tolerance on apple by directly binding to the promoter of genes involved in lignin biosynthesis and stress signal transduction pathways [16]. Increased monolignol contents and/or increased S/G ratios were determined in transgenic sweet potato plants hosting the *IbCAD1* gene and might enhance the antioxidation capacity against low-temperature stress and pathogen attack [136].

7.2. Biotic Resistance

Lignin accumulation is a useful and reliable defense indicator for biotic resistance, and some outstanding attempts to study it have been made in horticultural crops via lignin engineering. Overexpression of an MYB transcription factor, CmMYB19, from chrysanthemum enhanced lignin accumulation and impeded the reproduction of aphids on the host [137]. LrNAC35, an NAC member from *Lilium regale*, could directly activate the expression of lignin-related genes in petunia (*Petunia hybrida*) and contributed to reduced susceptibility to cucumber mosaic virus and tobacco mosaic virus attack by increasing lignin deposition in the cell walls [138]. Inhibition of microRNA397b in *Malus hupehensis* promoted *MhLAC7* expression and resulted in elevated lignin accumulation, conferring increased tolerance to *Botryosphaeria dothidea* infection [139].

7.3. Fruit Development and Ripening

Lignin accumulation is essentially required for fruit development and can affect fruit quality formation to a great extent. Ripening Inducing Factor (FaRIF), an NAC transcription factor highly expressed in strawberry receptacles, can influence lignin deposition and accumulation to regulate fruit ripening and quality formation [140]. PpMYB36, encoding a pear MYB transcription factor, was demonstrated to contribute to lignin accumulation and russet coloration in pear fruit [141]. PbrSAUR13, a member of the small auxin-up RNA gene family in *Pyrus bretschneideri*, enhanced lignin synthesis and stone cell accumulation in the fruit, whereas PbrSAUR52 had the opposite effect [142]. CsMYB330 can recognize and bind AC elements in the Cs4CL1 promoter to positively regulate fruit juice sac lignification, whereas CsMYB308 might be a transcriptional repressor [143].

7.4. Lodging Resistance and Stem Strength

Stem lodging is a critical problem adversely influencing crop productivity throughout the world. It involves stem breaking and bending, a state in which the plant cannot return to its vertical stance anymore. It is widely recognized that stem lodging interferes with crop photosynthesis and ultimately impacts yield and quality formation. It has been demonstrated that lignin deposition enhances plant cell walls, provides mechanical strength, and is closely related to resistance to lodging. The role of lignin in stem strength and lodging resistance was extensively investigated in herbaceous peony [144,145]. Ca treatment could increase lignin accumulation and reinforce the mechanical strength in peony stems, whereas ethyl glycol tetraacetic acid (EGTA), an effective Ca^{2+} chelator, had a negative role [146,147]. PlMYB83 could directly activate the promoter of PlMYB43, PlCOMT2, and PlLAC4 to manipulate lignin production, secondary wall thickness, and stem strength [148].

7.5. Postharvest Storage Regulation

Over the period of postharvest storage, horticultural products are usually prone to low-temperature-induced stress, ultimately resulting in morphological changes, lignin accumulation, and reduced quality. A feasible and effective way to protect postharvest storage products can be achieved through modulating lignin accumulation and distribution. On the one hand, lignin production can be decreased to maintain fruit quality when exogenous active substances are applied. On the other hand, worsening fruit appearance and integrity resulting from oxidative stress can be alleviated by enhancing or engineering lignin synthesis. Exogenous L-cysteine and γ-aminobutyric acid could alleviate the degree of lignification and delay senescence to promote the quality of postharvest loquat fruit [149]. Tuberous roots of transgenic sweet potato lines overexpressing IbLfp, encoding a lignin-forming peroxidase, showed improved tolerance to cold storage during low-temperature storage due to enhanced lignin accumulation [150].

8. Prospects

Lignin is an important secondary metabolite and one of the main components making up plant cell walls. Lignin can provide mechanical support for horticultural crops and the ability to respond to external environmental changes. At the same time, lignin is an important dietary fiber, being widely present in horticultural crops such as vegetables and fruits, and has an important regulatory effect on human health. However, the excessive presence of lignin can affect the texture, taste, and quality of horticultural products. Therefore, it is necessary to maintain the lignin content of horticultural crops within a certain range. At present, most research on the quality of horticultural plants mainly focuses on nutritional quality indicators such as vitamins, carotenoids, and anthocyanins, and there is a lack of research on other nutrient substances. The pathway of lignin synthesis in most horticultural crops is not clear enough, and most of the mined transcription regulatory factors are based on the NAC-MYB regulatory network. There are still few reports on other regulatory factors modulating the distribution and accumulation of lignin in horticultural plants, and the functions of these lignin-related genes still need to be further verified through biotechnology, molecular technology, and biochemical technology.

Excessive lignin can affect the taste and quality of crops, while insufficient lignin can lead to poor crop resistance, reduced yield, and even death. Studies in ornamental crops and non-edible organs have shown that lignin plays an important role in plant lodging resistance, stress resistance, and disease resistance. However, for other edible horticultural crops, excessive accumulation of lignin can lead to lignification and a decreased quality of horticultural products, resulting in irreversible economic losses. Therefore, it is necessary to reasonably regulate lignin metabolism and improve the quality of horticultural crops. This can be approached from the following aspects: firstly, research on lignin strengthening in ornamental plants can be carried out to reinforce stem strength, enhancing their ability to resist against lodging, biotic and abiotic stresses, and diseases; secondly, through the regulation of lignin metabolism in specific organs by exogenous active substances, the production of horticultural products with a low lignin content can be achieved; thirdly, lignin synthesis and regulatory factors specifically expressed in horticultural products can be explored, and horticultural germplasm resources with low lignification could be created through gene editing and other technologies.

Author Contributions: Conceptualization, G.-L.W. and A.-S.X.; acquisition of data for the work, G.-L.W. and J.-Q.W.; writing—original draft preparation, G.-L.W. and Y.-Y.C.; writing—review and editing, Y.-J.X. and C.-L.Z.; supervision, X.-Q.R. and A.-S.X.; analysis of data for the work, Z.-Z.H.; funding acquisition, G.-L.W. and A.-S.X. All authors have read and agreed to the published version of the manuscript.

Funding: This research was funded by the National Natural Science Foundation of China (32102369, 32372681), Natural Science Foundation of Jiangsu Province (BK20211366), New Century Excellent Talent of the Ministry of Education (NCET-11-0670) and Jiangsu Natural Science Foundation for Distinguished Young Scholars (BK20130027).

Data Availability Statement: Data are contained within the article.

Conflicts of Interest: The authors declare no conflicting interest regarding the publication of this work.

References

1. Liu, Q.; Luo, L.; Zheng, L. Lignins: Biosynthesis and Biological Functions in Plants. *Int. J. Mol. Sci.* **2018**, *19*, 335. [CrossRef] [PubMed]
2. Dong, N.-Q.; Lin, H.-X. Contribution of phenylpropanoid metabolism to plant development and plant–environment interactions. *J. Integr. Plant Biol.* **2021**, *63*, 180–209. [CrossRef]
3. Zhao, Q. Lignification: Flexibility, Biosynthesis and Regulation. *Trends Plant Sci.* **2016**, *21*, 713–721. [CrossRef] [PubMed]
4. Abraham, B.; Syamnath, V.L.; Arun, K.B.; Fathima Zahra, P.M.; Anjusha, P.; Kothakotta, A.; Chen, Y.-H.; Ponnusamy, V.K.; Nisha, P. Lignin-based nanomaterials for food and pharmaceutical applications: Recent trends and future outlook. *Sci. Total Environ.* **2023**, *881*, 163316. [CrossRef]
5. Wang, Y.; Fan, C.; Hu, H.; Li, Y.; Sun, D.; Wang, Y.; Peng, L. Genetic modification of plant cell walls to enhance biomass yield and biofuel production in bioenergy crops. *Biotechnol. Adv.* **2016**, *34*, 997–1017. [CrossRef]
6. Shu, F.; Jiang, B.; Yuan, Y.; Li, M.; Wu, W.; Jin, Y.; Xiao, H. Biological Activities and Emerging Roles of Lignin and Lignin-Based Products–A Review. *Biomacromolecules* **2021**, *22*, 4905–4918. [CrossRef]
7. Renault, H.; Werck-Reichhart, D.; Weng, J.-K. Harnessing lignin evolution for biotechnological applications. *Curr. Opin. Biotechnol.* **2019**, *56*, 105–111. [CrossRef]
8. Vanholme, R.; De Meester, B.; Ralph, J.; Boerjan, W. Lignin biosynthesis and its integration into metabolism. *Curr. Opin. Biotechnol.* **2019**, *56*, 230–239. [CrossRef]
9. Dixon, R.A.; Barros, J. Lignin biosynthesis: Old roads revisited and new roads explored. *Open Biol.* **2019**, *9*, 190215. [CrossRef]

10. Yao, T.; Feng, K.; Xie, M.; Barros, J.; Tschaplinski, T.J.; Tuskan, G.A.; Muchero, W.; Chen, J.-G. Phylogenetic Occurrence of the Phenylpropanoid Pathway and Lignin Biosynthesis in Plants. *Front. Plant Sci.* **2021**, *12*, 704697. [CrossRef]
11. Zhao, Q.; Dixon, R.A. Transcriptional networks for lignin biosynthesis: More complex than we thought? *Trends Plant Sci.* **2011**, *16*, 227–233. [CrossRef] [PubMed]
12. Nakano, Y.; Yamaguchi, M.; Endo, H.; Rejab, N.A.; Ohtani, M. NAC-MYB-based transcriptional regulation of secondary cell wall biosynthesis in land plants. *Front. Plant Sci.* **2015**, *6*, 288. [CrossRef] [PubMed]
13. Li, F.; Zhang, Y.; Tian, C.; Wang, X.; Zhou, L.; Jiang, J.; Wang, L.; Chen, F.; Chen, S. Molecular module of CmMYB15-like-Cm4CL2 regulating lignin biosynthesis of chrysanthemum (*Chrysanthemum morifolium*) in response to aphid (*Macrosiphoniella sanborni*) feeding. *New Phytol.* **2023**, *237*, 1776–1793. [CrossRef] [PubMed]
14. Xue, C.; Yao, J.-L.; Xue, Y.-S.; Su, G.-Q.; Wang, L.; Lin, L.-K.; Allan, A.C.; Zhang, S.-L.; Wu, J. PbrMYB169 positively regulates lignification of stone cells in pear fruit. *J. Exp. Bot.* **2019**, *70*, 1801–1814. [CrossRef] [PubMed]
15. Xue, Y.; Shan, Y.; Yao, J.-L.; Wang, R.; Xu, S.; Liu, D.; Ye, Z.; Lin, J.; Li, X.; Xue, C.; et al. The transcription factor PbrMYB24 regulates lignin and cellulose biosynthesis in stone cells of pear fruits. *Plant Physiol.* **2023**, *192*, 1997–2014. [CrossRef]
16. Chen, K.; Song, M.; Guo, Y.; Liu, L.; Xue, H.; Dai, H.; Zhang, Z. MdMYB46 could enhance salt and osmotic stress tolerance in apple by directly activating stress-responsive signals. *Plant Biotechnol. J.* **2019**, *17*, 2341–2355. [CrossRef] [PubMed]
17. Shi, Y.; Man, J.; Huang, Y.; Zhang, J.; Zhang, Z.; Yin, G.; Wang, X.; Liu, S.; Chen, Y.; Wang, X.; et al. Overexpression of PnMYB2 from Panax notoginseng induces cellulose and lignin biosynthesis during cell wall formation. *Planta* **2022**, *255*, 107. [CrossRef]
18. Li, X.; Wang, N.; She, W.; Guo, Z.; Pan, H.; Yu, Y.; Ye, J.; Pan, D.; Pan, T. Identification and Functional Analysis of the CgNAC043 Gene Involved in Lignin Synthesis from Citrusgrandis "San Hong". *Plants* **2022**, *11*, 403. [CrossRef]
19. Chen, K.; Guo, Y.; Song, M.; Liu, L.; Xue, H.; Dai, H.; Zhang, Z. Dual role of MdSND1 in the biosynthesis of lignin and in signal transduction in response to salt and osmotic stress in apple. *Hortic. Res.* **2020**, *7*, 204. [CrossRef]
20. Wang, R.; Xue, Y.; Fan, J.; Yao, J.-L.; Qin, M.; Lin, T.; Lian, Q.; Zhang, M.; Li, X.; Li, J.; et al. A systems genetics approach reveals PbrNSC as a regulator of lignin and cellulose biosynthesis in stone cells of pear fruit. *Genome Biol.* **2021**, *22*, 313. [CrossRef]
21. Zhao, W.; Ding, L.; Liu, J.; Zhang, X.; Li, S.; Zhao, K.; Guan, Y.; Song, A.; Wang, H.; Chen, S.; et al. Regulation of lignin biosynthesis by an atypical bHLH protein CmHLB in Chrysanthemum. *J. Exp. Bot.* **2022**, *73*, 2403–2419. [CrossRef] [PubMed]
22. Ge, H.; Xu, H.; Li, X.; Chen, J. The MADS-box gene EjAGL15 positively regulates lignin deposition in the flesh of loquat fruit during its storage. *Front. Plant Sci.* **2023**, *14*, 1166262. [CrossRef]
23. Yan, C.; Hu, Z.; Nie, Z.; Li, J.; Yao, X.; Yin, H. CcBLH6, a bell-like homeodomain-containing transcription factor, regulates the fruit lignification pattern. *Planta* **2021**, *253*, 90. [CrossRef]
24. Guillaumie, S.; Mzid, R.; Méchin, V.; Léon, C.; Hichri, I.; Destrac-Irvine, A.; Trossat-Magnin, C.; Delrot, S.; Lauvergeat, V. The grapevine transcription factor WRKY2 influences the lignin pathway and xylem development in tobacco. *Plant Mol. Biol.* **2010**, *72*, 215–234. [CrossRef] [PubMed]
25. Singh, S.K.; Shree, A.; Verma, S.; Singh, K.; Kumar, K.; Srivastava, V.; Singh, R.; Saxena, S.; Singh, A.P.; Pandey, A.; et al. The nuclear effector ArPEC25 from the necrotrophic fungus Ascochyta rabiei targets the chickpea transcription factor CaβLIM1a and negatively modulates lignin biosynthesis, increasing host susceptibility. *Plant Cell* **2023**, *35*, 1134–1159. [CrossRef] [PubMed]
26. Liu, Y.; Liu, Q.; Li, X.; Zhang, Z.; Ai, S.; Liu, C.; Ma, F.; Li, C. MdERF114 enhances the resistance of apple roots to Fusarium solani by regulating the transcription of MdPRX63. *Plant Physiol.* **2023**, *192*, 2015–2029. [CrossRef]
27. Zeng, J.-K.; Li, X.; Zhang, J.; Ge, H.; Yin, X.-R.; Chen, K.-S. Regulation of loquat fruit low temperature response and lignification involves interaction of heat shock factors and genes associated with lignin biosynthesis. *Plant Cell Environ.* **2016**, *39*, 1780–1789. [CrossRef]
28. Liu, X.; Wu, C.; Su, D.; Yang, Y.; Xian, Z.; Yu, C.; Li, Z.; Hao, Y.; Chen, R. The SlHB8 Acts as a Negative Regulator in Stem Development and Lignin Biosynthesis. *Int. J. Mol. Sci.* **2021**, *22*, 13343. [CrossRef]
29. Li, X.; Huang, H.; Rizwan, H.M.; Wang, N.; Jiang, J.; She, W.; Zheng, G.; Pan, H.; Guo, Z.; Pan, D.; et al. Transcriptome Analysis Reveals Candidate Lignin-Related Genes and Transcription Factors during Fruit Development in Pomelo (*Citrus maxima*). *Genes* **2022**, *13*, 845. [CrossRef]
30. Zhang, Q.; Wang, L.; Wang, Z.; Zhang, R.; Liu, P.; Liu, M.; Liu, Z.; Zhao, Z.; Wang, L.; Chen, X.; et al. The regulation of cell wall lignification and lignin biosynthesis during pigmentation of winter jujube. *Hortic. Res.* **2021**, *8*, 238. [CrossRef]
31. Wang, Q.; Hu, J.; Yang, T.; Chang, S. Anatomy and lignin deposition of stone cell in Camellia oleifera shell during the young stage. *Protoplasma* **2021**, *258*, 361–370. [CrossRef] [PubMed]
32. Du, T.; Qin, Z.; Zhou, Y.; Zhang, L.; Wang, Q.; Li, Z.; Hou, F. Comparative Transcriptome Analysis Reveals the Effect of Lignin on Storage Roots Formation in Two Sweetpotato (*Ipomoea batatas* (L.) Lam.) Cultivars. *Genes* **2023**, *14*, 1263. [CrossRef] [PubMed]
33. Zhang, X.; Ran, D.; Wu, P.; Cao, Z.; Xu, F.; Xia, N.; Gao, H.; Jiang, Y.; Yang, C.; He, N.; et al. Transcriptome and metabolite profiling to identify genes associated with rhizome lignification and the function of ZoCSE in ginger (*Zingiber officinale*). *Funct. Plant Biol.* **2022**, *49*, 689–703. [CrossRef]
34. Estrada, B.; Bernal, M.A.; Díaz, J.; Pomar, F.; Merino, F. Fruit Development in Capsicum annuum: Changes in Capsaicin, Lignin, Free Phenolics, and Peroxidase Patterns. *J. Agric. Food Chem.* **2000**, *48*, 6234–6239. [CrossRef]
35. Wang, G.-L.; Huang, Y.; Zhang, X.-Y.; Xu, Z.-S.; Wang, F.; Xiong, A.-S. Transcriptome-based identification of genes revealed differential expression profiles and lignin accumulation during root development in cultivated and wild carrots. *Plant Cell Rep.* **2016**, *35*, 1743–1755. [CrossRef]

36. Liu, J.; Huang, Q.; Kang, P.; Liang, L.; Chen, J. Lignin Accumulation in Three Pumelo Cultivars in Association with Sucrose and Energy Depletion. *Biomolecules* **2019**, *9*, 701. [CrossRef] [PubMed]
37. Li, X.; Meng, Z.; Malik, A.U.; Zhang, S.; Wang, Q. Maintaining the quality of postharvest broccoli by inhibiting ethylene accumulation using diacetyl. *Front. Nutr.* **2022**, *9*, 1055651. [CrossRef]
38. Qian, C.; Ji, Z.; Sun, Y.; Zhang, M.; Kan, J.; Xiao, L.; Liu, J.; Jin, C.; Yang, W.; Qi, X. Lignin Biosynthesis in Postharvest Water Bamboo (*Zizania latifolia*) Shoots during Cold Storage Is Regulated by RBOH-Mediated Reactive Oxygen Species Signaling. *J. Agric. Food Chem.* **2023**, *71*, 3201–3209. [CrossRef]
39. Yang, B.; Han, Y.; Wu, W.; Fang, X.; Chen, H.; Gao, H. Impact of melatonin application on lignification in water bamboo shoot during storage. *Food Chem. X* **2022**, *13*, 100254. [CrossRef]
40. Li, X.; Wang, J.; Qu, Y.; Li, Y.; Humaira, Y.; Muhammad, S.; Pu, H.; Yu, L.; Li, H. Comparison of storage and lignin accumulation characteristics between two types of snow pea. *PLoS ONE* **2022**, *17*, e0268776. [CrossRef]
41. Yang, T.-D.; Chen, Y.-L.; Zeng, F.-K.; Ye, M.-Q.; Wang, L.; Luo, Z.; Qi, Y.-W.; Chen, F.-P. Effects of modified atmosphere packaging on the postharvest quality of mulberry leaf vegetable. *Sci. Rep.* **2022**, *12*, 10893. [CrossRef]
42. Dai, Y.; Wang, S.; Huang, W.; Li, Z.; Zhang, S.; Zhang, H.; Li, G.; Fang, Z.; Sun, R.; Li, F.; et al. Transcriptome Analysis of Chinese Cabbage Provides Insights into the Basis of Understanding the Lignin Affected by Low Temperature. *Genes* **2022**, *13*, 2084. [CrossRef]
43. Hou, D.; Lu, H.; Zhao, Z.; Pei, J.; Yang, H.; Wu, A.; Yu, X.; Lin, X. Integrative transcriptomic and metabolomic data provide insights into gene networks associated with lignification in postharvest Lei bamboo shoots under low temperature. *Food Chem.* **2022**, *368*, 130822. [CrossRef] [PubMed]
44. Xiao, L.; Jiang, X.; Deng, Y.; Xu, K.; Duan, X.; Wan, K.; Tang, X. Study on Characteristics and Lignification Mechanism of Postharvest Banana Fruit during Chilling Injury. *Foods* **2023**, *12*, 1097. [CrossRef] [PubMed]
45. Wang, Q.; Wu, X.; Yuan, C.; Lou, Z.; Li, Y. Effect of Saturated Steam Heat Treatment on Physical and Chemical Properties of Bamboo. *Molecules* **2020**, *25*, 1999. [CrossRef] [PubMed]
46. Lima, R.B.; dos Santos, T.B.; Vieira, L.G.E.; Ferrarese, M.d.L.L.; Ferrarese-Filho, O.; Donatti, L.; Boeger, M.R.T.; Petkowicz, C.L.d.O. Heat stress causes alterations in the cell-wall polymers and anatomy of coffee leaves (*Coffea arabica* L.). *Carbohydr. Polym.* **2013**, *93*, 135–143. [CrossRef]
47. Gu, H.; Wang, Y.; Xie, H.; Qiu, C.; Zhang, S.; Xiao, J.; Li, H.; Chen, L.; Li, X.; Ding, Z. Drought stress triggers proteomic changes involving lignin, flavonoids and fatty acids in tea plants. *Sci. Rep.* **2020**, *10*, 15504. [CrossRef]
48. Živanović, B.; Milić Komić, S.; Nikolić, N.; Mutavdžić, D.; Srećković, T.; Veljović Jovanović, S.; Prokić, L. Differential Response of Two Tomato Genotypes, Wild Type cv. Ailsa Craig and Its ABA-Deficient Mutant flacca to Short-Termed Drought Cycles. *Plants* **2021**, *10*, 2308. [CrossRef]
49. Liu, W.; Jiang, Y.; Jin, Y.; Wang, C.; Yang, J.; Qi, H. Drought-induced ABA, H_2O_2 and JA positively regulate CmCAD genes and lignin synthesis in melon stems. *BMC Plant Biol.* **2021**, *21*, 83. [CrossRef]
50. Liu, W.; Jiang, Y.; Wang, C.; Zhao, L.; Jin, Y.; Xing, Q.; Li, M.; Lv, T.; Qi, H. Lignin synthesized by CmCAD2 and CmCAD3 in oriental melon (*Cucumis melo* L.) seedlings contributes to drought tolerance. *Plant Mol. Biol.* **2020**, *103*, 689–704. [CrossRef]
51. Becerra-Moreno, A.; Redondo-Gil, M.; Benavides, J.; Nair, V.; Cisneros-Zevallos, L.; Jacobo-Velázquez, D.A. Combined effect of water loss and wounding stress on gene activation of metabolic pathways associated with phenolic biosynthesis in carrot. *Front. Plant Sci.* **2015**, *6*, 837. [CrossRef]
52. Garrido, Y.; Tudela, J.A.; Marín, A.; Mestre, T.; Martínez, V.; Gil, M.I. Physiological, phytochemical and structural changes of multi-leaf lettuce caused by salt stress. *J. Sci. Food Agric.* **2014**, *94*, 1592–1599. [CrossRef] [PubMed]
53. De Lima, R.B.; dos Santos, T.B.; Vieira, L.G.E.; de Lourdes Lúcio Ferrarese, M.; Ferrarese-Filho, O.; Donatti, L.; Boeger, M.R.T.; de Oliveira Petkowicz, C.L. Salt stress alters the cell wall polysaccharides and anatomy of coffee (*Coffea arabica* L.) leaf cells. *Carbohydr. Polym.* **2014**, *112*, 686–694. [CrossRef]
54. Sánchez-Aguayo, I.; Rodríguez-Galán, J.M.; García, R.; Torreblanca, J.; Pardo, J.M. Salt stress enhances xylem development and expression of S-adenosyl-l-methionine synthase in lignifying tissues of tomato plants. *Planta* **2004**, *220*, 278–285. [CrossRef] [PubMed]
55. Rather, B.A.; Mir, I.R.; Masood, A.; Anjum, N.A.; Khan, N.A. Ethylene-nitrogen synergism induces tolerance to copper stress by modulating antioxidant system and nitrogen metabolism and improves photosynthetic capacity in mustard. *Environ. Sci. Pollut. Res.* **2022**, *29*, 49029–49049. [CrossRef] [PubMed]
56. Kováčik, J.; Klejdus, B. Dynamics of phenolic acids and lignin accumulation in metal-treated Matricaria chamomilla roots. *Plant Cell Rep.* **2008**, *27*, 605–615. [CrossRef] [PubMed]
57. Gao, F.; Zhang, X.; Zhang, J.; Li, J.; Niu, T.; Tang, C.; Wang, C.; Xie, J. Zinc oxide nanoparticles improve lettuce (*Lactuca sativa* L.) plant tolerance to cadmium by stimulating antioxidant defense, enhancing lignin content and reducing the metal accumulation and translocation. *Front. Plant Sci.* **2022**, *13*, 1015745. [CrossRef]
58. Liu, X.; Zhu, Q.; Liu, W.; Zhang, J. 24-Epibrassinolide confers zinc stress tolerance in watermelon seedlings through modulating antioxidative capacities and lignin accumulation. *PeerJ.* **2023**, *11*, e15330. [CrossRef]
59. Yang, Y.; He, Y.; Lv, S.; Zhu, H.; Wang, T.; Wang, G.; Hong, N.; Wang, L. The PcMYB44-mediated miR397-PcLACs module regulates defence-induced lignification in pear resistance to fungal disease. *Mol. Plant Pathol.* **2023**, *24*, 1107–1125. [CrossRef]

60. Kumar, P.; Singh, B.; Rajak, S.; Pandey, S.; Pati, P.K. Dynamics of reactive oxygen species and lignin biosynthesis during leaf spot disease of *Withania somnifera* (L.) Dunal. *Plant Biol.* **2023**, *25*, 757–770. [CrossRef]
61. Chen, C.; Cai, N.; Wan, C.; Huang, Q.; Chen, J. Cell wall modification and lignin biosynthesis involved in disease resistance against Diaporthe citri in harvested pummelo fruit elicited by carvacrol. *J. Sci. Food Agric.* **2022**, *102*, 3140–3149. [CrossRef] [PubMed]
62. Yu, Y.; Yu, Y.; Cui, N.; Ma, L.; Tao, R.; Ma, Z.; Meng, X.; Fan, H. Lignin biosynthesis regulated by CsCSE1 is required for Cucumis sativus defence to Podosphaera xanthii. *Plant Physiol. Biochem.* **2022**, *186*, 88–98. [CrossRef] [PubMed]
63. Zhang, Z.; Long, Y.; Yin, X.; Wang, W.; Li, W.; Chen, T.; Chen, J.; Chen, X.; Wang, B.; Ma, J. Metabolome and Transcriptome Analysis of Sulfur-Induced Kiwifruit Stem Laccase Gene Involved in Syringyl Lignin Synthesis against Bacterial Canker. *J. Agric. Food Chem.* **2023**, *71*, 13566–13576. [CrossRef] [PubMed]
64. Zúñiga, E.; Luque, J.; Martos, S. Lignin biosynthesis as a key mechanism to repress Polystigma amygdalinum, the causal agent of the red leaf blotch disease in almond. *J. Plant Physiol.* **2019**, *236*, 96–104. [CrossRef] [PubMed]
65. Hussain, K.; Jaweed, T.H.; Kamble, A.C. Modulation of phenylpropanoid and lignin biosynthetic pathway is crucial for conferring resistance in pigeon pea against Fusarium wilt. *Gene* **2023**, *851*, 146994. [CrossRef]
66. Liu, J.-X.; Feng, K.; Wang, G.-L.; Xu, Z.-S.; Wang, F.; Xiong, A.-S. Elevated CO_2 induces alteration in lignin accumulation in celery (*Apium graveolens* L.). *Plant Physiol. Biochem.* **2018**, *127*, 310–319. [CrossRef]
67. Wang, Y.-H.; Wu, X.-J.; Sun, S.; Xing, G.-M.; Wang, G.-L.; Que, F.; Khadr, A.; Feng, K.; Li, T.; Xu, Z.-S.; et al. DcC4H and DcPER Are Important in Dynamic Changes of Lignin Content in Carrot Roots under Elevated Carbon Dioxide Stress. *J. Agric. Food Chem.* **2018**, *66*, 8209–8220. [CrossRef]
68. Shabbaj, I.I.; AbdElgawad, H.; Tammar, A.; Alsiary, W.A.; Madany, M.M.Y. Future climate CO_2 can harness ROS homeostasis and improve cell wall fortification to alleviate the hazardous effect of Phelipanche infection in pea seedlings. *Plant Physiol. Biochem.* **2021**, *166*, 1131–1141. [CrossRef]
69. Maldonado, R.; Molina-Garcia, A.D.; Sanchez-Ballesta, M.T.; Escribano, M.I.; Merodio, C. High CO_2 Atmosphere Modulating the Phenolic Response Associated with Cell Adhesion and Hardening of Annona cherimola Fruit Stored at Chilling Temperature. *J. Agric. Food Chem.* **2002**, *50*, 7564–7569. [CrossRef]
70. Hao, Y.; Chen, F.; Wu, G.; Gao, W. Impact of Postharvest Nitric Oxide Treatment on Lignin Biosynthesis-Related Genes in Wax Apple (*Syzygium samarangense*) Fruit. *J. Agric. Food Chem.* **2016**, *64*, 8483–8490. [CrossRef]
71. Sun, M.; Yang, X.-L.; Zhu, Z.-P.; Xu, Q.-Y.; Wu, K.-X.; Kang, Y.-J.; Wang, H.; Xiong, A.-S. Comparative transcriptome analysis provides insight into nitric oxide suppressing lignin accumulation of postharvest okra (*Abelmoschus esculentus* L.) during cold storage. *Plant Physiol. Biochem.* **2021**, *167*, 49–67. [CrossRef] [PubMed]
72. Qi, X.; Ji, Z.; Lin, C.; Li, S.; Liu, J.; Kan, J.; Zhang, M.; Jin, C.; Qian, C. Nitric oxide alleviates lignification and softening of water bamboo (*Zizania latifolia*) shoots during postharvest storage. *Food Chem.* **2020**, *332*, 127416. [CrossRef] [PubMed]
73. Sun, M.; Yang, T.; Qiao, X.-H.; Zhao, P.; Zhu, Z.-P.; Wang, G.-L.; Xu, L.-L.; Xiong, A.-S. Nitric oxide regulates the lignification and carotenoid biosynthesis of postharvest carrot (*Daucus carota* L.). *Postharvest Biol. Technol.* **2024**, *207*, 112593. [CrossRef]
74. Corti Monzón, G.; Pinedo, M.; Di Rienzo, J.; Novo-Uzal, E.; Pomar, F.; Lamattina, L.; de la Canal, L. Nitric oxide is required for determining root architecture and lignin composition in sunflower. Supporting evidence from microarray analyses. *Nitric Oxide* **2014**, *39*, 20–28. [CrossRef] [PubMed]
75. Ren, Y.; Xue, Y.; Tian, D.; Zhang, L.; Xiao, G.; He, J. Improvement of Postharvest Anthracnose Resistance in Mango Fruit by Nitric Oxide and the Possible Mechanisms Involved. *J. Agric. Food Chem.* **2020**, *68*, 15460–15467. [CrossRef]
76. Wang, B.; Li, Z.; Han, Z.; Xue, S.; Bi, Y.; Prusky, D. Effects of nitric oxide treatment on lignin biosynthesis and texture properties at wound sites of muskmelons. *Food Chem.* **2021**, *362*, 130193. [CrossRef]
77. Han, M.-h.; Yang, N.; Wan, Q.-w.; Teng, R.-m.; Duan, A.-q.; Wang, Y.-h.; Zhuang, J. Exogenous melatonin positively regulates lignin biosynthesis in Camellia sinensis. *Int. J. Biol. Macromol.* **2021**, *179*, 485–499. [CrossRef]
78. Zhao, D.; Luan, Y.; Shi, W.; Tang, Y.; Huang, X.; Tao, J. Melatonin enhances stem strength by increasing lignin content and secondary cell wall thickness in herbaceous peony. *J. Exp. Bot.* **2022**, *73*, 5974–5991. [CrossRef]
79. Qu, G.; Wu, W.; Ba, L.; Ma, C.; Ji, N.; Cao, S. Melatonin Enhances the Postharvest Disease Resistance of Blueberries Fruit by Modulating the Jasmonic Acid Signaling Pathway and Phenylpropanoid Metabolites. *Front. Chem.* **2022**, *10*, 957581. [CrossRef]
80. Boonsiriwit, A.; Lee, M.; Kim, M.; Itkor, P.; Lee, Y.S. Exogenous Melatonin Reduces Lignification and Retains Quality of Green Asparagus (*Asparagus officinalis* L.). *Foods* **2021**, *10*, 2111. [CrossRef]
81. Liu, S.; Zha, Z.; Chen, S.; Tang, R.; Zhao, Y.; Lin, Q.; Duan, Y.; Wang, K. Hydrogen-rich water alleviates chilling injury-induced lignification of kiwifruit by inhibiting peroxidase activity and improving antioxidant system. *J. Sci. Food Agric.* **2023**, *103*, 2675–2680. [CrossRef] [PubMed]
82. Chen, C.; Nie, Z.; Wan, C.; Gan, Z.; Chen, J. Suppression on postharvest juice sac granulation and cell wall modification by chitosan treatment in harvested pummelo (*Citrus grandis* L. Osbeck) stored at room temperature. *Food Chem.* **2021**, *336*, 127636. [CrossRef] [PubMed]
83. Jogaiah, S.; Satapute, P.; De Britto, S.; Konappa, N.; Udayashankar, A.C. Exogenous priming of chitosan induces upregulation of phytohormones and resistance against cucumber powdery mildew disease is correlated with localized biosynthesis of defense enzymes. *Int. J. Biol. Macromol.* **2020**, *162*, 1825–1838. [CrossRef] [PubMed]

84. Chen, X.; Li, N.; Liu, C.; Wang, H.; Li, Y.; Xie, Y.; Ma, F.; Liang, J.; Li, C. Exogenous GABA improves the resistance of apple seedlings to long-term drought stress by enhancing GABA shunt and secondary cell wall biosynthesis. *Tree Physiol.* **2022**, *42*, 2563–2577. [CrossRef] [PubMed]
85. Jiao, Y.; Gong, X.; Qi, K.; Xie, Z.; Wang, Y.; Yuan, K.; Pan, Q.; Zhang, S.; Shiratake, K.; Khanizadeh, S.; et al. Transcriptome analysis provides new ideas for studying the regulation of glucose-induced lignin biosynthesis in pear calli. *BMC Plant Biol.* **2022**, *22*, 310. [CrossRef]
86. Zhao, D.; Xu, C.; Luan, Y.; Shi, W.; Tang, Y.; Tao, J. Silicon enhances stem strength by promoting lignin accumulation in herbaceous peony (*Paeonia lactiflora* Pall.). *Int. J. Biol. Macromol.* **2021**, *190*, 769–779. [CrossRef]
87. Wang, G.-L.; Que, F.; Xu, Z.-S.; Wang, F.; Xiong, A.-S. Exogenous gibberellin enhances secondary xylem development and lignification in carrot taproot. *Protoplasma* **2017**, *254*, 839–848. [CrossRef]
88. Wang, G.-L.; An, Y.-H.; Wang, Y.-H.; Liu, J.-X.; Wang, J.-Z.; Sun, M.; Xiong, A.-S. Gibberellin-Induced Alterations to the Expression of Cell Wall-Related Genes in the Xylem of Carrot Root. *J. Plant Growth Regul.* **2021**, *40*, 787–797. [CrossRef]
89. Singh, V.; Sergeeva, L.; Ligterink, W.; Aloni, R.; Zemach, H.; Doron-Faigenboim, A.; Yang, J.; Zhang, P.; Shabtai, S.; Firon, N. Gibberellin Promotes Sweetpotato Root Vascular Lignification and Reduces Storage-Root Formation. *Front. Plant Sci.* **2019**, *10*, 1320. [CrossRef]
90. García-Rojas, M.; Meneses, M.; Oviedo, K.; Carrasco, C.; Defilippi, B.; González-Agüero, M.; León, G.; Hinrichsen, P. Exogenous gibberellic acid application induces the overexpression of key genes for pedicel lignification and an increase in berry drop in table grape. *Plant Physiol. Biochem.* **2018**, *126*, 32–38. [CrossRef]
91. López Núñez-Flores, M.J.; Gutiérrez, J.; Gómez-Ros, L.V.; Novo Uzal, E.; Sottomayor, M.; Ros Barceló, A. Downregulation of the Basic Peroxidase Isoenzyme from Zinnia elegans by Gibberellic Acid. *J. Integr. Plant Biol.* **2010**, *52*, 244–251. [CrossRef] [PubMed]
92. Kong, Q.; Mostafa, H.H.A.; Yang, W.; Wang, J.; Nuerawuti, M.; Wang, Y.; Song, J.; Zhang, X.; Ma, L.; Wang, H.; et al. Comparative transcriptome profiling reveals that brassinosteroid-mediated lignification plays an important role in garlic adaption to salt stress. *Plant Physiol. Biochem.* **2021**, *158*, 34–42. [CrossRef]
93. Han, Y.; Yang, R.; Zhang, X.; Wang, Q.; Wang, B.; Zheng, X.; Li, Y.; Prusky, D.; Bi, Y. Brassinosteroid Accelerates Wound Healing of Potato Tubers by Activation of Reactive Oxygen Metabolism and Phenylpropanoid Metabolism. *Foods* **2022**, *11*, 906. [CrossRef] [PubMed]
94. Liu, X.; Zhu, Q.; Liu, W.; Zhang, J. Exogenous Brassinosteroid Enhances Zinc tolerance by activating the Phenylpropanoid Biosynthesis pathway in *Citrullus lanatus* L. *Plant Signal Behav.* **2023**, *18*, 2186640. [CrossRef] [PubMed]
95. An, Y.-H.; Zhou, H.; Yuan, Y.-H.; Li, L.; Sun, J.; Shu, S.; Guo, S.-R. 24-Epibrassinolide-induced alterations in the root cell walls of *Cucumis sativus* L. under $Ca(NO_3)_2$ stress. *Protoplasma* **2018**, *255*, 841–850. [CrossRef] [PubMed]
96. Ren, R.; Yang, X.; Song, A.; Li, C.; Yang, H.; Kang, Y. Control of Phytophthora melonis damping-off treated with 24-epibrassinolide and a histological study of cucumber hypocotyl. *Protoplasma* **2020**, *257*, 1519–1529. [CrossRef]
97. Li, Y.; Hua, J.; Hou, X.; Qi, N.; Li, C.; Wang, C.; Yao, Y.; Huang, D.; Zhang, H.; Liao, W. Brassinosteroids is involved in methane-induced adventitious root formation via inducing cell wall relaxation in marigold. *BMC Plant Biol.* **2023**, *23*, 2. [CrossRef]
98. Naing, A.H.; Soe, M.T.; Yeum, J.H.; Kim, C.K. Ethylene Acts as a Negative Regulator of the Stem-Bending Mechanism of Different Cut Snapdragon Cultivars. *Front. Plant Sci.* **2021**, *12*, 745038. [CrossRef] [PubMed]
99. Soe, M.T.; Naing, A.H.; Kim, S.R.; Kim, C.K. Characterizing the effects of different chemicals on stem bending of cut snapdragon flower. *Plant Methods* **2022**, *18*, 4. [CrossRef]
100. Figueroa, N.E.; Gatica-Meléndez, C.; Figueroa, C.R. Ethylene application at the immature stage of Fragaria chiloensis fruit represses the anthocyanin biosynthesis with a concomitant accumulation of lignin. *Food Chem.* **2021**, *358*, 129913. [CrossRef]
101. Nascimento, V.L.; Pereira, A.M.; Siqueira, J.A.; Pereira, A.S.; Silva, V.F.; Costa, L.C.; Ribeiro, D.M.; Zsögön, A.; Nunes-Nesi, A.; Araújo, W.L. Exogenous ethylene reduces growth via alterations in central metabolism and cell wall composition in tomato (*Solanum lycopersicum*). *J. Plant Physiol.* **2021**, *263*, 153460. [CrossRef] [PubMed]
102. Xie, G.; Feng, Y.; Chen, Y.; Zhang, M. Effects of 1-Methylcyclopropene (1-MCP) and Ethylene on Postharvest Lignification of Common Beans (*Phaseolus vulgaris* L). *ACS Omega* **2020**, *5*, 8659–8666. [CrossRef] [PubMed]
103. Zhu, X.; Ye, L.; Ding, X.; Gao, Q.; Xiao, S.; Tan, Q.; Huang, J.; Chen, W.; Li, X. Transcriptomic analysis reveals key factors in fruit ripening and rubbery texture caused by 1-MCP in papaya. *BMC Plant Biol.* **2019**, *19*, 309. [CrossRef] [PubMed]
104. Xu, S.; Sun, M.; Yao, J.-L.; Liu, X.; Xue, Y.; Yang, G.; Zhu, R.; Jiang, W.; Wang, R.; Xue, C.; et al. Auxin inhibits lignin and cellulose biosynthesis in stone cells of pear fruit via the PbrARF13-PbrNSC-PbrMYB132 transcriptional regulatory cascade. *Plant Biotechnol. J.* **2023**, *21*, 1408–1425. [CrossRef] [PubMed]
105. Yao, X.; Li, H.; Nie, J.; Liu, H.; Guo, Y.; Lv, L.; Yang, Z.; Sui, X. Disruption of the amino acid transporter CsAAP2 inhibits auxin-mediated root development in cucumber. *New Phytol.* **2023**, *239*, 639–659. [CrossRef]
106. Ma, Q.; Ding, Y.; Chang, J.; Sun, X.; Zhang, L.; Wei, Q.; Cheng, Y.; Chen, L.; Xu, J.; Deng, X. Comprehensive insights on how 2,4-dichlorophenoxyacetic acid retards senescence in post-harvest citrus fruits using transcriptomic and proteomic approaches. *J. Exp. Bot.* **2014**, *65*, 61–74. [CrossRef]
107. Shi, M.; Liu, X.; Zhang, H.; He, Z.; Yang, H.; Chen, J.; Feng, J.; Yang, W.; Jiang, Y.; Yao, J.-L.; et al. The IAA- and ABA-responsive transcription factor CgMYB58 upregulates lignin biosynthesis and triggers juice sac granulation in pummelo. *Hortic. Res.* **2020**, *7*, 139. [CrossRef]

108. Nunes da Silva, M.; Vasconcelos, M.W.; Pinto, V.; Balestra, G.M.; Mazzaglia, A.; Gomez-Cadenas, A.; Carvalho, S.M.P. Role of methyl jasmonate and salicylic acid in kiwifruit plants further subjected to Psa infection: Biochemical and genetic responses. *Plant Physiol. Biochem.* **2021**, *162*, 258–266. [CrossRef]
109. Zhang, M.; Shi, Y.; Liu, Z.; Zhang, Y.; Yin, X.; Liang, Z.; Huang, Y.; Grierson, D.; Chen, K. An EjbHLH14-EjHB1-EjPRX12 module is involved in methyl jasmonate alleviation of chilling-induced lignin deposition in loquat fruit. *J. Exp. Bot.* **2022**, *73*, 1668–1682. [CrossRef]
110. Li, S.; Xiao, L.; Chen, M.; Cao, Q.; Luo, Z.; Kang, N.; Jia, M.; Chen, J.; Xiang, M. The involvement of the phenylpropanoid and jasmonate pathways in methyl jasmonate-induced soft rot resistance in kiwifruit (*Actinidia chinensis*). *Front. Plant Sci.* **2022**, *13*, 1097733. [CrossRef]
111. Zhu, Y.; Wang, Y.; Jiang, H.; Liu, W.; Zhang, S.; Hou, X.; Zhang, S.; Wang, N.; Zhang, R.; Zhang, Z.; et al. Transcriptome analysis reveals that PbMYB61 and PbMYB308 are involved in the regulation of lignin biosynthesis in pear fruit stone cells. *Plant J.* **2023**, *116*, 217–233. [CrossRef] [PubMed]
112. Fu, S.; Shao, J.; Roy, A.; Brlansky, R.H.; Zhou, C.; Hartung, J.S. Transcriptomic analyses reveal physiological changes in sweet orange roots affected by citrus blight. *BMC Genom.* **2019**, *20*, 969. [CrossRef] [PubMed]
113. Feng, H.; Xu, L.; Wang, Y.; Tang, M.; Zhu, X.; Zhang, W.; Sun, X.; Nie, S.; Muleke, E.M.m.; Liu, L. Identification of critical genes associated with lignin biosynthesis in radish (*Raphanus sativus* L.) by de novo transcriptome sequencing. *Mol. Genet. Genom.* **2017**, *292*, 1151–1163. [CrossRef]
114. Gao, X.; Guo, P.; Wang, Z.; Chen, C.; Ren, Z. Transcriptome profiling reveals response genes for downy mildew resistance in cucumber. *Planta* **2021**, *253*, 112. [CrossRef]
115. Ma, J.; Li, X.; He, M.; Li, Y.; Lu, W.; Li, M.; Sun, B.; Zheng, Y. A Joint Transcriptomic and Metabolomic Analysis Reveals the Regulation of Shading on Lignin Biosynthesis in Asparagus. *Int. J. Mol. Sci.* **2023**, *24*, 1539. [CrossRef]
116. Ren, L.; Zhang, T.; Wu, H.; Ge, X.; Wan, H.; Chen, S.; Li, Z.; Ma, D.; Wang, A. Blocking IbmiR319a Impacts Plant Architecture and Reduces Drought Tolerance in Sweet Potato. *Genes* **2022**, *13*, 404. [CrossRef]
117. Hu, Y.; Cheng, H.; Zhang, Y.; Zhang, J.; Niu, S.; Wang, X.; Li, W.; Zhang, J.; Yao, Y. The MdMYB16/MdMYB1-miR7125-MdCCR module regulates the homeostasis between anthocyanin and lignin biosynthesis during light induction in apple. *New Phytol.* **2021**, *231*, 1105–1122. [CrossRef]
118. Sharma, N.K.; Yadav, S.; Gupta, S.K.; Irulappan, V.; Francis, A.; Senthil-Kumar, M.; Chattopadhyay, D. MicroRNA397 regulates tolerance to drought and fungal infection by regulating lignin deposition in chickpea root. *Plant Cell Environ.* **2023**, *46*, 3501–3517. [CrossRef]
119. Xue, C.; Yao, J.-L.; Qin, M.-F.; Zhang, M.-Y.; Allan, A.C.; Wang, D.-F.; Wu, J. PbrmiR397a regulates lignification during stone cell development in pear fruit. *Plant Biotechnol. J.* **2019**, *17*, 103–117. [CrossRef]
120. Jeena, G.S.; Joshi, A.; Shukla, R.K. Bm-miR172c-5p Regulates Lignin Biosynthesis and Secondary Xylem Thickness by Altering the Ferulate 5 Hydroxylase Gene in Bacopa monnieri. *Plant Cell Physiol.* **2021**, *62*, 894–912. [CrossRef]
121. Yang, X.; Kang, Y.; Liu, Y.; Shi, M.; Zhang, W.; Fan, Y.; Yao, Y.; Li, H.; Qin, S. Integrated analysis of miRNA-mRNA regulatory networks of potato (*Solanum tuberosum* L.) in response to cadmium stress. *Ecotoxicol. Environ. Saf.* **2021**, *224*, 112682. [CrossRef] [PubMed]
122. Kang, Y.; Yang, X.; Liu, Y.; Shi, M.; Zhang, W.; Fan, Y.; Yao, Y.; Zhang, J.; Qin, S. Integration of mRNA and miRNA analysis reveals the molecular mechanism of potato (*Solanum tuberosum* L.) response to alkali stress. *Int. J. Biol. Macromol.* **2021**, *182*, 938–949. [CrossRef]
123. Yang, C.; Liang, Y.; Qiu, D.; Zeng, H.; Yuan, J.; Yang, X. Lignin metabolism involves Botrytis cinerea BcGs1- induced defense response in tomato. *BMC Plant Biol.* **2018**, *18*, 103. [CrossRef] [PubMed]
124. Wang, P.; Wu, X.; Shi, Z.; Tao, S.; Liu, Z.; Qi, K.; Xie, Z.; Qiao, X.; Gu, C.; Yin, H.; et al. A large-scale proteogenomic atlas of pear. *Mol. Plant* **2023**, *16*, 599–615. [CrossRef] [PubMed]
125. Yang, C.; Xu, L.; Zhang, N.; Islam, F.; Song, W.; Hu, L.; Liu, D.; Xie, X.; Zhou, W. iTRAQ-based proteomics of sunflower cultivars differing in resistance to parasitic weed Orobanche cumana. *Proteomics* **2017**, *17*, 1700009. [CrossRef]
126. Li, S.; Su, X.; Jin, Q.; Li, G.; Sun, Y.; Abdullah, M.; Cai, Y.; Lin, Y. iTRAQ-Based Identification of Proteins Related to Lignin Synthesis in the Pear Pollinated with Pollen from Different Varieties. *Molecules* **2018**, *23*, 548. [CrossRef]
127. Li, Y.; Liu, S.; Zhang, D.; Liu, A.; Zhu, W.; Zhang, J.; Yang, B. Integrative Omic Analysis Reveals the Dynamic Change in Phenylpropanoid Metabolism in Morus alba under Different Stress. *Plants* **2023**, *12*, 3265. [CrossRef]
128. Wang, Y.; Zhang, X.; Yang, S.; Yuan, Y. Lignin Involvement in Programmed Changes in Peach-Fruit Texture Indicated by Metabolite and Transcriptome Analyses. *J. Agric. Food Chem.* **2018**, *66*, 12627–12640. [CrossRef]
129. Gong, X.; Qi, K.; Chen, J.; Zhao, L.; Xie, Z.; Yan, X.; Khanizadeh, S.; Zhang, S.; Tao, S. Multi-omics analyses reveal stone cell distribution pattern in pear fruit. *Plant J.* **2023**, *113*, 626–642. [CrossRef]
130. Huang, H.-Y.; Ren, Q.-Q.; Lai, Y.-H.; Peng, M.-Y.; Zhang, J.; Yang, L.-T.; Huang, Z.-R.; Chen, L.-S. Metabolomics combined with physiology and transcriptomics reveals how Citrus grandis leaves cope with copper-toxicity. *Ecotoxicol. Environ. Saf.* **2021**, *223*, 112579. [CrossRef]
131. Song, F.; Li, Z.; Wang, C.; Jiang, Y.; Wang, Z.; He, L.; Ma, X.; Zhang, Y.; Song, X.; Liu, J.; et al. CsMYB15 positively regulates Cs4CL2-mediated lignin biosynthesis during juice sac granulation in navel orange. *Front. Plant Sci.* **2023**, *14*, 1223820. [CrossRef] [PubMed]

132. Zhang, J.; Yin, X.-R.; Li, H.; Xu, M.; Zhang, M.-X.; Li, S.-J.; Liu, X.-F.; Shi, Y.-N.; Grierson, D.; Chen, K.-S. ETHYLENE RESPONSE FACTOR39-MYB8 complex regulates low-temperature-induced lignification of loquat fruit. *J. Exp. Bot.* **2020**, *71*, 3172–3184. [CrossRef] [PubMed]
133. Tu, M.; Wang, X.; Yin, W.; Wang, Y.; Li, Y.; Zhang, G.; Li, Z.; Song, J.; Wang, X. Grapevine VlbZIP30 improves drought resistance by directly activating VvNAC17 and promoting lignin biosynthesis through the regulation of three peroxidase genes. *Hortic. Res.* **2020**, *7*, 150. [CrossRef]
134. Zhao, D.; Luan, Y.; Shi, W.; Zhang, X.; Meng, J.; Tao, J. A Paeonia ostii caffeoyl-CoA O-methyltransferase confers drought stress tolerance by promoting lignin synthesis and ROS scavenging. *Plant Sci.* **2021**, *303*, 110765. [CrossRef]
135. Duan, A.-Q.; Tao, J.-P.; Jia, L.-L.; Tan, G.-F.; Liu, J.-X.; Li, T.; Chen, L.-Z.; Su, X.-J.; Feng, K.; Xu, Z.-S.; et al. AgNAC1, a celery transcription factor, related to regulation on lignin biosynthesis and salt tolerance. *Genomics* **2020**, *112*, 5254–5264. [CrossRef]
136. Lee, C.-J.; Kim, S.-E.; Park, S.-U.; Lim, Y.-H.; Choi, H.-Y.; Kim, W.-G.; Ji, C.Y.; Kim, H.S.; Kwak, S.-S. Tuberous roots of transgenic sweetpotato overexpressing IbCAD1 have enhanced low-temperature storage phenotypes. *Plant Physiol. Biochem.* **2021**, *166*, 549–557. [CrossRef]
137. Wang, Y.; Sheng, L.; Zhang, H.; Du, X.; An, C.; Xia, X.; Chen, F.; Jiang, J.; Chen, S. CmMYB19 Over-Expression Improves Aphid Tolerance in Chrysanthemum by Promoting Lignin Synthesis. *Int. J. Mol. Sci.* **2017**, *18*, 619. [CrossRef]
138. Sun, D.; Zhang, X.; Zhang, Q.; Ji, X.; Jia, Y.; Wang, H.; Niu, L.; Zhang, Y. Comparative transcriptome profiling uncovers a Lilium regale NAC transcription factor, LrNAC35, contributing to defence response against cucumber mosaic virus and tobacco mosaic virus. *Mol. Plant Pathol.* **2019**, *20*, 1662–1681. [CrossRef]
139. Yu, X.; Gong, H.; Cao, L.; Hou, Y.; Qu, S. MicroRNA397b negatively regulates resistance of Malus hupehensis to Botryosphaeria dothidea by modulating MhLAC7 involved in lignin biosynthesis. *Plant Sci.* **2020**, *292*, 110390. [CrossRef]
140. Martín-Pizarro, C.; Vallarino, J.G.; Osorio, S.; Meco, V.; Urrutia, M.; Pillet, J.; Casañal, A.; Merchante, C.; Amaya, I.; Willmitzer, L.; et al. The NAC transcription factor FaRIF controls fruit ripening in strawberry. *Plant Cell* **2021**, *33*, 1574–1593. [CrossRef]
141. Ma, C.; Wang, X.; Yu, M.; Zheng, X.; Sun, Z.; Liu, X.; Tian, Y.; Wang, C. PpMYB36 Encodes a MYB-Type Transcription Factor That Is Involved in Russet Skin Coloration in Pear (*Pyrus pyrifolia*). *Front. Plant Sci.* **2021**, *12*, 776816. [CrossRef] [PubMed]
142. Wang, M.; Manzoor, M.A.; Wang, X.; Feng, X.; Zhao, Y.; He, J.; Cai, Y. Comparative Genomic Analysis of SAUR Gene Family, Cloning and Functional Characterization of Two Genes (PbrSAUR13 and PbrSAUR52) in *Pyrus bretschneideri*. *Int. J. Mol. Sci.* **2022**, *23*, 7054. [CrossRef] [PubMed]
143. Jia, N.; Liu, J.; Sun, Y.; Tan, P.; Cao, H.; Xie, Y.; Wen, B.; Gu, T.; Liu, J.; Li, M.; et al. Citrus sinensis MYB transcription factors CsMYB330 and CsMYB308 regulate fruit juice sac lignification through fine-tuning expression of the Cs4CL1 gene. *Plant Sci.* **2018**, *277*, 334–343. [CrossRef] [PubMed]
144. Zhao, D.; Luan, Y.; Xia, X.; Shi, W.; Tang, Y.; Tao, J. Lignin provides mechanical support to herbaceous peony (*Paeonia lactiflora* Pall.) stems. *Hortic. Res.* **2020**, *7*, 213. [CrossRef] [PubMed]
145. Yang, Y.; Huang, Y.; Ren, A.; Wan, Y.; Liu, Y. Xylem development and phloem conductivity in relation to the stem mechanical strength of *Paeonia lactiflora*. *J. Plant Physiol.* **2023**, *283*, 153963. [CrossRef] [PubMed]
146. Zhao, D.; Tang, Y.; Xia, X.; Sun, J.; Meng, J.; Shang, J.; Tao, J. Integration of Transcriptome, Proteome, and Metabolome Provides Insights into How Calcium Enhances the Mechanical Strength of Herbaceous Peony Inflorescence Stems. *Cells* **2019**, *8*, 102. [CrossRef] [PubMed]
147. Tang, Y.; Zhao, D.; Meng, J.; Tao, J. EGTA reduces the inflorescence stem mechanical strength of herbaceous peony by modifying secondary wall biosynthesis. *Hortic. Res.* **2019**, *6*, 36. [CrossRef]
148. Tang, Y.; Lu, L.; Sheng, Z.; Zhao, D.; Tao, J. An R2R3-MYB network modulates stem strength by regulating lignin biosynthesis and secondary cell wall thickening in herbaceous peony. *Plant J.* **2023**, *113*, 1237–1258. [CrossRef]
149. Zhang, H.; Pu, J.; Liu, H.; Wang, M.; Du, Y.; Tang, X.; Luo, X.; Wang, Y.; Deng, Q. Effects of L-Cysteine and γ-Aminobutyric Acid Treatment on Postharvest Quality and Antioxidant Activity of Loquat Fruit during Storage. *Int. J. Mol. Sci.* **2023**, *24*, 10541. [CrossRef]
150. Lee, C.-J.; Park, S.-U.; Kim, S.-E.; Lim, Y.-H.; Ji, C.Y.; Kim, Y.-H.; Kim, H.S.; Kwak, S.-S. Overexpression of IbLfp in sweetpotato enhances the low-temperature storage ability of tuberous roots. *Plant Physiol. Biochem.* **2021**, *167*, 577–585. [CrossRef]

agronomy

MDPI

Perspective

Recent Progress in Genetic Transformation and Gene Editing Technology in Cucurbit Crops

Jing Feng [1,†], Naonao Wang [2,†], Yang Li [1], Huihui Wang [1], Wenna Zhang [2], Huasen Wang [1] and Sen Chai [1,*]

1 Engineering Laboratory of Genetic Improvement of Horticultural Crops of Shandong Province, College of Horticulture, Qingdao Agricultural University, Qingdao 266109, China
2 Beijing Key Laboratory of Growth and Developmental Regulation for Protected Vegetable Crops, China Agricultural University, Beijing 100193, China
* Correspondence: chaisen@qau.edu.cn
† These authors contributed equally to this work.

Abstract: Cucurbits (Cucurbitaceae) include major horticultural crops with high nutritional and economic value that also serve as model plants for studying plant development and crop improvement. Conventional breeding methods have made important contributions to the production of cucurbit crops but have led to a breeding bottleneck because of the narrow genetic bases and low variation rates of these crops. With the development of molecular techniques, innovations in germplasm development through transgenesis and gene editing have led to breakthroughs in horticultural crop breeding. Although the development of genetic transformation and gene editing techniques for cucurbit crops has lagged behind that for other major crops, great progress has been made in recent years. Here, we summarize recent advances in improving the genetic transformation efficiency of cucurbit crops, including the screening of germplasm and the application of physical treatments, morphogenic genes, and selection markers. In addition, we review the application of gene editing technology to cucurbit crops, including CRISPR (clustered regularly interspaced short palindromic repeat)/Cas9 (CRISPR-associated nuclease 9)-mediated gene knockout and base editing. This work provides a reference for improving genetic transformation efficiency and gene editing technology for cucurbit crops.

Keywords: cucurbit crops; genetic transformation; gene editing; biotechnology; plant regeneration; breeding; genomics

Citation: Feng, J.; Wang, N.; Li, Y.; Wang, H.; Zhang, W.; Wang, H.; Chai, S. Recent Progress in Genetic Transformation and Gene Editing Technology in Cucurbit Crops. *Agronomy* **2023**, *13*, 755. https://doi.org/10.3390/agronomy13030755

Academic Editor: Yang Zhu

Received: 15 February 2023
Revised: 2 March 2023
Accepted: 2 March 2023
Published: 5 March 2023

1. Introduction

Cucurbits (Cucurbitaceae) are a major group of horticultural crops that are cultivated worldwide, including numerous species with extremely high economic and nutritional value. Cucurbit crops such as melon (*Cucumis melo*), pumpkin (*Cucurbita moschata* Duchesne), watermelon (*Citrullus lanatus*), and cucumber (*Cucumis sativus*) are very popular among consumers. Cucurbit plants are also used as model materials for studying plant development and quality improvement because of their variable shapes and abundance of flavor compounds [1]. In the past few decades, conventional genetic breeding technologies (cross breeding and mutation breeding) have played important roles in breeding selection of cucurbit varieties with high yields and quality. However, the low genetic diversity and variation rates in these crops limit the breeding of complex genetic characters and hamper innovation. With the development of modern biotechnology techniques, innovation via transgenesis and gene editing has become an important focus in the development of horticultural crops and has given rise to new breeding approaches such as "de novo domestication", "rapid breeding", "haploid breeding", and "breaking of genetic linkage" [2,3]. Notably, the recent development of transgenic and gene editing techniques for multiple cucurbit species has greatly facilitated the study of the molecular mechanisms underlying gene function in these crops and should provide new methods for cultivating cucurbit crops with economically desirable traits.

Genetic transformation is a supporting technology for genetic engineering and includes the cloning, delivery, integration, and expression of target genes and the regeneration, screening, and identification of transgene-positive plants. Vector delivery is the first step in transformation, via methods such as protoplast transfection, Agrobacterium (*Agrobacterium tumefaciens*)-mediated transformation, and particle bombardment. Agrobacterium-mediated transformation is currently the most widely used genetic transformation method because of its high efficiency. Agrobacterium-mediated transformation involves the stable integration of T-DNA from a Ti plasmid containing the target gene into plant chromosomal DNA via infection. The transformation process relies on the efficiency of the plant regeneration system. Much progress has been made in the in vitro culture of cucurbit crops via common methods such as organ culture, somatic embryogenesis, anther culture, and protoplast culture [4–7]. The most common transformation method for cucurbit crops uses cotyledons as explants for direct or indirect plant regeneration; indirect regeneration is mainly used for watermelon and direct regeneration for cucumber and melons. Trulson et al. established the first system for Agrobacterium-mediated transformation of cucurbit crops using hypocotyls as explants to achieve the stable expression of the *Neomycin Phosphotransferase II (NPTII)* gene in cucumber [8]. In 1990, Fang et al. successfully transferred the *NPTII* gene into melons [9]. This was followed by the successful transformation of cucurbit crops such as watermelon, followed by pumpkin, opening the door to the genetic transformation of numerous cucurbit crops.

In recent years, the maturity of transgenic technology enables wide application of gene editing technology in many plants. In gene editing, an engineered endonuclease identifies specific DNA sequences or guide RNAs via its DNA-binding domain (DBD) and cuts the target DNA sequence precisely and efficiently, thus editing specific DNA sequences [10]. Researchers have developed many specific gene editing systems, including a class of nucleases that identify their targets via protein-DNA interactions, e.g., mega nucleases (MNs), zinc finger nucleases (ZFNs), and transcription-activator-like effector nucleases (TALENs) [11,12]. The other class of nucleases includes CRISPR/Cas (clustered regularly interspaced short palindromic repeats/CRISPR-associated) and CRISPR/Cpf1 (a class 2/type V CRISPR RNA-guided endonuclease), which identify their targets through RNA–DNA base pairing [13]. CRISPR technology is regarded as the optimal method for acquiring genome edited crops given its low cost, high flexibility, and high reliability [14–16]. Gene editing technologies have been applied to cucurbit crops relatively recently, particularly CRISPR, which has been successfully used in cucumber, watermelon, melon, and pumpkin.

Although transformation systems have been established for some cucurbit crops, these systems suffer from low genetic transformation efficiency and poorly developed technology, and there are still gaps between these systems and those for other major crops. Researchers have tried to improve the genetic transformation efficiency of cucurbit crops in various ways and have optimized the genetic transformation steps to improve the transformation efficiency of multiple plant varieties. In addition, researchers have achieved gene knockout and base editing in cucurbit crops using CRISPR technology, and the genes responsible for many important agronomic traits have been explored. Here, we summarize the methods used to improve the genetic transformation efficiency of cucurbit crops and the application of gene editing technology to them.

2. Methods to Improve the Genetic Transformation Efficiency of Cucurbit Crops

Agrobacterium-mediated transformation using cotyledons as explants is currently the major technique adopted for many cucurbit crops. The efficiency of genetic transformation is influenced by many factors, including the regeneration efficiency of the plant, infection method, Agrobacterium strain, screening method, and use of exogenous phytohormones (Figure 1). Researchers have recently optimized the transformation process based on previous work, including genotype screening, Agrobacterium infection methods, application of morphogenic genes, and diversification of screening markers, thus improving the transformation efficiency (Figure 1) [17–31].

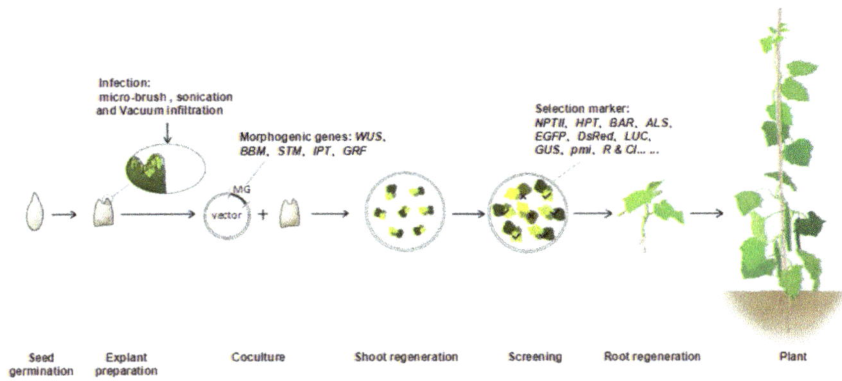

Figure 1. Efficient methods for the genetic transformation of cucurbit crops.

Seed germination: The sterilized seeds were sown on germination medium and cultured for 24–72 h until the cotyledons swelled and the hypocotyls extended to about 5–7 mm (different plant standards). Explant preparation: The cotyledons of the seeds were divided into two parts along the gap, and 2/3 of the adaxial end was taken as the explants. Physical treatments such as sonication, micro-brush, and vacuum infiltration were applied to assist the infection according to the experimental requirements. Coculture: The transformation vector with morphogenic genes was transformed into Agrobacteria and cocultured with explants for 3–4 days. Shoot regeneration: Explants after coculture were placed on the corresponding induction medium to induce shoots. Screening: Transformed positive shoots were identified using different screening markers. Root regeneration: Rooting was induced on positive shoots. Plant: Positive plants were obtained.

2.1. Germplasm Influences Transformation Efficiency

Differences in the genotypes of the target germplasm lead to significant differences in regeneration efficiency, representing an important factor influencing genetic transformation efficiency (Table 1). Identifying suitable germplasm is a direct method for improving the efficiency of genetic transformation. Germplasm screening has often been employed in watermelon, cucumber, and melon, less so in pumpkins, and rarely in other cucurbit crops. Different watermelon germplasms show significant differences in transformation efficiency. The transformation efficiency reached 8–17% in watermelon varieties such as Arka manik, Sugar baby, Arka muthu, IIHR-14, Feeling, and YL but only approximately 1–2% in China baby and Quality [17–20]. In general, the genetic transformation efficiency of cucumber ranges from 1% to 23%. The genetic transformation efficiency significantly differs among varieties, e.g., up to 23% in Jinyan no.7 but only approximately 1% in Eunchim and 404 [21,22]. The varieties Cu2, Poinsett 76, and Xintaimici are widely used by researchers and show stable genetic transformation efficiencies of 5–20% under different conditions [23–26]. Genetic transformation has been reported in more than 20 varieties of melon, such as Kirkagaç 637, CM-15, Charentais, and Silver Light, but their transformation efficiency is less than 7% in most cases. Fewer melon varieties with high transformation efficiency are currently available [27,32–35]. Finally, although it is difficult to transform pumpkins, successful transformation was recently achieved. Using JingXinZhen No. 4, Xin et al. achieved a transformation efficiency of up to 3.56% [22].

Table 1. Details of some successful reports on Agrobacterium-mediated transformation of cucurbit crops.

Species	Germplasm	Stability	Agrobacterium Strains	Transformation Efficiency (%)	T1	Reference
Watermelon	Feeling, China baby, Quality	Kanamycin	LBA4404	8.5–10%, 1–2.1%, 0–1.3% (PCR)	Reported	[18]
Watermelon	Feeling, China baby, Quality	Kanamycin	LBA4404	7%, 2.22%, 1.25% (PCR and Southern blot)	NR	[17]
Watermelon	the female parent line of Zhengkang No. 6	Kanamycin/cefotaxime	LBA4404	3 plantlets (PCR and Southern blot)	Reported	[36]
Watermelon	PI179878	Hygromycin B	EHA105	1.67% (shoots)	NR	[37]
Watermelon	ZG94	Bialaphos resistance	EHA105	23% (PCR)	Reported	[38]
Watermelon	Arka manik, Sugar baby, Arka muthu, IIHR-14	BASTA	EHA105	17.33%, 12.33%, 16%, 14.66% (GUS, PCR, Southern hybridization)	NR	[19]
Watermelon	TC	NR	EHA105	6.5% (GFP, PCR)	NR	[29]
Watermelon	YL	Glufosinate-ammonium	EHA105	12.5% (PCR)	NR	[20]
Cucumber	Eunchim	Phospinotricin	EHA101	1.7% (Northern blot)	NR	[21]
Cucumber	Poinsett 76	Phosphinothricin	EHA105 LBA4404	21.0%, 8.5% (shoots)	Reported	[24]
Cucumber	Eunsung	Paromomycin	EHA101	4.01% (Southern blot)	Reported	[39]
Cucumber	Xintaimici	Kanamycin	LBA4404	4.8% (PCR, Southern blot, Western blot)	NR	[40]
Cucumber	Shinhokusei 1	Kanamycin	EHA105	11.9 ± 3.5% (plants)	Reported	[41]
Cucumber	Ilan	Kanamycin	EHA105	3 plants (PCR)	Reported	[42]
Cucumber	Cu2	GFP	EHA105	1.32% (plantlets)	Reported	[43]
Cucumber	Xintaimici	Kanamycin	GV3101	8.10% (PCR)	NR	[26]
Cucumber	CCMC	GUS/Hygromycin B	EHA105	0.24% (PCR)	NR	[30]
Cucumber	Cu2, Xintaimici, 404, Eu1	GFP	EHA105	5.18%, 2.20%, 1.97%, 2.46% (GFP)	Reported	[22]
Melon	Silver Light	Kanamycin	LBA4404	Unspecified (2 transgenic lines)	Reported	[44]
Melon	Hetao		Transformation via the pollen-tube pathway	4.3% (PCR)	Reported	[45]
Melon	Védrantis	Kanamycin	GV2260	0.35–3.0% (PCR and GUS histochemical analysis)	NR	[46]
Melon	Hybrid line M01–3		Ovary-injection transformation	2.7% (PCR)	Reported	[47]

Table 1. *Cont.*

Species	Germplasm	Stability	Agrobacterium Strains	Transformation Efficiency (%)	T1	Reference
Melon	Cantaloupe F39 Honeydew 150	Kanamycin/GUS	EHA105	0.3%, 0.5% (PCR, GUS and Southern blot)	NR	[48]
Melon	Geumnodajieunchun	Kanamycin/geneticin	LBA4404	2.9%, 7.1% (PCR)	NR	[49]
Melon	Silver Light	Kanamycin	LBA4404	11 transgenic lines (PCR and Southern blot)	Reported	[50]
Melon	Silver Light	Kanamycin	LBA4404	0.8% (PCR)	Reported	[33]
Melon	CM-23	Kanamycin	EHA105	4% (PCR, Southern hybridization)	NR	[51]
Melon	M-15	Kanamycin	EHA105	13% (PCR)	Reported	[32]
Melon	CM-15	Kanamycin	EHA105	4 plants (PCR)	Reported	[34]
Melon	Charentais, Galia, Piel de sapo, Blanco	Kanamycin	EHA105	1.3%, 1.6%, 3.8%, 2.5% (plants)	Reported	[27]
Melon	Charentais	Hygromycin B	EHA105	5.68% (shoots)	NR	[35]
Melon	m1	GFP	EHA105	3.95% (GFP)	Reported	[22]
Pumpkin	JingXinZhen No. 4	GFP	EHA105	3.56% (GFP)	Reported	[22]

2.2. *Auxiliary Physical Treatments Significantly Enhance Agrobacterium Infection*

Various strains of Agrobacterium, an efficient natural carrier of transgenes, are widely used for the genetic transformation of plants because of their easy transformation and low cost. The most commonly used Agrobacterium strains for the genetic transformation of cucurbit crops include EHA105, GV3101, and LBA4404; different strains show somewhat different transformation efficiencies in various germplasms [26–28].

In addition to the effect of strain type, the key to successful transformation lies in the transmission process during infection. For cucurbit crops, the most commonly used explants are leaves, whose regenerable cells are located at the U-shaped cut end that is produced when generating explants. However, it is difficult for Agrobacteria to reach this region under routine infection conditions [43]. To achieve effective Agrobacterium infection, various methods are used, such as sonication, micro-brush (KITA, Nanotek Brush) treatment, and vacuum infiltration to promote the penetration of Agrobacteria into cells deep in the explant. Sonication generates many minor wounds on explants to facilitate the entry of Agrobacteria and improve the genetic transformation efficiency. Sonication improved the transformation efficiency of watermelon from the original rate of 8% to 17% [19]. Similarly, slightly scratching cucumber explants to allow better access by Agrobacteria significantly increased the entry of the transgene, as revealed by β-glucuronidase (*GUS*) staining [30]. Vacuum infiltration is another effective way to increase the probability of Agrobacterium entry. This method was first applied in the genetic transformation of soybean (*Glycine max*), Arabidopsis (*Arabidopsis thaliana*), and wheat (*Triticum aestivum*) and has also been successfully utilized in cucurbit crops [52]. A variety of cucumber germplasms were successfully transformed using vacuum infiltration. The "optimal infiltration intensity" strategy, which integrates multiple auxiliary methods, was recently proposed. The researchers used sonication and micro-brush techniques to wound cotyledonary node explants and created vacuum conditions with a sterile syringe, significantly improving the transformation efficiency of cucumber, melon, and pumpkin [22].

2.3. *Morphogenic Genes Significantly Increase Transformation Efficiency*

The genetic transformation efficiency of cucurbit crops can be improved by introducing some auxiliary physical methods. However, exploring and optimizing genetic transformation conditions requires a lot of time and effort, and the same genetic transformation system cannot be universally applied to different genotypes. Many genes capable of promoting or reprogramming cell fate can improve the genetic transformation efficiency of explants. These genes are known as morphogenic genes, including the wounding-associated *Wound-Induced Dedifferentiation 1–4* (*WIND1–4*); epigenetic modification genes, including *Polycomb Repressive Complex1/2* (*PRC1/2*) and *Arabidopsis Trithorax4* (*ATX4*); growth regulatory genes, including *Indole3-Acetic Acids* (*IAAs*), *YUCCAS* (*YUCs*), and *Arabidopsis Response Regulator7/15* (*ARR7/15*); and developmental regulatory genes, including *Wuschel* (*WUS*), *Shoot Meristemless* (*STM*), and *Wuschel Related Homeobox5/11* (*WOX5/11*) [53–55]. Among morphogenic genes, the *WUS-STM* and *Growth-Regulating Factor* and *GRF-Interacting Factor* (*GRF-GIF*) complexes are commonly used to improve genetic transformation efficiency. For example, co-expressing *WUS* and *STM* greatly improved the in vitro transformation efficiency of monocotyledonous plants such as rice (*Oryza sativa*), sorghum (*Sorghum bicolor*), and some maize (*Zea mays*) inbred lines [56]. Co-expressing *WUS* and *Isopentenyl Transferase* (*IPT*) induced meristem growth and budding in *Nicotiana benthamiana* seedlings [57]. In addition, exogenously expressing *AtGRF5* promoted budding and improved the genetic transformation efficiency of soybean, oilseed rape (*Brassica napus*), and sunflower (*Helianthus annuus*) [58]. Expressing fusion protein of *TaGRF4* and *OsGIF1* improved the regeneration speed and efficiency of wheat and rice and expanded the scope of varieties that could be used for genetic transformation [59].

Expressing morphogenic genes can improve the genetic transformation efficiency of watermelon. Among these genes, *AtGRF5*, *TaGRF4-OsGIF1*, *ZmWUS*, *ZmWUS–ZmBBM*, *ZmWUS–IPT*, and *AtGRF5* showed the best effects, increasing the genetic transformation efficiency of watermelon cultivar WW150 approximately 40-fold to 24.73% [28]. In another study, overexpressing the endogenous genes *ClGRF4–ClGIF1* achieved a transformation efficiency of up to 47.02% in watermelon cultivar TC, representing an approximately 9-fold increase compared with the control (with a transformation efficiency of 5.23%). Mutating the *miR396* target site in *ClGRF4* further increased the transformation efficiency to 67.27% and had significant effects in eight other watermelon varieties, such as YL, M08, 148, and 97103 [29]. The role of morphogenic genes in increasing transformation efficiency has been verified in watermelon, for which indirect regeneration is mainly used, but whether these genes can be used for other cucurbit crops such as cucumber and melon (for which direct regeneration is mainly used) remains to be studied.

2.4. Diversification of the Screening Markers Used in Genetic Transformation

Introducing screening marker genes during transformation can facilitate the identification of positive transgenic events and can help determine whether the transformation was successful. Three major categories of screening markers are currently used in the genetic transformation of cucurbit crops. The first category is resistance genes, e.g., *NPTII*, *Hygromycin phosphotransferase (HPT)*, *Phosphinothricin N-acetyltransferase (BAR)*, and *Acetolactate Synthase (ALS)*. The buds of explants harboring a resistance gene can grow normally under selection, whereas buds lacking the resistance gene cannot grow. The second category is reporter genes, e.g., *Enhanced Green Fluorescent Protein* (*eGFP*), *Discosoma striata red Fluorescent Protein* (*DsRed*), *luciferase* (*LUC*), and *β-glucuronidase* (*GUS*). Under specific physical/chemical conditions, buds harboring a reporter gene exhibit a different appearance from buds that have not been transformed. The third category is metabolism-related genes—e.g., *Phosphomannose isomerase* (*pmi*) and *maize R, C1*, and *B transcription factor genes*—which allow transformed regenerated buds to grow under specific nutritional conditions [31,60].

Among the resistance genes, *NPTII* has been used as a selection marker, but its effectiveness differs significantly between germplasms. Cotyledon explants of cucumber variety CMCC are extremely sensitive to hygromycin B, with good screening effects even at low concentrations [30]. However, escape is likely to occur when resistance genes are used; that is, buds free of any resistance gene can sometimes grow under selection [27]. In contrast, the use of reporter genes results in no escapes. Under specific physical/chemical conditions, reporter genes such as *eGFP*, *DsRed*, *LUC*, and *GUS* enable transgene-positive buds to exhibit a color different from that of non-transformed buds, thus achieving effective screening and avoiding the disadvantage of escapes. However, reporter genes are not perfect. They may be interfered with by the plants themselves, making it difficult to identify transgenic material under specific conditions. Moreover, high levels of reporter gene expression can damage the plant, leading to abnormal phenotypes. In addition to resistance genes and reporter genes, researchers have also developed metabolism-related genes, which allow transgene-positive buds to grow under specific nutritional conditions. For example, expression of the *pmi* gene enables the conversion of mannose-6-phosphate into fructose-6-phosphate, allowing regenerated buds to grow normally using mannose as a carbon source. Using the mannose selection system in cucumber, effective transgene-positive bud screening was achieved using a medium containing 10 g/L mannose and 10 g/L saccharose, leading to reduced damage and escape compared with the use of reporter genes and resistance genes [61].

2.5. Other Factors

Genetic transformation involves multiple factors and multiple steps. Therefore, various issues are likely to occur during transformation, which differ somewhat in different species. For example, cucumber is rich in endogenous ethylene, but the accumulation of ethylene is unfavorable for cell growth and differentiation and in severe cases leads to the aging of cells. $AgNO_3$ can be added to the medium during transformation to inhibit ethylene production and promote cell differentiation and regeneration. However, Ag^+ is a heavy metal that is somewhat toxic to plants, and the long-term use of $AgNO_3$ can lead to plant deformity [43].

Browning, vitrification, and Agrobacterium contamination are likely to occur during melon transformation. Browning occurs when cells experience metabolic changes upon the activation of polyphenol oxidase in explants. This process can occur because of the germplasm selected, physiological conditions, culture conditions, and damage [62]. Antioxidants such as vitamin C, sodium thiosulfate, and cysteine can be added during melon transformation to inhibit browning. Vitrification is a condition in which the stems and leaves of explants are water-soaked and fragile. Vitrified seedlings tend to grow slowly and show weak differentiation ability. Vitrification may be related to phytohormone levels, culture conditions, and relative humidity. Vitrification of seedlings during transformation can be reduced by increasing the concentrations of agar and saccharose in the medium as appropriate, strengthening ventilation and air exchange, and controlling relative humidity levels. Finally, since Agrobacterium mediation is generally required for transformation, Agrobacterium contamination/overgrowth may occur during tissue culture. For Agrobacterium contamination, cefotaxime, plant preservative mixture (PPM), and Timentin can be added to the medium to inhibit the reproduction of the Agrobacteria and improve the survival rates of plants [43].

3. Application of Gene Editing Technology in Cucurbit Crops

Gene editing technology is an important strategy for gene function studies and molecular breeding. A gene editing system, which is introduced into a plant through genetic transformation, uses engineered endonucleases to edit the specific nucleotide sequences of the plant genome precisely and efficiently. CRISPR technology has become a mainstream gene editing technology used for cucurbit crops thanks to its efficiency, precision, and easy operation. After gene editing, the carrier fragments can be eliminated from the plants through hybridization [63]. CRISPR/Cas9-based knockout and base editing of genes have been successfully achieved in various cucurbit crops (Figure 2 and Table 2).

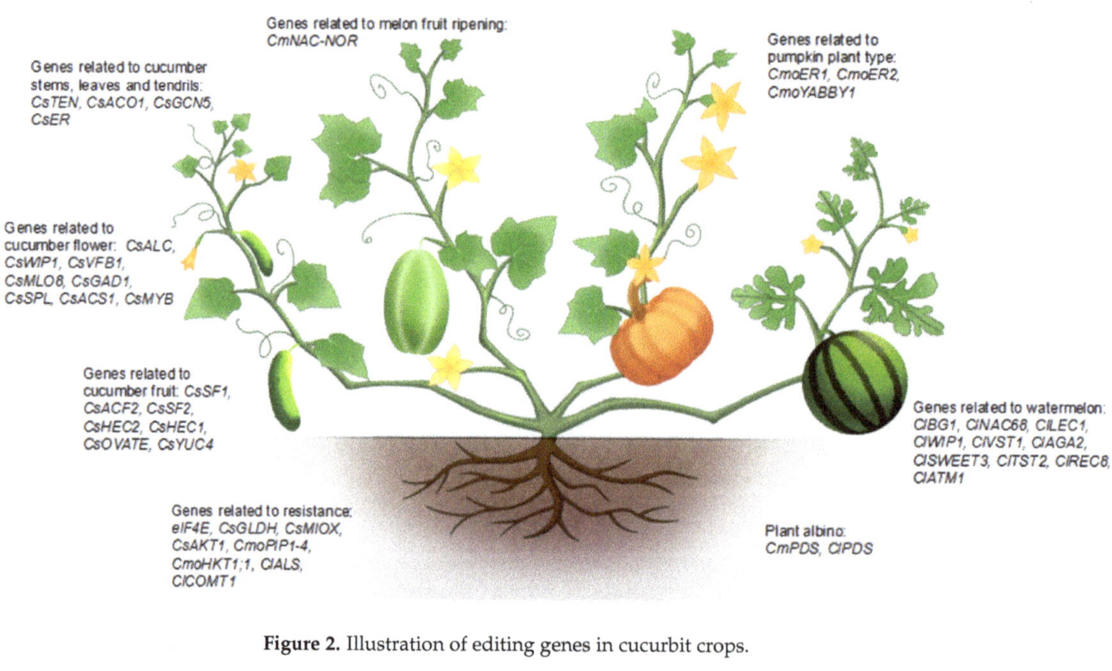

Figure 2. Illustration of editing genes in cucurbit crops.

Table 2. Details of editing genes of cucurbit crops.

Species	Gene	ID	Gene Function or Phenotype	References
Watermelon	*ClALS*	Cla019277	Herbicide resistance	[38]
Watermelon	*ClPDS*	Cla97C07G142100	Albino plants	[28]
Watermelon	*ClPDS*	Cla010898	Albino plants	[37]
Watermelon	*ClBG1*	Cla97C08G153160	Regulation of seed size and seed germination	[64]
Watermelon	*ClNAC68*	Cla97C03G059250	Regulation of fruit sugar content and seed development	[65]
Watermelon	*ClLEC1*	Cla97C10G188900	Reproductive development	[29]
Watermelon	*ClWIP1*	Cla008537	Sex determination	[66]
Watermelon	*ClVST1*	Cla97C02G031010	Vacuolar sugar transporter	[67]
Watermelon	*ClAGA2, ClSWEET3, ClTST2*	Cla97C04G070460, Cla97C01G000640, Cla97C02G036390	Carbohydrate partitioning	[68]
Watermelon	*ClREC8*	Cla97C07G132920	Seedless watermelon	[69]
Watermelon	*ClCOMT1*	Cla97C10G188660	Tolerance to abiotic stresses	[70]
Watermelon	*ClATM1*	Cla010576	Male sterility	[71]
Cucumber	*CsALC*	Csa2G356640.1	Pollen tube emergence	[72]
Cucumber	*eIF4E*	XM_004147349	Antivirus	[42]
Cucumber	*CsWIP1, CsVFB1, CsMLO8, CsGAD1*	Csa4M290830, Csa4M641640, Csa5M623470, Csa5M348050	Gynoecious inbred lines	[43]
Cucumber	*CsSPL*	Csa3M850670	Male and female fertility and ovule development	[73]
Cucumber	*CsSF1, CsACF2*	Csa2G174140, Csa1G580750	Cucurbit-specific RING-type E3 ligase, Rate-limiting enzyme for ethylene biosynthesis	[74]
Cucumber	*CsSF2*	Csa2G337260	Fruit elongation	[75]
Cucumber	*CsTEN, CsACO1*	Csa5P644520.1, Csa6G160180	Identity and mobility of tendrils	[76]
Cucumber	*CsHEC2*	Csa2G285890	Fruit wart	[77]
Cucumber	*CsACS1, CsMYB*	CsaV3_6G044400, CsaV3_6G044410	Female floral development	[78]
Cucumber	*CsHEC1, CsOVATE, CsYUC4*	Csa4G639900, Csa4G038760, Csa2G379350	Fruit neck length	[79]
Cucumber	*CsGCN5*	Csa6G527060	Gene transcription and plant development	[30]
Cucumber	*CsER*	CsaV3_4G036080	Compact plant architecture	[22]
Cucumber	*CsGLDH, CsMIOX*	Csa4M236360.1, Csa2M000640	Ascorbic acid biosynthesis	[80]
Cucumber	*CsAKT1*	CsaV3_1G029650	Salt tolerance	[81]
Melon	*CmPDS*	MELO3C017772.2	Albino plants	[35]
Melon	*CmNAC-NOR*	MELO3C016540.2	Fruit ripening	[82]
Melon	*CmNAC-NOR*	MELO3C016540	Fruit coloring and ripening	[83]
Melon	*eIF4E*	MELO3C002698.2	Antivirus	[84]
Pumpkin	*CmoER1, CmoER2*	CmoCh09G003660, CmoCh01G017570	Compact plant architecture	[22]
Pumpkin	*CmoPIP1-4*	CmoCh04G011950	Salt stress	[85]
Pumpkin	*CmoHKT1;1*	CmoCh10G003830	High-affinity K+ transporter1	[86]
Pumpkin	*CmoYABBY1*	CmoCh15G012090	Architecture regulation	[1]

3.1. CRISPR/Cas9-Based Gene Knockout

CRISPR/Cas9 is a microbial adaptive immune mechanism that achieves precise gene editing based on the RNA-guided Cas9 nuclease within the Type II prokaryotic CRISPR/Cas system [87]. The present CRISPR-Cas9 systems are modifications of bacterial CRISPR/Cas9, comprising modified Cas9 endonuclease and a single guide (sg)RNA [88]. Under the guidance of a specific sgRNA, Cas9 specifically identifies the protospacer adjacent motif (PAM, NGG) in its target sequence and produces blunt-ended double-strand breaks approximately 3 bp upstream of the PAM to trigger the HDR or NHEJ pathway, thus achieving precise editing of plant genes [89].

In 2016, CRISPR/Cas9 was successfully applied to cucurbit crops for the first time. The researchers used CRISPR/Cas9 to mutate the DNA sequence at the target site of *eukaryotic translation initiation factor 4E* (*eIF4E*) at its N′ and C′ termini in cucumber. Two homozygous *eIF4E* knockout mutants exhibited immunity to *Cucumber vein yellowing virus* (*Ipomovirus*) infection and resistance to the *potyviruses Zucchini yellow mosaic virus* and *Papaya ring spot mosaic virus-W* [42]. Subsequently, Hu et al. used the cucumber U6 promoter to drive sgRNA to knock out *WIP Domain Protein 1* (*CsWIP1*) using CRISPR/Cas9 technology and successfully created a gynoecious cucumber line [43]. Thereafter, Liu et al. explored the role of *Sporocyteless* (*CsSPL*) in ovule development and reproduction [73]. Since then, CRISPR/Cas9 has been widely used in cucumber. Researchers have successfully used CRISPR/Cas9 in breakthrough studies on the functions of genes for sex, fruit length, tendril formation, and stress resistance in cucumber [22,30,75–81].

In addition to the successful application of CRISPR/Cas9 in cucumber, gene knockout using CRISPR/Cas9 has also been reported in watermelon. Tian et al. knocked out the *phytoene desaturase* (*ClPDS*) gene in watermelon and successfully obtained albino regenerated buds, demonstrating the feasibility of applying CRISPR/Cas9 technology to watermelon [37]. In 2019, researchers used CRISPR/Cas9 to knock out the *ClWIP1* gene and obtained a gynoecious line to explore the molecular mechanism that determines the sex of watermelon [66]. Researchers also used CRISPR/Cas9 to successfully knock out genes (e.g., a key component of the cohesin complex in meiosis *(ClREC8), caffeic acid O-methyltransferase* (*ClCOMT1*), *β-glucosidase* (*ClBG1*), *NAC transcription factor68* (*ClNAC68*), *vacuolar sugar transporter (ClVST1), alkaline alpha-galactosidase (ClAGA2), Citrullus lanatusSugars Will Eventually Be Exported Transporter 3 (ClSWEET3), Tonoplast Sugar Transporter (ClTST2), Citrullus lanatus Abnormal Tapetum 1 (ClATM1)*, and *nuclear transcription factor Y subunit B-9-like* (*ClLEC1*)) and achieved good results, increasing seed fertility, stress resistance, seed size, and sugar transport in watermelon [29,64–66,68–71]

Compared with that in cucumber and watermelon, the application of gene editing technology in other cucurbit crops is lagging. For melons, researchers successfully used CRISPR/Cas9 to knock out the *CmPDS* gene in 2019 and obtained albino regenerated buds [35]. Researchers subsequently used CRISPR/Cas9 to knock out *eIF4E, ERECTA* (*CmER*), and the NAC transcription factor gene *Nonripening* (*CmNAC-NOR*) and revealed their functions in virus resistance, plant architecture, and fruit ripening, respectively [22,82–84]. In pumpkin, researchers used *Agrobacterium rhizogenes* as an infection tool to develop an efficient CRISPR/Cas9-based root transformation system to analyze stem–root communication in cucurbit crops and successfully knocked out *high-affinity K + transporter 1* (*CmoHKT1;1*) and *Plasma membrane intrinsic proteins* (*CmoPIP1–4*) [85,86]. Besides the root transformation system, Xin et al. obtained a knockout mutant of *CmoER* and achieved stable gene editing in pumpkin [22]. This team recently used CRISPR/Cas9 technology to achieve domain B knockout of *CmoYABBY1*, thus strengthening the translation of this gene and inhibiting stem growth proportionally in a dose-dependent manner, creating a pumpkin plant with a dominant bushy trait (Figure 2) [1].

3.2. Precise Editing via Cytosine Base Editors

In addition to the gene knockout achieved in cucurbit crops, precise single-nucleotide editing of genes in watermelon has been achieved using cytosine base editors (CBEs). CBEs consist of a Cas9 nickase (nCas9) with a D10A mutation, which deactivates RuvC (one of the two Cas9 nuclease domains), fused with a cytidine deaminase and a uracil DNA glycosylase (UDG) inhibitor (UGI). CBEs induce a C:G > T:A base change. Tian et al. used a novel Cas9 variant fused with CBE3 to achieve a single-nucleotide conversion from C to T in the Pro190 (CCG) codon of the *ALS* gene of watermelon without merging the DNA templates of the donor, thereby creating a non-GM herbicide-resistant watermelon variety [38]. It is worth noting that the first successful application of CRISPR/Cas9 technology in watermelon was also achieved by the same team [37].

4. Discussion and Future Prospects

In this review, we systematically summarized recent innovations and applications of genetic transformation and gene editing technology in cucurbit crops. Significant improvements have been made, but there are still many outstanding issues in the genetic transformation of cucurbit crops.

4.1. Establish Genotype-Free Transformation Systems

Genotypes play a pivotal role in transformation. Extensive studies have been performed on cucurbit crops, and the germplasms examined have varied widely among research teams. In contrast to other species, no widely used germplasm is currently available for the stable transformation of cucurbit crops. A uniform genotype would help researchers explore the functions of genes and analyze their genetic relationships. Therefore, testing the genetic transformation efficiency of the existing germplasms and identifying the most suitable receptor germplasms could help improve the genetic transformation efficiency of cucurbit crops. Certainly, for future molecular breeding, striving to establish a transformation system that is not limited by genotype is our ultimate goal, e.g., establishment of a genotype-free genetic transformation system using morphogenic genes. The use of morphogenic genes can significantly improve genetic transformation efficiency, as verified in multiple plant varieties [28,29]. However, the strong expression of morphogenic genes can affect the growth and development of plants, making it crucial to eliminate the abnormal phenotypes of regenerated seedlings when morphogenic genes are used. Various studies have used inducible *35s-MdBBM-GR* vectors or the heat-induced Cre/*loxP* gene editing system during specific periods to reduce the damage caused by morphogenic genes [90,91]. In the future, innovative ways to solve this problem should continue to be explored.

4.2. Further Development of Gene Editing Technologies in Cucurbit Crops

The development prospects of gene editing technology are very broad. For example, the CRISPR-Cas9 system in tissue-specific or induced promoter control can help with spatial or temporal genomic alterations [92,93]. CRISPR-Cas9 also enables rapid functional identification of genes by generating gRNA libraries. As a result, CRISPR-Cas9 has become an efficient, simple, and fast tool for growing agriculturally improved crops that can both add better traits and remove undesirable traits [94]. Certainly, one of the best applications of the CRISPR-Cas9 system is the use of label-free "cisgene" plants with improved agronomic traits. These plants will eventually be immune to current transgene regulation because CRISPR-associated RNA-guided nucleases (RGENs) or ribonucleoproteins (RNPs) induce genomic mutations without any foreign DNA [95]. However, the development of gene editing technology in cucurbit crops is lagging, with successful results (including gene knockout) achieved only through CRISPR/Cas9 and base editing [37,38]. Efficient editing systems, such as "precise editing" via prime editing and gene editing systems optimized for the characteristics of cucurbit crops, remain to be further explored.

In conclusion, some progress has been made in the genetic transformation and gene editing of cucurbit crops, but many issues remain to be resolved.

Author Contributions: Conceptualization, J.F., N.W. and S.C.; formal analysis, J.F. and N.W.; investigation, J.F. and N.W.; resources, J.F., N.W. and H.W. (Huihui Wang); data curation, Y.L.; writing—original draft preparation, J.F., N.W. and S.C.; writing—review and editing, W.Z., N.W. and S.C.; visualization, J.F. and N.W.; supervision, H.W. (Huasen Wang) and S.C.; project administration, S.C.; funding acquisition, S.C. All authors have read and agreed to the published version of the manuscript.

Funding: This work was supported by the National Natural Science Foundation of China (32102404 to S.C.), This work was also supported by the Natural Science Foundation of Shandong Province (ZR2020QC157 to S.C.).

Data Availability Statement: Not applicable.

Conflicts of Interest: The authors declare no conflict of interest.

References

1. Wang, S.; Wang, K.; Li, Z.; Li, Y.; He, J.; Li, H.; Wang, B.; Xin, T.; Tian, H.; Tian, J.; et al. Architecture design of cucurbit crops for enhanced productivity by a natural allele. *Nat. Plants* **2022**, *8*, 1394–1407. [CrossRef] [PubMed]
2. Nanasato, Y.; Tabei, Y. A method of transformation and current progress in transgenic research on cucumbers and Cucurbita species. *Plant Biotechnol. (Tokyo Jpn.)* **2020**, *37*, 141–146. [CrossRef]
3. Li, H.; Yang, X.; Shang, Y.; Zhang, Z.; Huang, S. Vegetable biology and breeding in the genomics era. *Sci. China Life Sci.* **2023**, *66*, 226–250. [CrossRef]
4. Debeaujon, I.; Branchard, M. Somatic embryogenesis in Cucurbitaceae. *Plant Cell Tiss. Org. Cult.* **1993**, *34*, 91–100. [CrossRef]
5. Hooghvorst, I.; Nogués, S. Opportunities and Challenges in Doubled Haploids and Haploid Inducer-Mediated Genome-Editing Systems in Cucurbits. *Agronomy* **2020**, *10*, 1441. [CrossRef]
6. Jarl, C.I.; Bokelmann, G.S.; De Haas, J.M. Protoplast regeneration and fusion in Cucumis: Melon × cucumber. *Plant Cell Tiss. Org. Cult.* **1995**, *43*, 259–265. [CrossRef]
7. Kurtar, E.S.; Seymen, M. Another Culture in Cucurbita Species. *Methods Mol. Biol.* **2021**, *2289*, 111–121. [CrossRef]
8. Trulson, A.J.; Simpson, R.B.; Shahin, E.A. Transformation of cucumber (*Cucumis sativus* L.) plants with Agrobacterium rhizogenes. *Theor. Appl. Genet.* **1986**, *73*, 11–15. [CrossRef]
9. Fang, G.; Grumet, R. Agrobacterium tumefaciens mediated transformation and regeneration of muskmelon plants. *Plant Cell Rep.* **1990**, *9*, 160–164. [CrossRef]
10. Wyman, C.; Kanaar, R. DNA double-strand break repair: All's well that ends well. *Annu. Rev. Genet.* **2006**, *40*, 363–383. [CrossRef]
11. Puchta, H.; Fauser, F. Synthetic nucleases for genome engineering in plants: Prospects for a bright future. *Plant J.* **2014**, *78*, 727–741. [CrossRef]
12. Voytas, D.F.; Gao, C. Precision Genome Engineering and Agriculture: Opportunities and Regulatory Challenges. *PLoS Biol.* **2014**, *12*, e1001877. [CrossRef] [PubMed]
13. Zetsche, B.; Gootenberg, J.S.; Abudayyeh, O.O.; Slaymaker, I.M.; Makarova, K.S.; Essletzbichler, P.; Volz, S.E.; Joung, J.; van der Oost, J.; Regev, A.; et al. Cpf1 Is a Single RNA-Guided Endonuclease of a Class 2 CRISPR-Cas System. *Cell* **2015**, *163*, 759–771. [CrossRef] [PubMed]
14. Liu, Y.; Zhang, C.; Wang, X.; Li, X.; You, C. CRISPR/Cas9 technology and its application in horticultural crops. *Hortic. Plant J.* **2022**, *8*, 395–407. [CrossRef]
15. Mao, Y.; Botella, J.R.; Liu, Y.; Zhu, J.-K. Gene editing in plants: Progress and challenges. *Natl. Sci. Rev.* **2019**, *6*, 421–437. [CrossRef]
16. Chen, K.; Wang, Y.; Zhang, R.; Zhang, H.; Gao, C. CRISPR/Cas Genome Editing and Precision Plant Breeding in Agriculture. *Annu. Rev. Plant Biol.* **2019**, *70*, 667–697. [CrossRef] [PubMed]
17. Huang, Y.-C.; Chiang, C.-H.; Li, C.-M.; Yu, T.-A. Transgenic watermelon lines expressing the nucleocapsid gene of Watermelon silver mottle virus and the role of thiamine in reducing hyperhydricity in regenerated shoots. *Plant Cell Tissue Organ Cult. (PCTOC)* **2010**, *106*, 21–29. [CrossRef]
18. Yu, T.A.; Chiang, C.H.; Wu, H.W.; Li, C.M.; Yang, C.F.; Chen, J.H.; Chen, Y.W.; Yeh, S.D. Generation of transgenic watermelon resistant to Zucchini yellow mosaic virus and Papaya ringspot virus type W. *Plant Cell Rep.* **2011**, *30*, 359–371. [CrossRef]
19. Vasudevan, V.; Sathish, D.; Ajithan, C.; Sathish, S.; Manickavasagam, M. Efficient Agrobacterium-mediated in planta genetic transformation of watermelon [Citrullus lanatus Thunb.]. *Plant Biotechnol. Rep.* **2021**, *15*, 447–457. [CrossRef]
20. Cao, L.; Wei, W.; Shen, J.; Xu, Z.; Li, Z. Study on the optimization of transformation systems in watermelon. *Veg. Res.* **2022**, *2*, 1–9. [CrossRef]
21. Kim, H.-A.; Min, S.-R.; Choi, D.-W.; Choi, P.-S.; Hong, S.-G. Development of transgenic cucumber expressing TPSP gene and morphological alterations. *J. Plant Biotechnol.* **2010**, *37*, 72–76. [CrossRef]
22. Xin, T.; Tian, H.; Ma, Y.; Wang, S.; Yang, L.; Li, X.; Zhang, M.; Chen, C.; Wang, H.; Li, H.; et al. Targeted creating new mutants with compact plant architecture using CRISPR/Cas9 genome editing by an optimized genetic transformation procedure in cucurbit plants. *Hortic. Res.* **2022**, *9*, uhab086. [CrossRef]
23. Vasudevan, A.; Selvaraj, N.; Ganapathi, A.N.; Choi, C.W. Agrobacterium-mediated Genetic Transformation in Cucumber (*Cucumis sativus* L.). *Am. J. Biochem. Biotechnol.* **2007**, *3*, 24–32. [CrossRef]

24. Selvaraj, N.; Kasthurirengan, S.; Vasudevan, A.; Manickavasagam, M.; Choi, C.W.; Ganapathi, A. Evaluation of green fluorescent protein as a reporter gene and phosphinothricin as the selective agent for achieving a higher recovery of transformants in cucumber (Cucumis sativus L. cv. Poinsett76) via Agrobacterium tumefaciens. *Vitr. Cell. Dev. Biol.-Plant* **2010**, *46*, 329–337. [CrossRef]
25. Wang, Y.; Wang, J.; Shi, B.; Yu, T.; Qi, J.; Meyerowitz, E.M.; Jiao, Y. The Stem Cell Niche in Leaf Axils Is Established by Auxin and Cytokinin in Arabidopsis. *Plant Cell* **2014**, *26*, 2055–2067. [CrossRef] [PubMed]
26. Du, C.; Chai, L.a.; Liu, C.; Si, Y.; Fan, H. Improved Agrobacterium tumefaciens-mediated transformation using antibiotics and acetosyringone selection in cucumber. *Plant Biotechnol. Rep.* **2022**, *16*, 17–27. [CrossRef]
27. García-Almodóvar, R.C.; Gosalvez, B.; Aranda, M.A.; Burgos, L. Production of transgenic diploid Cucumis melo plants. *Plant Cell Tissue Organ Cult. (PCTOC)* **2017**, *130*, 323–333. [CrossRef]
28. Pan, W.; Cheng, Z.; Han, Z.; Yang, H.; Zhang, W.; Zhang, H. Efficient genetic transformation and CRISPR/Cas9-mediated genome editing of watermelon assisted by genes encoding developmental regulators. *J. Zhejiang Univ. Sci. B* **2022**, *23*, 339–344. [CrossRef]
29. Feng, Q.; Xiao, L.; He, Y.; Liu, M.; Wang, J.; Tian, S.; Zhang, X.; Yuan, L. Highly efficient, genotype-independent transformation and gene editing in watermelon (*Citrullus lanatus*) using a chimeric ClGRF4-GIF1 gene. *J. Integr. Plant Biol.* **2021**, *63*, 2038–2042. [CrossRef]
30. Zhao, Z.; Qi, Y.; Yang, Z.; Cheng, L.; Sharif, R.; Raza, A.; Chen, P.; Hou, D.; Li, Y. Exploring the Agrobacterium-mediated transformation with CRISPR/Cas9 in cucumber (*Cucumis sativus* L.). *Mol. Biol. Rep.* **2022**, *49*, 11481–11490. [CrossRef]
31. Sundar, I.K.; Sakthivel, N. Advances in selectable marker genes for plant transformation. *J. Plant Physiol.* **2008**, *165*, 1698–1716. [CrossRef]
32. Zhang, H.J.; Gao, P.; Wang, X.Z.; Luan, F.S. An efficient regeneration protocol for Agrobacterium-mediated transformation of melon (*Cucumis melo* L.). *Genet Mol. Res.* **2014**, *13*, 54–63. [CrossRef] [PubMed]
33. Bezirganoglu, I.; Hwang, S.Y.; Shaw, J.F.; Fang, T.J. Efficient production of transgenic melon via Agrobacterium-mediated transformation. *Genet Mol. Res.* **2014**, *13*, 3218–3227. [CrossRef] [PubMed]
34. Zhang, H.; Luan, F. Overexpression of the CmACS-3 gene in melon causes abnormal pollen development. *Genet Mol. Res.* **2015**, *14*, 10433–10443. [CrossRef]
35. Hooghvorst, I.; López-Cristoffanini, C.; Nogués, S. Efficient knockout of phytoene desaturase gene using CRISPR/Cas9 in melon. *Sci. Rep.* **2019**, *9*, 17077. [CrossRef]
36. Liu, L.; Gu, Q.; Ijaz, R.; Zhang, J.; Ye, Z. Generation of transgenic watermelon resistance to Cucumber mosaic virus facilitated by an effective Agrobacterium-mediated transformation method. *Sci. Hortic.* **2016**, *205*, 32–38. [CrossRef]
37. Tian, S.; Jiang, L.; Gao, Q.; Zhang, J.; Zong, M.; Zhang, H.; Ren, Y.; Guo, S.; Gong, G.; Liu, F.; et al. Efficient CRISPR/Cas9-based gene knockout in watermelon. *Plant Cell Rep.* **2017**, *36*, 399–406. [CrossRef] [PubMed]
38. Tian, S.; Jiang, L.; Cui, X.; Zhang, J.; Guo, S.; Li, M.; Zhang, H.; Ren, Y.; Gong, G.; Zong, M.; et al. Engineering herbicide-resistant watermelon variety through CRISPR/Cas9-mediated base-editing. *Plant Cell Rep.* **2018**, *37*, 1353–1356. [CrossRef]
39. Jang, H.-A.; Kim, H.-A.; Kwon, S.-Y.; Choi, D.-W.; Choi, P.-S. The use of cotyledonary-node explants in Agrobacterium tumefaciensmediated transformation of cucumber (*Cucumis sativus* L.). *J. Plant Biotechnol.* **2011**, *38*, 198–202. [CrossRef]
40. Wang, J.; Zhang, S.; Wang, X.; Wang, L.; Xu, H.; Wang, X.; Shi, Q.; Wei, M.; Yang, F. Agrobacterium-mediated transformation of cucumber (*Cucumis sativus* L.) using a sense mitogen-activated protein kinase gene (CsNMAPK). *Plant Cell Tissue Organ Cult. (PCTOC)* **2012**, *113*, 269–277. [CrossRef]
41. Nanasato, Y.; Okuzaki, A.; Tabei, Y. Improving the transformation efficiency of Cucurbita species: Factors and strategy for practical application. *Plant Biotechnol.* **2013**, *30*, 287–294. [CrossRef]
42. Chandrasekaran, J.; Brumin, M.; Wolf, D.; Leibman, D.; Klap, C.; Pearlsman, M.; Sherman, A.; Arazi, T.; Gal-On, A. Development of broad virus resistance in non-transgenic cucumber using CRISPR/Cas9 technology. *Mol. Plant Pathol.* **2016**, *17*, 1140–1153. [CrossRef]
43. Hu, B.; Li, D.; Liu, X.; Qi, J.; Gao, D.; Zhao, S.; Huang, S.; Sun, J.; Yang, L. Engineering Non-transgenic Gynoecious Cucumber Using an Improved Transformation Protocol and Optimized CRISPR/Cas9 System. *Mol. Plant* **2017**, *10*, 1575–1578. [CrossRef] [PubMed]
44. Wu, H.W.; Yu, T.A.; Raja, J.A.J.; Christopher, S.J.; Wang, S.L.; Yeh, S.D. Double-Virus Resistance of Transgenic Oriental Melon Conferred by Untranslatable Chimeric Construct Carrying Partial Coat Protein Genes of Two Viruses. *Plant Dis.* **2010**, *94*, 1341–1347. [CrossRef]
45. Hao, J.; Niu, Y.; Yang, B.; Gao, F.; Zhang, L.; Wang, J.; Hasi, A. Transformation of a marker-free and vector-free antisense ACC oxidase gene cassette into melon via the pollen-tube pathway. *Biotechnol. Lett.* **2011**, *33*, 55–61. [CrossRef] [PubMed]
46. Chovelon, V.; Restier, V.; Giovinazzo, N.; Dogimont, C.; Aarrouf, J. Histological study of organogenesis in Cucumis melo L. after genetic transformation: Why is it difficult to obtain transgenic plants? *Plant Cell Rep.* **2011**, *30*, 2001–2011. [CrossRef] [PubMed]
47. Shan, W.; Zhao, C.; Fan, J.; Cong, H.; Liang, S.; Yu, X. Antisense suppression of alcohol acetyltransferase gene in ripening melon fruit alters volatile composition. *Sci. Hortic.* **2012**, *139*, 96–101. [CrossRef]
48. Ren, Y.; Bang, H.; Curtis, I.S.; Gould, J.; Patil, B.S.; Crosby, K.M. Agrobacterium-mediated transformation and shoot regeneration in elite breeding lines of western shipper cantaloupe and honeydew melons (*Cucumis melo* L.). *Plant Cell Tissue Organ Cult. (PCTOC)* **2011**, *108*, 147–158. [CrossRef]

49. Choi, J.Y.; Shin, J.S.; Chung, Y.S.; Hyung, N.-I. An efficient selection and regeneration protocol for Agrobacterium-mediated transformation of oriental melon (*Cucumis melo* L. var. makuwa). *Plant Cell Tissue Organ Cult. (PCTOC)* **2012**, *110*, 133–140. [CrossRef]
50. Bezirganoglu, I.; Hwang, S.-Y.; Fang, T.J.; Shaw, J.-F. Transgenic lines of melon (*Cucumis melo* L. var. *makuwa* cv. 'Silver Light') expressing antifungal protein and chitinase genes exhibit enhanced resistance to fungal pathogens. *Plant Cell Tissue Organ Cult. (PCTOC)* **2013**, *112*, 227–237. [CrossRef]
51. Zhang, H.; Gao, P.; Wang, X.; Luan, F. An improved method of Agrobacterium tumefaciens-mediated genetic transformation system of melon (*Cucumis melo* L.). *J. Plant Biochem. Biotechnol.* **2013**, *23*, 278–283. [CrossRef]
52. Zhang, Z.; Li, X.; Ma, S.; Shan, N.; Zhang, X.; Sui, X.; Zhang, Z.; Sui, X. A Protocol for Agrobacterium-mediated Transformation of Cucumber (*Cucumis sativus* L.) from cotyledon explants. *Protoc. Exch.* **2017**. [CrossRef]
53. Mendez-Hernandez, H.A.; Ledezma-Rodriguez, M.; Avilez-Montalvo, R.N.; Juarez-Gomez, Y.L.; Skeete, A.; Avilez-Montalvo, J.; De-la-Pena, C.; Loyola-Vargas, V.M. Signaling Overview of Plant Somatic Embryogenesis. *Front. Plant Sci.* **2019**, *10*, 77. [CrossRef]
54. Zhang, X.; Xu, G.; Cheng, C.; Lei, L.; Sun, J.; Xu, Y.; Deng, C.; Dai, Z.; Yang, Z.; Chen, X.; et al. Establishment of an Agrobacterium-mediated genetic transformation and CRISPR/Cas9-mediated targeted mutagenesis in Hemp (*Cannabis Sativa* L.). *Plant Biotechnol. J.* **2021**, *19*, 1979–1987. [CrossRef] [PubMed]
55. Maren, N.A.; Duan, H.; Da, K.; Yencho, G.C.; Ranney, T.G.; Liu, W. Genotype-independent plant transformation. *Hortic. Res.* **2022**, *9*, uhac047. [CrossRef]
56. Lowe, K.; Wu, E.; Wang, N.; Hoerster, G.; Hastings, C.; Cho, M.J.; Scelonge, C.; Lenderts, B.; Chamberlin, M.; Cushatt, J.; et al. Morphogenic Regulators Baby boom and Wuschel Improve Monocot Transformation. *Plant Cell* **2016**, *28*, 1998–2015. [CrossRef]
57. Maher, M.F.; Nasti, R.A.; Vollbrecht, M.; Starker, C.G.; Clark, M.D.; Voytas, D.F. Plant gene editing through de novo induction of meristems. *Nat. Biotechnol.* **2020**, *38*, 84–89. [CrossRef] [PubMed]
58. Kong, J.; Martin-Ortigosa, S.; Finer, J.; Orchard, N.; Gunadi, A.; Batts, L.A.; Thakare, D.; Rush, B.; Schmitz, O.; Stuiver, M.; et al. Overexpression of the Transcription Factor GROWTH-REGULATING FACTOR5 Improves Transformation of Dicot and Monocot Species. *Front. Plant Sci.* **2020**, *11*, 572319. [CrossRef]
59. Debernardi, J.M.; Tricoli, D.M.; Ercoli, M.F.; Hayta, S.; Ronald, P.; Palatnik, J.F.; Dubcovsky, J. A GRF-GIF chimeric protein improves the regeneration efficiency of transgenic plants. *Nat Biotechnol* **2020**, *38*, 1274–1279. [CrossRef] [PubMed]
60. Miki, B.; McHugh, S. Selectable marker genes in transgenic plants: Applications, alternatives and biosafety. *J. Biotechnol.* **2004**, *107*, 193–232. [CrossRef]
61. He, Z.; Duan, Z.; Liang, W.; Chen, F.; Yao, W.; Liang, H.; Yue, C.; Sun, Z.; Chen, F.; Dai, J. Mannose selection system used for cucumber transformation. *Plant Cell Rep.* **2006**, *25*, 953–958. [CrossRef]
62. Hesami, M.; Tohidfar, M.; Alizadeh, M.; Daneshvar, M.H. Effects of sodium nitroprusside on callus browning of Ficus religiosa: An important medicinal plant. *J. For. Res.* **2020**, *31*, 789–796. [CrossRef]
63. Soda, N.; Verma, L.; Giri, J. CRISPR-Cas9 based plant genome editing: Significance, opportunities and recent advances. *Plant Physiol. Biochem.* **2018**, *131*, 2–11. [CrossRef] [PubMed]
64. Wang, Y.; Wang, J.; Guo, S.; Tian, S.; Zhang, J.; Ren, Y.; Li, M.; Gong, G.; Zhang, H.; Xu, Y. CRISPR/Cas9-mediated mutagenesis of ClBG1 decreased seed size and promoted seed germination in watermelon. *Hortic. Res.* **2021**, *8*, 70. [CrossRef]
65. Wang, J.; Wang, Y.; Zhang, J.; Ren, Y.; Li, M.; Tian, S.; Yu, Y.; Zuo, Y.; Gong, G.; Zhang, H.; et al. The NAC transcription factor ClNAC68 positively regulates sugar content and seed development in watermelon by repressing ClINV and ClGH3.6. *Hortic. Res.* **2021**, *8*, 214. [CrossRef] [PubMed]
66. Zhang, J.; Guo, S.; Ji, G.; Zhao, H.; Sun, H.; Ren, Y.; Tian, S.; Li, M.; Gong, G.; Zhang, H.; et al. A unique chromosome translocation disrupting ClWIP1 leads to gynoecy in watermelon. *Plant J. Cell Mol. Biol.* **2020**, *101*, 265–277. [CrossRef] [PubMed]
67. Ren, Y.; Sun, H.; Zong, M.; Guo, S.; Ren, Z.; Zhao, J.; Li, M.; Zhang, J.; Tian, S.; Wang, J.; et al. Localization shift of a sugar transporter contributes to phloem unloading in sweet watermelons. *New Phytol.* **2020**, *227*, 1858–1871. [CrossRef]
68. Ren, Y.; Li, M.; Guo, S.; Sun, H.; Zhao, J.; Zhang, J.; Liu, G.; He, H.; Tian, S.; Yu, Y.; et al. Evolutionary gain of oligosaccharide hydrolysis and sugar transport enhanced carbohydrate partitioning in sweet watermelon fruits. *Plant Cell* **2021**, *33*, 1554–1573. [CrossRef]
69. Cao, L.; Li, C.; Li, H.; Wang, Z.; Jiang, Y.; Guo, Y.; Sun, P.; Chen, X.; Li, Q.; Tian, H.; et al. Disruption of REC8 in Meiosis I led to watermelon seedless. *Plant Sci.* **2022**, *323*, 111394. [CrossRef] [PubMed]
70. Chang, J.; Guo, Y.; Yan, J.; Zhang, Z.; Yuan, L.; Wei, C.; Zhang, Y.; Ma, J.; Yang, J.; Zhang, X.; et al. The role of watermelon caffeic acid O-methyltransferase (ClCOMT1) in melatonin biosynthesis and abiotic stress tolerance. *Hortic. Res.* **2021**, *8*, 210. [CrossRef] [PubMed]
71. Zhang, R.; Chang, J.; Li, J.; Lan, G.; Xuan, C.; Li, H.; Ma, J.; Zhang, Y.; Yang, J.; Tian, S.; et al. Disruption of the bHLH transcription factor Abnormal Tapetum 1 causes male sterility in watermelon. *Hortic. Res.* **2021**, *8*, 258. [CrossRef] [PubMed]
72. Cheng, Z.; Liu, X.; Yan, S.; Liu, B.; Zhong, Y.; Song, W.; Chen, J.; Wang, Z.; Che, G.; Liu, L.; et al. Pollen tube emergence is mediated by ovary-expressed ALCATRAZ in cucumber. *Nat. Commun.* **2023**, *14*, 258. [CrossRef] [PubMed]
73. Liu, X.; Ning, K.; Che, G.; Yan, S.; Han, L.; Gu, R.; Li, Z.; Weng, Y.; Zhang, X. CsSPL functions as an adaptor between HD-ZIP III and CsWUS transcription factors regulating anther and ovule development in *Cucumis sativus* (cucumber). *Plant J.* **2018**, *94*, 535–547. [CrossRef] [PubMed]

74. Xin, T.; Zhang, Z.; Li, S.; Zhang, S.; Li, Q.; Zhang, Z.H.; Huang, S.; Yang, X. Genetic Regulation of Ethylene Dosage for Cucumber Fruit Elongation. *Plant Cell* **2019**, *31*, 1063–1076. [CrossRef]
75. Zhang, Z.; Wang, B.; Wang, S.; Lin, T.; Yang, L.; Zhao, Z.; Zhang, Z.; Huang, S.; Yang, X. Genome-wide Target Mapping Shows Histone Deacetylase Complex1 Regulates Cell Proliferation in Cucumber Fruit. *Plant Physiol.* **2020**, *182*, 167–184. [CrossRef]
76. Yang, X.; Yan, J.; Zhang, Z.; Lin, T.; Xin, T.; Wang, B.; Wang, S.; Zhao, J.; Zhang, Z.; Lucas, W.J.; et al. Regulation of plant architecture by a new histone acetyltransferase targeting gene bodies. *Nat. Plants* **2020**, *6*, 809–822. [CrossRef]
77. Wang, Z.; Wang, L.; Han, L.; Cheng, Z.; Liu, X.; Wang, S.; Liu, L.; Chen, J.; Song, W.; Zhao, J.; et al. HECATE2 acts with GLABROUS3 and Tu to boost cytokinin biosynthesis and regulate cucumber fruit wart formation. *Plant Physiol.* **2021**, *187*, 1619–1635. [CrossRef]
78. Zhang, H.; Li, S.; Yang, L.; Cai, G.; Chen, H.; Gao, D.; Lin, T.; Cui, Q.; Wang, D.; Li, Z.; et al. Gain-of-function of the 1-aminocyclopropane-1-carboxylate synthase gene ACS1G induces female flower development in cucumber gynoecy. *Plant Cell* **2021**, *33*, 306–321. [CrossRef]
79. Wang, Z.; Zhou, Z.; Wang, L.; Yan, S.; Cheng, Z.; Liu, X.; Han, L.; Chen, G.; Wang, S.; Song, W.; et al. The CsHEC1-CsOVATE module contributes to fruit neck length variation via modulating auxin biosynthesis in cucumber. *Proc. Natl. Acad. Sci. USA* **2022**, *119*, e2209717119. [CrossRef]
80. Liu, P.; Li, Q.; Wang, H.; Lu, T.; Li, Y.; Sui, X.; Yu, H.; Jiang, W. Ethylene signaling plays an important role in UV-B-induced ascorbic acid accumulation in *Cucumis sativus* leaves. *Authorea* **2022**. [CrossRef]
81. Peng, Y.; Chen, L.; Zhu, L.; Cui, L.; Yang, L.; Wu, H.; Bie, Z. CsAKT1 is a key gene for the CeO2 nanoparticle's improved cucumber salt tolerance: A validation from CRISPR-Cas9 lines. *Environ. Sci. Nano* **2022**, *9*, 4367–4381. [CrossRef]
82. Liu, B.; Santo Domingo, M.; Mayobre, C.; Martín-Hernández, A.M.; Pujol, M.; Garcia-Mas, J. Knock-Out of CmNAC-NOR Affects Melon Climacteric Fruit Ripening. *Front. Plant Sci.* **2022**, *13*, 878037. [CrossRef]
83. Wang, J.; Tian, S.; Yu, Y.; Ren, Y.; Guo, S.; Zhang, J.; Li, M.; Zhang, H.; Gong, G.; Wang, M.; et al. Natural variation in the NAC transcription factor NONRIPENING contributes to melon fruit ripening. *J. Integr. Plant Biol.* **2022**, *64*, 1448–1461. [CrossRef] [PubMed]
84. Pechar, G.S.; Donaire, L.; Gosalvez, B.; García-Almodovar, C.; Sánchez-Pina, M.A.; Truniger, V.; Aranda, M.A. Editing melon eIF4E associates with virus resistance and male sterility. *Plant Biotechnol. J.* **2022**, *20*, 2006–2022. [CrossRef]
85. Sohail, H.; Noor, I.; Nawaz, M.A.; Ma, M.; Shireen, F.; Huang, Y.; Yang, L.; Bie, Z. Genome-wide identification of plasma-membrane intrinsic proteins in pumpkin and functional characterization of CmoPIP1-4 under salinity stress. *Environ. Exp. Bot.* **2022**, *202*. [CrossRef]
86. Geng, S.; Sohail, H.; Cao, H.; Sun, J.; Chen, Z.; Zhou, L.; Wang, W.; Ye, R.; Yang, L.; Bie, Z. An efficient root transformation system for CRISPR/Cas9-based analyses of shoot-root communication in cucurbit crops. *Hortic. Res.* **2022**, *9*. [CrossRef] [PubMed]
87. Garneau, J.E.; Dupuis, M.-È.; Villion, M.; Romero, D.A.; Barrangou, R.; Boyaval, P.; Fremaux, C.; Horvath, P.; Magadán, A.H.; Moineau, S. The CRISPR/Cas bacterial immune system cleaves bacteriophage and plasmid DNA. *Nature* **2010**, *468*, 67–71. [CrossRef] [PubMed]
88. Sorek, R.; Lawrence, C.M.; Wiedenheft, B. CRISPR-mediated adaptive immune systems in bacteria and archaea. *Annu. Rev. Biochem.* **2013**, *82*, 237–266. [CrossRef] [PubMed]
89. Li, J.; Yu, X.; Zhang, C.; Li, N.; Zhao, J. The application of CRISPR/Cas technologies to Brassica crops: Current progress and future perspectives. *Abiotech* **2022**, *3*, 146–161. [CrossRef] [PubMed]
90. Chen, J.; Tomes, S.; Gleave, A.P.; Hall, W.; Luo, Z.; Xu, J.; Yao, J.L. Significant improvement of apple (Malus domestica Borkh.) transgenic plant production by pre-transformation with a Baby boom transcription factor. *Hortic. Res.* **2022**, *9*. [CrossRef]
91. Luo, W.; Tan, J.; Li, T.; Feng, Z.; Ding, Z.; Xie, X.; Chen, Y.; Chen, L.; Liu, Y.G.; Zhu, Q.; et al. Overexpression of maize GOLDEN2 in rice and maize calli improves regeneration by activating chloroplast development. *Sci. China Life Sci.* **2023**, *66*, 340–349. [CrossRef] [PubMed]
92. Hyun, Y.; Kim, J.; Cho, S.W.; Choi, Y.; Kim, J.-S.; Coupland, G. Site-directed mutagenesis in Arabidopsis thaliana using dividing tissue-targeted RGEN of the CRISPR/Cas system to generate heritable null alleles. *Planta* **2015**, *241*, 271–284. [CrossRef]
93. Wang, Z.-P.; Xing, H.-L.; Dong, L.; Zhang, H.-Y.; Han, C.-Y.; Wang, X.-C.; Chen, Q.-J. Egg cell-specific promoter-controlled CRISPR/Cas9 efficiently generates homozygous mutants for multiple target genes in Arabidopsis in a single generation. *Genome Biol.* **2015**, *16*, 144. [CrossRef] [PubMed]
94. Belhaj, K.; Chaparro-Garcia, A.; Kamoun, S.; Nekrasov, V. Plant genome editing made easy: Targeted mutagenesis in model and crop plants using the CRISPR/Cas system. *Plant Methods* **2013**, *9*, 39. [CrossRef]
95. Abbott, A. Europe's genetically edited plants stuck in legal limbo. *Nature* **2015**, *528*, 319–320. [CrossRef] [PubMed]

MDPI
St. Alban-Anlage 66
4052 Basel
Switzerland
www.mdpi.com

Agronomy Editorial Office
E-mail: agronomy@mdpi.com
www.mdpi.com/journal/agronomy

www.ingramcontent.com/pod-product-compliance
Lightning Source LLC
Chambersburg PA
CBHW061608210926
51850CB00009B/2253
* 9 7 8 3 7 2 5 8 0 9 3 1 8 *